Composite Materials: Science and Engineering

Composite Materials: Science and Engineering

Editor: Nicholas Thomson

NY RESEARCH PRESS

New York

Published by NY Research Press
118-35 Queens Blvd., Suite 400,
Forest Hills, NY 11375, USA
www.nyresearchpress.com

Composite Materials: Science and Engineering
Edited by Nicholas Thomson

International Standard Book Number: 978-1-63238-728-8 (Hardback)

Cataloging-in-Publication Data

Composite materials : science and engineering / edited by Nicholas Thomson.
p. cm.
Includes bibliographical references and index.
ISBN 978-1-63238-728-8
1. Composite materials. 2. Materials. 3. Materials science. 4. Engineering. I. Thomson, Nicholas.
TA418.9.C6 C66 2020
620.118--dc23

Printed in China.

Contents

Preface

The composite material is made up of the combination of two or more constituent materials that have distinct chemical and physical properties. When two different materials are combined they often produce a material that has different characteristics from either of the individual components. Each component remains separate within the finished structure. There are two categories of these constituent materials, matrix and reinforcement. The matrix material is used to support and surround the reinforcement materials by maintaining their relative positions. The reinforcements enhance the matrix properties by their special mechanical and physical properties. At least one portion of each constituent material should be added in the composition. Composite materials are lighter, stronger and less expensive than the traditionally used materials. They are often used in the construction of bridges, buildings, and other structures such as boat hulls and racing car bodies. This book is a valuable compilation of topics, ranging from the basic to the most complex advancements in the field of composite materials. Most of the topics introduced in it cover new techniques and applications of such materials. It is appropriate for students seeking detailed information in this area as well as for experts.

Significant researches are present in this book. Intensive efforts have been employed by authors to make this book an outstanding discourse. This book contains the enlightening chapters which have been written on the basis of significant researches done by the experts.

Finally, I would also like to thank all the members involved in this book for being a team and meeting all the deadlines for the submission of their respective works. I would also like to thank my friends and family for being supportive in my efforts.

Editor

1

Spectral Characteristics of Acoustic Emission in Carbon Fiber-Reinforced Composite Materials Subjected to Cyclic Loading

Vladimir L. Shkuratnik and Petr V. Nikolenko

Mining Institute of National University of Science and Technology MISiS, 119991 Moscow, Russia

Correspondence should be addressed to Petr V. Nikolenko; petrov-87@mail.ru

Academic Editor: Carlo Santulli

The article is devoted to the investigation of the spectral characteristics of acoustic emission signals that appear under various schemes of composite materials loading. The tests involved samples of composites reinforced with layers sheets of carbon fiber fabric and dispersed carbon fibers. Based on the results of laboratory tests, a comparison is made between the traditional parameters of acoustic emission and the complete spectrograms of the acoustic emission response developed with the use of a special algorithm. The relationship between the emission activity and the change in the spectral composition of emission hits is shown. For example, for some composites, the acoustic emission memory effect (Kaiser effect) manifests itself not only in the time domain but also in the spectral domain in a form of a sharp change in the amplitudes in the frequency range 130/150 kHz. Also, when the samples were loaded according to the Brazilian scheme, the presence of the so-called "inverse" Kaiser effect is observed, in which the memory carrier "remembers" the previously experienced level of tensile stresses and reproduces this information during subsequent unloading. Such effect manifests itself in the form of a sharp change in the amplitudes in the low-frequency region of the spectrum.

1. Introduction

The method of acoustic emission (AE) is widely known as an effective tool for assessing the degree of disturbance of various composite materials [1–4] both at the stage of testing samples in laboratory conditions and when monitoring the state of products and structures. Less well known is the use of AE for studying the stress-strain state of complex and structurally heterogeneous media (rocks, building structures, etc.). This application is based on the so-called acoustical-emission memory effect, also known as the Kaiser effect (KE). This effect is observed in rocks, metals, and composite materials and consists in nonreproducibility of acoustic emission parameters in the subsequent loading cycle in comparison with the previous, when instant recovery of these parameters take place at the moment of reaching the maximum stress level of the preceding cycle [5]. To this day, as the main informative parameters in the analysis of AE, time characteristics are used, such as activity, total activity, and count.

The first attempts of practical application of KE for the evaluation of stress fields in the rock mass refer to the second half of the 1970s. In this case, rock samples extracted from the massif by column drilling were used as a "memory carrier" [6, 7]. Tests of these samples under various conditions of mechanical loading with simultaneous measurement of AE informative parameters made it possible to estimate the stresses acting in the rock massif. However, the accuracy of such an estimate was inadequate in a number of cases, which was associated with a large scatter of mechanical properties and the disturbance of rocks of even one genotype. Further development of the method consisted in determining the parameters of formation and manifestation of KE in rock samples of different genotypes under different loading schemes [8–12].

In recent years, the method based on the use of KE has evolved by replacing the "memory carrier" in the form of rocks with difficult-to-predict properties for composite materials whose acoustic and mechanical properties are being set at the synthesis stage [13]. The use of composites placed in measuring boreholes significantly reduced the spread of stress estimates in the massif by eliminating the uncertainty factor caused by the random distribution of properties in the "memory carrier." An additional advantage of using materials with programmable properties is the possibility of using specific techniques and algorithms for processing the results of measurements [14]. At the forefront in this case comes the spectral analysis of AE signals, which makes it possible to obtain additional information on the defects growth in the investigated medium.

Attempts to use the analysis of acoustic emission spectra usually face a number of problems of an instrumental and methodical nature. Thus, when the maximum frequency of the spectrum f_{max} is used as the main informative parameter, it becomes necessary to use broadband acoustic transducers [15]. It is known that the sensitivity of such transducers is significantly lower in comparison to their resonant counterparts [16].

Another approach to the spectral processing of AE signals based on a detailed analysis of the spectra of individual acoustic hits was taken at different loading stages of the samples [17, 18]. The interpretation results in this case are characterized by low reliability and repeatability due to a significant random component in each separate AE hit. A combined analysis of all AE hits without additional processing is also uninformative, as it will be demonstrated further on.

Overcoming of the described problems is possible due to the complex spectral analysis of all waveforms of registered AE hits.

2. Materials and Methods

Studies were performed on cylindrical samples with a diameter of 45 mm and a height of 90 mm made from materials, synthesized under laboratory conditions. The matrix for all samples was the epoxy resin Poxy Systems L. Two types of filler were used: dispersed carbon fibers, with a length of 3 mm, a thickness of 15 μm, and a weight ratio to matrix 1/1000 (series of samples "A") and sheets of carbon fiber fabric located parallel to the axis of the cylinders (series of samples "B"). Before the entire polymerization of the matrix, all samples were subjected to evacuation at a pressure of 0.3 bar to remove air bubbles and maintain the consistency of the properties of synthesized materials. The fillers used in groups "A" and "B" composites have the same mechanical characteristics. The difference between composites is in the size and form of the initial inhomogeneities. Dispersed fibers create heterogeneities of a small size. In that case, crack growth under loading should be accompanied by the AE in the region of small wavelengths (high frequencies). When using carbon fiber sheets, the dimensions of the inhomogeneities are much larger and changes in the spectral composition should be expected in the low-frequency band. The polished sections of the samples of both series are shown in Figure 1.

Samples of both series were divided into two groups of 5 pieces in each. Samples of the first group underwent uniaxial multicyclic loading according to the classical scheme (along the axis of the cylinder). Samples of the second group were loaded by the so-called Brazilian scheme (along the diameter of the cylinder), suggesting the occurrence of significant tensile stresses [19]. In this case, the main parameters of the AE and the complete waveforms of all recorded pulses were recorded with the aid of the multichannel acoustic emission complex A-Line 32D with a set of resonant acoustic transducers GT-200. The bandwidth of the entire measuring path was 25–500 kHz. Each AE waveform was recorded separately using the threshold principle. The threshold value was set to 32 dB. The recording time of one waveform was 4096 samples at a sampling interval of 0.2 μs. Also, the time of occurrence of each AE hit was recorded with an accuracy of 1 μs. Scheme of the laboratory setup is shown in Figure 2.

3. Results and Discussion

The results of the experiment were the obtained dependences of the axial mechanical load σ and the acoustic emission activity $\dot{N}$ on the loading time t, as well as the package of individual oscillograms of each AE hit. All the obtained oscillograms were subjected to a FFT to transform them into a spectral form. The set of amplitude spectra obtained in this way was presented in the form of a so-called spectrogram—a three-dimensional image, where the AE hit number N is plotted along the horizontal axis, the frequency f is plotted along the vertical axis, and the amplitude is encoded by color. It should be noted that, in the building of a spectrogram, it was not the window Fourier transform of one long record, but the successive fast Fourier transform of the individual AE waveforms. For correct interpretation of the spectrograms obtained, it is required to superimpose $\sigma(t)$ on them. The peculiarity of the formation of a spectrogram is the sequential arrangement of the spectra in it as a function of the ordinal number. In this case, in the $\sigma(t)$ dependence, the time with the sampling period $t_s = 1$s is plotted along the horizontal axis. Taking into account the fact that several AE hits occur in one second, the spectrogram and the load graph are "out of synch". To solve this problem, there are two ways of correction of the spectrogram or correction of the $\sigma(t)$ dependence. Correction of the spectrogram is a laborious process and, subsequently, can complicate its further processing. The simplest solution is to transform $\sigma(t)$ into $\sigma(N)$.

When recording the load, the record starts from the first second. Thus, the sequence number of the record of the load values is also the second at which this load was registered. To transform $\sigma(t)$ into $\sigma(N)$ from the waveform files, information is taken about the exact time of occurrence of the AE hit t_{hit}. The values of t_{hit} are rounded to integer values. In this case, t_{hit} becomes simultaneously the index of the registered load σ. Thus, from the whole array of readings $\sigma(t)$ those that correspond to t_{hit} are selected. An example of the dependence $\dot{N}(t)$ obtained as a result of loading a sample from the first group and a spectrogram with superimposed $\sigma(N)$ is shown in Figure 3.

(a)

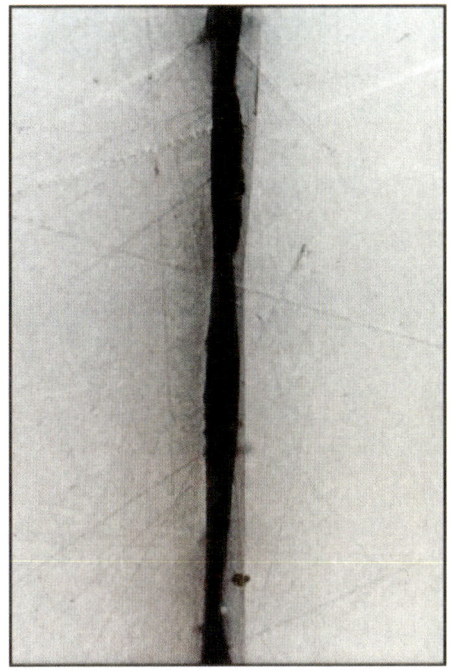

(b)

FIGURE 1: Polished sections of a composite materials of a series "A" (a) and a series "B" (b).

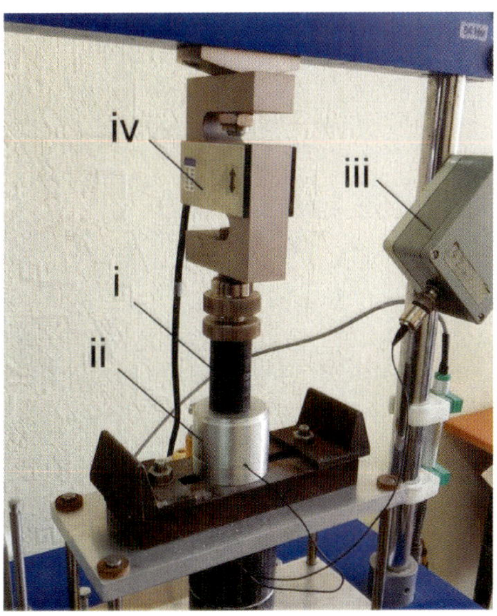

FIGURE 2: Laboratory setup with the tested sample (i), AE transducer (ii), preamplifier (iii), and stress gauge (iv).

From Figure 3(a) it can be seen that, on all loading cycles, starting from III, the jumps of AE activity are clearly revealed when the load reaches the maximum of the preceding cycle. It corresponds to KE manifestation. At the same time, no regularities in the spectral composition of the AE hits are observed on the obtained spectrogram (Figure 3(b)) without additional processing. In the spectrogram, only the region of high amplitudes along the horizontal line at 175 kHz is distinguished. This line is due to the resonant type of the frequency response of the transducer used, for which the resonant frequency is exactly $f_{res} = 175$ kHz.

The form of spectrogram shown in Figure 3 is primarily determined by the random nature of the individual AE signals. Reducing the impact of the "random" component is possible with the use of special processing. In this case, it is proposed to smooth the spectrogram with a moving average in a rectangular window in accordance with Equation (1):

$$m_{i,j} = \frac{1}{a \cdot b} \sum_{i=1}^{a} \sum_{j=1}^{b} A_{i,j}, \tag{1}$$

where a and b are the width and height of the smoothing window and $A_{i,j}$ are the elements of the original amplitude matrix of the AE hits.

It should be noted that when using the moving average, information on the absolute values of the amplitudes of the registered AE pulses is lost. Such a loss, however, does not have a significant effect on the results of the spectrograms interpretation, since in the final analysis only the relative amplitude variation at different stages of loading has a value.

When smoothing with a moving average, the problem arises in selecting the parameters of the smoothing window. In practice, specific numerical values for obtaining the best result are selected by the operator. However, if you use too large windows, there is edge effects-data loss at the edges of the smoothed area. In this case, a window with parameters $a = 15$ and $b = 5$ was chosen for smoothing with the number of registered AE pulses $N = 1150$ and the number of discretes along the frequency axis-2048. The result of the transformation of the spectrogram for a sample of the first group is shown in Figure 4.

As a result of processing the spectrogram, it is possible to single out a number of regularities on it. Firstly, in the range

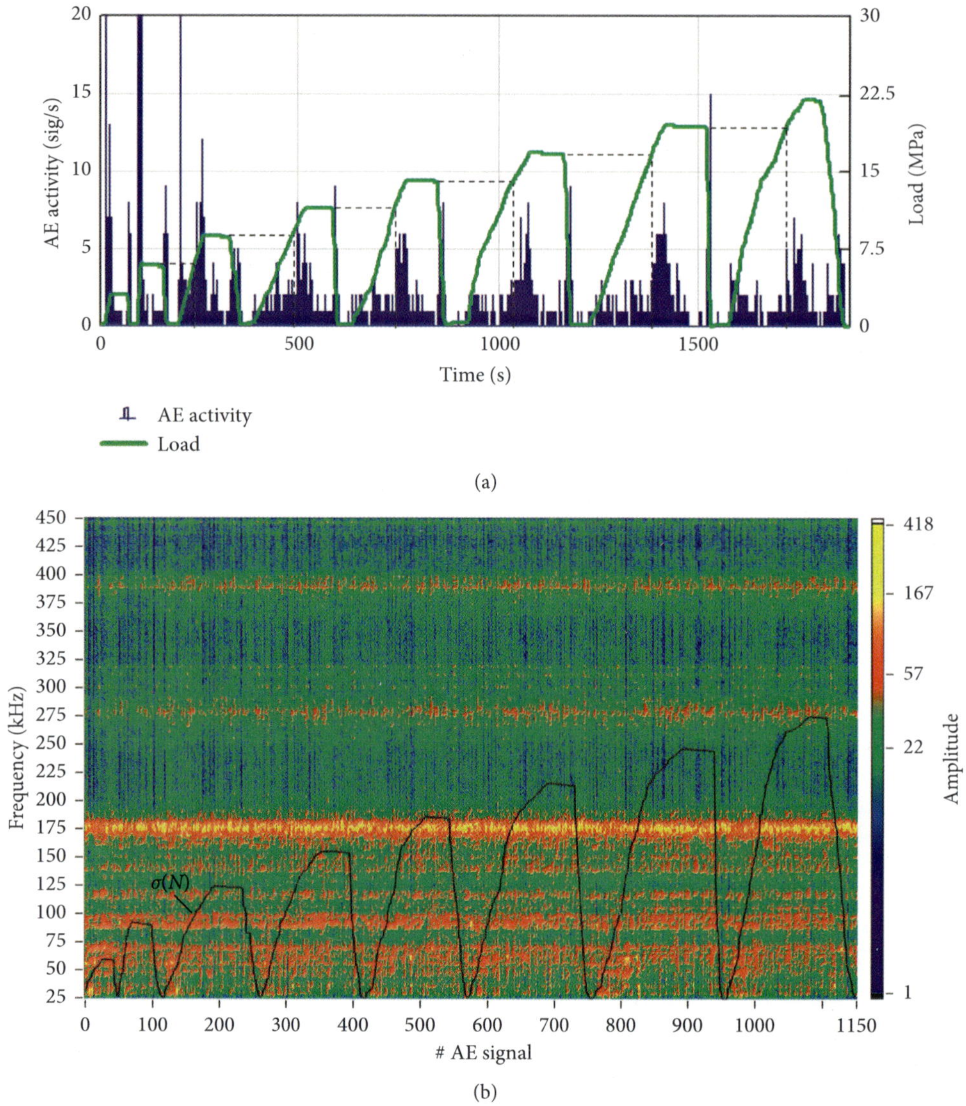

FIGURE 3: AE activity upon loading the sample from the first group of the "A" series (a) and an example of the obtained spectrogram with superposition of the $\sigma(N)$ (b).

up to 50 kHz, the role of low frequencies decreases with the increasing load. In this case, the low-frequency amplitudes return to their original levels when the load is removed. Such a change may be due to the closure of some of the cracks formed. In this case, with a change in the size of the defects, the spectral composition of the acoustic pulses corresponding to them also changes [3].

Along with low-frequency components, significant changes are also observed in the range of medium frequencies (130/150 kHz). In this range, the amplitude change occurs abruptly at the moment when the value of σ reaches the maximum level of the previous cycle and most clearly this change begins with the IV loading cycle, i.e., from the load level 0.1 of strength limit. Thus, the Kaiser effect in the material is manifested not only in the form of acoustic emission activity burst but also in the form of a sharp change in the spectral characteristics of the AE signals.

The results of loading a sample of the "B" series are shown in Figure 5.

Figure 5(a) demonstrates the presence of an acoustic emission memory effect in a composite reinforced with layers of carbon fiber fabric. However, in contrast to the samples of group "A", in spectral analysis of AE (Figure 5(b)), no significant change in the frequency band of 130/150 kHz is observed. This regularity can be explained by the absence of cracks of small size (Figure 1) capable of causing high-frequency AE hits in the "B" series. In this case, there is only a repeatable cycle-to-cycle amplitude reduction at frequencies up to 50 kHz with increasing load.

The results of loading the samples of group 2 according to the Brazilian scheme are illustrated in Figure 6. Unlike the first group, the load on the sample is represented as the mechanical force P (kN) produced by the loading device. Smoothing parameters were chosen similarly in group 1.

When analyzing the behavior of AE activity (Figure 6(a)), the manifestations of the Kaiser effect are clearly distinguished when the axial load exceeds the maximum of the preceding cycle P^*.The spectrogram (Figure 6(b)) shows a distinct and

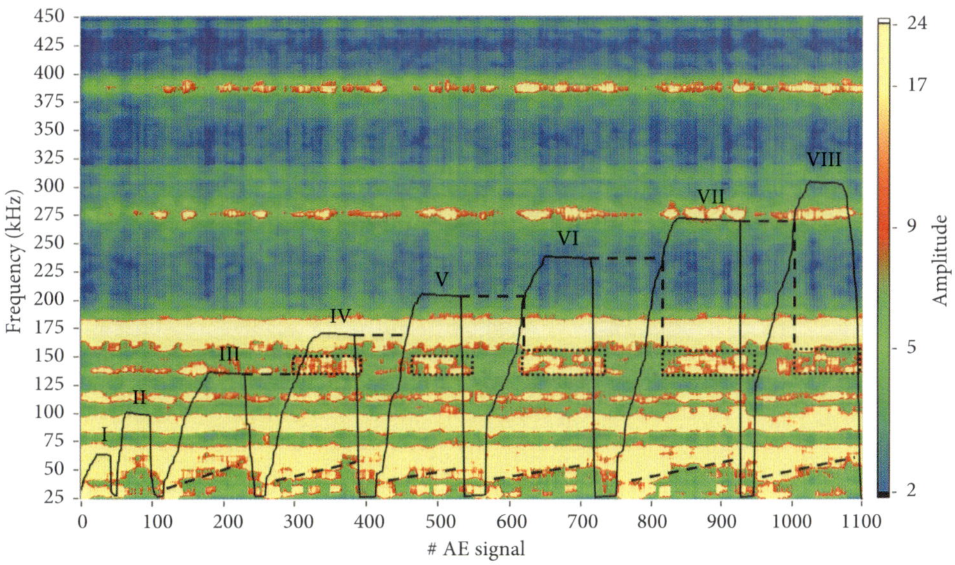

FIGURE 4: An example of a smoothed spectrogram for a sample of the first group of the "A" series with superposition of the $\sigma(N)$ dependence (Roman numerals indicate the numbers of the loading cycles).

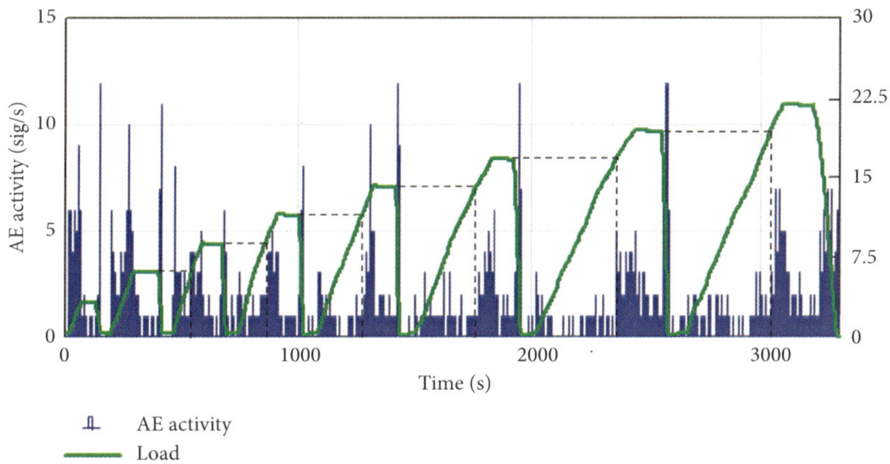

(a)

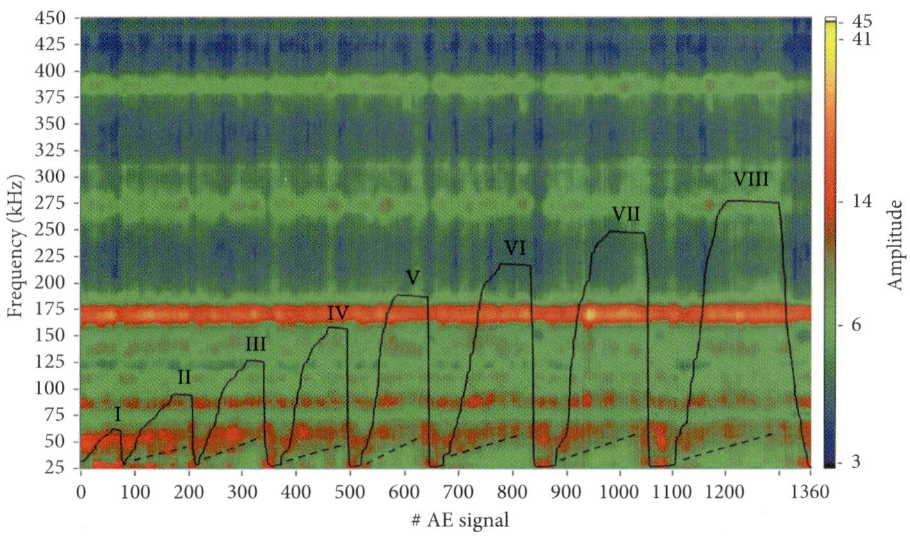

(b)

FIGURE 5: AE activity upon loading a sample from the first group of the "B" series (a) and an example of the obtained spectrogram with superposition of $\sigma(N)$ (b).

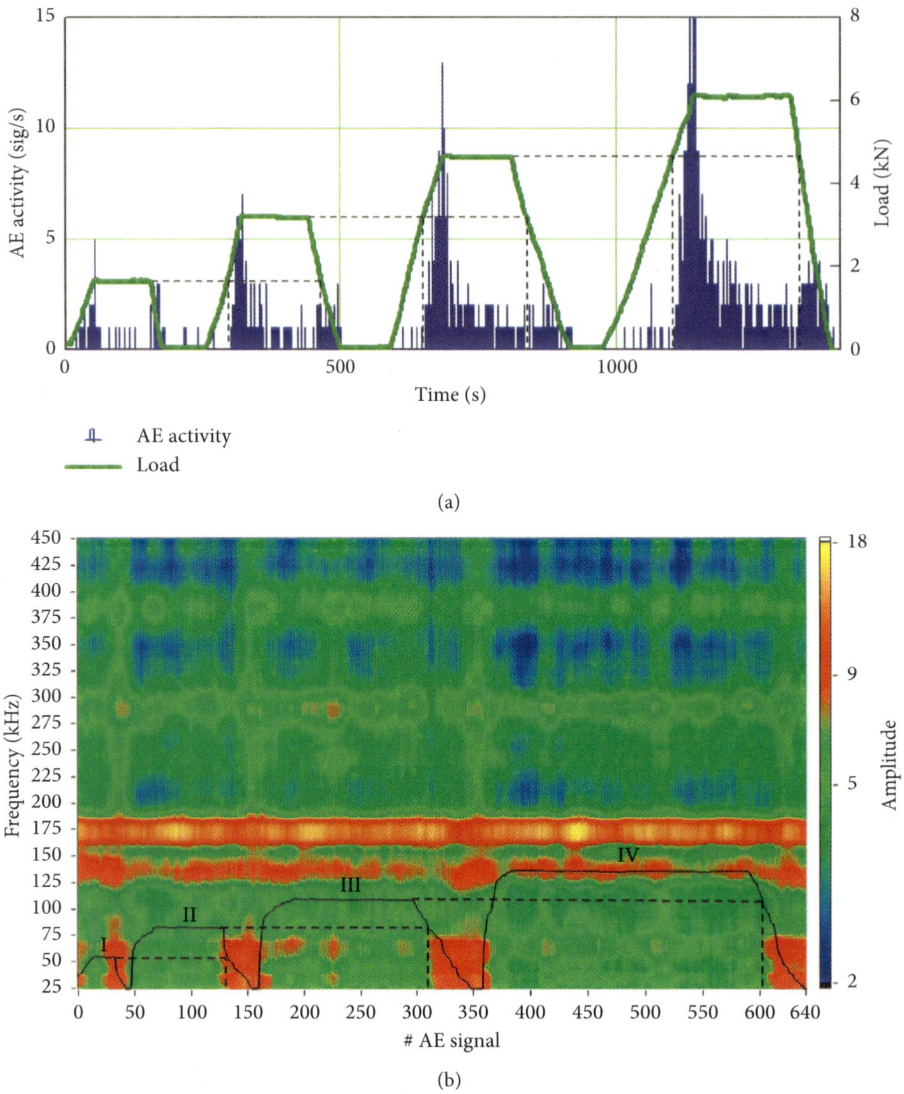

FIGURE 6: AE activity upon loading a sample from the second group of the "A" series (a) and an example of a processed spectrogram with superposition of $P(N)$ (b).

sharp change in amplitudes in the frequency range 25/70 kHz. The rise of amplitudes in this range occurs every time when the load decreases and passes through the P^* level. The low-frequency amplitudes remain high up to the intersection of the P^* level during the subsequent loading stage. At this point, they restore the original values. This behavior can be explained by the fact that, at the stage of unloading, acoustic emission has a predominantly frictional nature (i.e., it is related to friction between the cracks bridges). Formation of new and growth of existing cracks practically does not occur. In the process of removing the load, there are tangential stresses at the "matrix-reinforcing fiber" boundary [20], arising from the significant difference in the elastic moduli of epoxy resin ($E = 40$ MPa) and carbon fiber ($E = 50$ GPa), which leads to sliding breach at the "fiber-matrix" boundary. Tensile stresses capable of breaking the carbon fibers were not achieved during the loading process. Samples of the "B" series showed similar regularities.

When using AE spectral analysis, it is necessary to understand that, in practice, calculation of the exact characteristics of the signal source is practically impossible because of the need to calculate both the response function of the used transducers and the entire system as a whole. And if the approaches to calculating the response functions of the transducers are known [21, 22], finding the function of the whole system (including transducers, sample, and loading device) is extremely difficult. Despite the foregoing, the authors of the article are sure that the relative change in the spectral characteristics of AE signals in the process of loading can give important indirect information about the nature and degree of disturbance of composite materials.

4. Conclusions

The presented research is devoted to the analysis of the spectra of acoustic emission signals arising during cyclic loading of samples of composite materials. The following important conclusions are drawn:

(1) Analysis of acoustic emission spectra is most effective when all registered waveforms are included in the processing. In order to exclude the random component, it is necessary to use the smoothing of the spectrograms both along the time and frequency axis.

(2) The spectral composition of the AE hits during cyclic loading varies both in the low-frequency range (up to 50 kHz) and in the frequency range 130/150 kHz for composite samples reinforced with dispersed fibers. Moreover, a sharp change in amplitudes in the range of 130/150 kHz corresponds to the moment when the load reaches the maximum previously experienced value. Thus, the acoustical-emission memory effect (the Kaiser effect) manifests itself not only in the time domain but also in the spectral domain.

(3) The spectral composition of the AE hits under cyclic loading of composite samples reinforced with sheets of carbon fiber fabric varies only in the range of low (up to 50 kHz) frequencies. This can be attributed to the lack of cracks that are small in length and capable of elastic closure-opening during loading and unloading.

(4) When the composites are loaded according to the so-called Brazilian scheme, which presupposes the appearance of tensile stresses, a significant change in the spectral composition of the AE occurs in the low-frequency region. Of particular interest is the effect of an abrupt increase in the amplitudes in the frequency range up to 50 kHz during unloading when the previously experienced maximum load is reached. The growth of AE activity in this case is practically indistinguishable. Thus, the sample exhibits an "inverse" Kaiser effect—an effect in which the memory carrier "remembers" the previously experienced level of tensile stresses and reproduces this information during subsequent unloading. Such effect can be used to evaluate tensile stresses in complex and inhomogeneous media such as rocks and building structures.

Conflicts of Interest

The authors declare that they have no conflicts of interest.

Acknowledgments

This work was supported by the Russian Scientific Foundation (project no. 17-77-10009).

References

[1] K. G. Dassios, E. Z. Kordatos, D. G. Aggelis, and T. E. Matikas, "Nondestructive damage evaluation in ceramic matrix composites for aerospace applications," *The Scientific World Journal*, vol. 2013, Article ID 715945, 12 pages, 2013.

[2] P. Zhang, Y. N. Zhao, Q. F. Li, P. Wang, and T. H. Zhang, "Flexural toughness of steel fiber reinforced high performance concrete containing nano-SiO_2 and fly ash," *The Scientific World Journal*, vol. 2014, Article ID 403743, 11 pages, 2014.

[3] J. Blom, J. Wastiels, and D. G. Aggelis, "Application of acoustic emission on the characterization of fracture in textile reinforced cement laminates," *The Scientific World Journal*, vol. 2014, Article ID 178020, 7 pages, 2014.

[4] D. G. Aggelis, K. G. Dassios, E. Z. Kordatos, and T. E. Matikas, "Damage accumulation in cyclically-loaded glass-ceramic matrix composites monitored by acoustic emission," *Scientific World Journal*, vol. 2013, article 869467, 10 pages, 2013.

[5] J. Kaiser, "Erkenntnisse und Folgerungen aus der Messung von Geräuschen bei Zugbeanspruchimg von metallischen Werkstoffen," *Archiv fur das Eisenhuttenwesen*, vol. 24, no. 1-2, pp. 43–45, 1953.

[6] T. Kanagawa, M. Hayashi, and H. Nakasa, "Estimation of spatial geo-stress in rock samples using Kaiser effect," Rep. No. 375017, Central Research Institute of Electric Power Industry (CRIEPI), Abico, Japan, 1976.

[7] K. Kurita and N. Fujii, "Stress memory of crystalline rocks in acoustic emission," *Geophysical Research Letters*, vol. 6, no. 1, pp. 9–12, 1979.

[8] H. J. Wang, L. Tang, X. H. Ren, A. Y. Yang, and Y. Niu, "Mechanism of rock deformation memory effect in low stress region and its memory fading," *Rock and Soil Mechanics*, vol. 35, no. 4, pp. 1007–1014, 2014.

[9] Q. Meng, M. E. Zhang, L. Han, H. Pu, and Y. Chen, "Acoustic emission characteristics of red sandstone specimens under uniaxial cyclic loading and unloading compression," *Rock Mechanics and Rock Engineering*, vol. 51, no. 4, pp. 969–988, 2018.

[10] D. J. Holcomb and L. S. Costin, "Detecting damage surfaces in brittle materials using acoustic emission," *Journal of Applied Mechanics-ASME*, vol. 53, no. 3, pp. 536–544, 1986.

[11] D. R. Hughson and A. M. Crawford, "Kaiser effect gauging: the influence of confining stress on its response," in *Proceedings of 6th International Journal of Rock Mechanics and Mining Sciences and Geomechanics Abstracts*, vol. 2, pp. 981–985, A. A. Balkema, Montreal, Canada, October 1987.

[12] Y. Filimonov, A. Lavrov, and V. Shkuratnik, "Effect of confining stress on acoustic emission in ductile rock," *Strain*, vol. 41, no. 1, pp. 33–35, 2005.

[13] P. V. Nikolenko and V. L. Shkuratnik, "Acoustic emission in composites and application for stress monitoring in in rock masse," *Journal of Mining Science*, vol. 50, no. 6, pp. 1088–1093, 2014.

[14] P. V. Nikolenko and V. V. Nabatov, "Interference protection in geoacoustic control of critical stresses in rocks," *Gornyi Zhurnal*, vol. 2015, no. 9, pp. 33–36, 2015.

[15] N. B. Zhang, Q. X. Qi, Z. H. Ouyang, H. Y. Li, S. K. Zhao, and Z. J. Xu, "Experimental on acoustic emission characteristics of marble with different stress paths," *Meitan Xuebao/Journal of the China Coal Society*, vol. 39, no. 2, pp. 389–394, 2014.

[16] S. Assous, D. Gunn, C. Hopper, PD. Jackson, L. Linnett, and M. Lovel, "An approach for correcting magnitude and phase distortion in wideband piezoelectric transducer systems," in *Proceedings of conference OCEANS 2007—Europe*, Aberdeen, Scotland, June 2007.

[17] S. D. Goodfellow and R. P. Young, "A laboratory acoustic emission experiment under in situ conditions," *Geophysical Research Letters*, vol. 41, no. 10, pp. 3422–3430, 2014.

[18] M. C. He, J. L. Miao, and J. L. Feng, “Rock burst process of limestone and its acoustic emission characteristics under true-triaxial unloading conditions,” *International Journal of Rock Mechanics and Mining Sciences*, vol. 47, no. 2, pp. 286–298, 2010.

[19] A. Lavrov and A. Vervoort, “Effects of non-uniform radial load distribution on stress field in Brazilian tests,” *Journal of Rock Mechanics ant Tunneling Technology*, no. 2, pp. 79–92, 2001.

[20] T. Nozawa, K. Ozawa, and H. Tanigawa, “Re-defining failure envelopes for silicon carbide composites based on damage process analysis by acoustic emission,” *Fusion Engineering and Design*, vol. 88, no. 9-10, pp. 2543–2546, 2013.

[21] K. Ono, H. Cho, and T. Matsuo, “Bar and plate-wave characterization of AE sensors,” *JAE*, vol. 26, p. 72, 2008.

[22] K. Ono, H. Cho, and T. Matsuo, “New characterization methods of AE sensors,” *JAE*, vol. 28, p. 256, 2010.

2

Concrete with Improved Chloride Binding and Chloride Resistivity by Blended Cements

Katalin Kopecskó[1] and György L. Balázs[2]

[1] *Department of Engineering Geology and Geotechnics, Budapest University of Technology, Muegyetem Rkp. 3, Budapest 1111, Hungary*
[2] *Department of Construction Materials and Technologies, Budapest University of Technology, Muegyetem Rkp. 3, Budapest 1111, Hungary*

Correspondence should be addressed to György L. Balázs; balazs.gyorgy@epito.bme.hu

Academic Editor: Frederic Dumur

Durability and service life of concrete structures can be endangered by chloride ions. Two phenomena help to keep control of chloride effects. On one hand cements are able to bind chloride ions by their aluminate clinker phases or by the clinker substituting materials. On the other hand resistivity of concrete against chloride penetration can be improved by careful selection of concrete constituents and production. Detailed results of two series of extensive experimental studies are presented herein. Chloride ion binding capacity of tested cements in decreasing sequence was the following: (1) CEM III/B 32,5 N-S; (2) CEM III/A 32,5 N; (3) CEM II/B 32,5 R; (4) CEM II/B-M (V-L) 32,5 R; (5) CEM I 42,5 N. Test results indicated that the increasing substitution of clinkers by GGBS improves the chloride resistivity in concrete made with the same water to cement ratio. The application of air entraining agent increases considerably the values of D_{nssm}. Based on the migration coefficients (D_{nssm}) the following sequence of efficiency was found (from the best): CEM III/B 32,5 N > CEM V/A (S-V) 32,5 N > CEM III/A 32,5 N > CEM II/B-S 42,5 R > CEM II/A-S 42,5 N > CEM I 42,5 N.

1. Introduction

One of the most important objectives of durability design is to control the penetration of chloride ions up to the level of steel reinforcement. The increase of concrete cover seems to be a possible solution. However, this should not be the optimal solution since the dead load of the structure increases by the increase of concrete cover in addition to the increased probability of spalling of concrete cover for high thicknesses. Another solution is the application of high performance concrete with low permeability and improved resistivity against chloride diffusion.

Empirical and physical prediction models are available for chloride ingress and corrosion initiation in concrete structures [1]. When chlorides penetrate through the concrete cover and reach the steel reinforcement, the corrosion will initiate when a certain chloride concentration (the threshold value) is reached [2–5]. Studies on the effects of chloride ions deal either with binding of chloride ions or with the transport processes (migration or penetration) of chloride in concrete. In order to understand the beginning and development of corrosion process of steel reinforcements, we have to understand both effects in detail [6–9]. Interaction of chloride ions and cements can be studied by testing directly the interaction of chloride ions and the clinker minerals. It is well-established that both C_3A and C_4AF are able to bind chloride ions by the formation of Friedel's salt ($C_3A{\cdot}CaCl_2{\cdot}H_{10}$), or its iron analogue ($C_3F{\cdot}CaCl_2{\cdot}H_{10}$), if chlorides originated from $CaCl_2$ are dissolved in the mixing water [10].

In the hardened cement paste two main chloride binding mechanisms are considered (i) through physical adsorption and (ii) through chemical reactions [11]. The CSH phase is considered to physically bind chlorides due to its high specific surface values [12–15]. The AFm compounds are generally known to be able to bind chloride ions through chemical substitution [12, 16–19]. When considering the chloride binding capacity of a commercial Portland cement, it is concluded in [20] that Friedel's salt is the only hydrated phase formed that incorporated chlorides into its structure, up to an external chloride concentration of 3 moles. This statement is in agreement with our further results [21]. Even the composition of the Portland cements could be different, because of the addition of latent hydraulic or pozzolanic components [22]. The clinker substituting constituents (slag, fly ash, and trass) are influencing the chloride binding ability. One of our previous studies indicated that hardened slag itself can also bind chloride ions. The presence of Friedel's salt was proved by XRD. Tests were carried out to study the chloride ion binding of blast furnace slag without cement clinkers (steam-cured samples for 8 hrs at 80°C). We have shown that the improved chloride ion binding capacity of the slag cements is owing to the slag content even if its hydration was not in the typical environment of cement hydration with high pH value [21].

Steam curing is a generally used method in the production of precast concrete elements. Increased temperature accelerates the hydration resulting in high early strength. However the accelerated curing affects detrimentally both the long-term strength and the durability. Hooton and Titherington [23] concluded that steam-cured concrete containing silica fume or blends of both silica fume and ground granulated blast furnace slag exhibit improved chloride penetration resistance compared to those of Portland cement concrete. Dhir and his coauthors [24] reported the results of chloride binding measurements of GGBS pastes, as well as chloride diffusion and permeability measurements of GGBS concrete mixes. They found that, in concrete designed to have a minimum volume of voids, the chloride ion binding capacity of the cement matrix becomes the dominant factor in how resistant the concrete is to chloride permeation. Luo and his colleges [25] experimentally studied both the chloride diffusion coefficient and the chloride binding capacity of Portland cement or blended cement made of Portland cement and 70 m% GGBS replacement. They found that (i) chloride diffusion coefficient decreased and (ii) chloride ion binding capacity improved in samples of blended cement.

Nevertheless, to determine the extent of *the chloride ion binding* by the cements made with different amount of GGBS or other clinker substituting materials is highly important (see details in Section 2). Clinker substituting constituents are also known as supplementary cementing materials (SCMs) with latent hydraulic or pozzolanic behaviour. One of the experimental parameters was the circumstances of the hardening: both naturally hardened and accelerated hardened (steam-cured) cement paste specimens were subjected to salt-treatment. The reason was to study the difference between the chloride ion binding capacities in case of cast-in situ (naturally hardened) or prefabricated concrete. Cements with increasing slag contents have improved capacity for chloride binding.

We have studied then *the chloride diffusion* as a function of the slag content or other substituting materials of cements as well as the presence of air entraining agents (see details Section 3). Concrete specimens were prepared by different water to cement ratios, in order to study the resistivity of concrete against chloride ions. The amount of cement was kept constant in all concrete mixtures. Test results of concrete without air entraining agents were compared to test results of concrete made with air entraining agents. Test results indicated that the increasing substitution of clinkers by GGBS improves the chloride resistivity in concrete made with the same water to cement ratio. The application of air entraining agent reduces the chloride resistivity, because the permeability increases by the increased porosity.

2. Blended Cements for Enhanced Chloride Binding Capacity

2.1. Purpose of Research. Purpose of this part of research was to answer the following questions: (a) what is the effect of the different amount of clinker substituting materials on the chloride ion binding of cements and (b) what is the influence of the curing temperature (natural hardening at 20°C, or steam curing at 80°C) on the hydration processes and on chloride ion binding capacities of cements? We followed the methods in [26] and the experimental observations were considerably extended.

2.2. Experimental Studies. Five types of cements were selected with different amounts of clinker substituting constituents. Sulphate contents of the standardised cement products were not modified (hence the influence of various sulphate contents of the cements was not studied). Cements involved into this comparative study were for reference CEM I 42,5 N ordinary Portland cement, two other Portland cements (CEM II/B-M (V-L) 32,5 R and CEM II/B-S 32,5 R), and two blast furnace slag cements (CEM III/A 32,5 N and CEM III/B 32,5 N-S). Cement clinkers and other main constituents different from clinkers were ground together during the production of cement. According to the standard MSZ EN 197-1:2000 [27] main constituents are specially selected inorganic materials in a proportion exceeding 5% by mass related to the sum of all main and minor additional constituents. Amount of the main constituents different from clinkers in case of cements involved in our study is summarized in Table 1. The composition of clinker minerals was the same in case of the cements as in CEM I 42,5 R ordinary Portland cement (reference). Composition of clinker minerals in ordinary Portland cements and composition of clinker minerals in the blended cements were the same. Samples of cement pastes were prepared by the required amount of water to reach the standard (semiplastic) consistence according to the Hungarian standard MSZ EN 196-3:2005 [28] (Table 1). The chemical compositions of tested cements are given in Table 2.

Tests were carried out to study the influence of steam curing and salt-treatment of cements paste specimens (Table 3).

The samples of series 1 and 2 were naturally hardened and after demoulding these samples were kept at room temperature (22 ± 3°C) in 100% r.h. The samples of series of 3 and 4 were steam-cured for 3 hrs at 80°C; then samples were kept in the same condition as series of 1 and 2. Salt-treatment meant to keep the specimens in 10% NaCl solution, between 28 and 38 days (24 hrs in the salt solution, followed by 24 hrs of drying cyclically). Salt-treated samples were kept then at room temperature (22 ± 3°C) and 100% r.h. Tests of hardened samples were carried out at different ages (90 and 180 days).

TABLE 1: Cement types involved into the study, percentages of clinker substituting constituents, and the applied water-cement ratio of standard (semiplastic) consistence.

Cements	Type and amount of main constituents in cements	Water to cement ratio
CEM I 42,5 N (OPC)	0 m%	0.273
CEM II/B-M (V-L) 32,5 R	18 m% V, 10 m% L	0.270
CEM II/B-S 32,5 R	26 m% GGBS	0.272
CEM III/A 32,5 N	40 m% GGBS	0.285
CEM III/B 32,5 N-S	62 m% GGBS	0.302

Notes. V: fly ash, L: limestone powder, and GGBS: ground granulated blast furnace slag.

Hydration and chloride binding were studied by thermal test (TG/DTG/DTA) using Derivatograph Q-1500 D and X-ray diffraction (XRD) using Philips PW 3710 diffractometer. Derivatograph Q-1500 D is simultaneous thermoanalytical test method where thermogravimetric (TG) and differential thermoanalytical (DTA) curves are produced during the same measurement. The 1st derivative of thermogravimetric curve (DTG) is also obtained. Reference material was alumina; heating rate was 10°C/min up to about 1000°C. Tests were carried out in air atmosphere. The other method was X-ray diffraction (XRD). X-ray powder diffraction analyses were carried out by a Philips PW 3710 diffractometer. Parameter values of measurements were as follows: generator tension was 40 kV, generator current was 30 mA, and tube anode was Cu (long fine focus), with monochromator. For the analysis the Total Access Diffraction Database PDF-2 (PLUS 42) software was used. Investigation of both thermoanalytical and powder X-ray diffraction measurements made it possible to carry out detailed analysis of phase modifications.

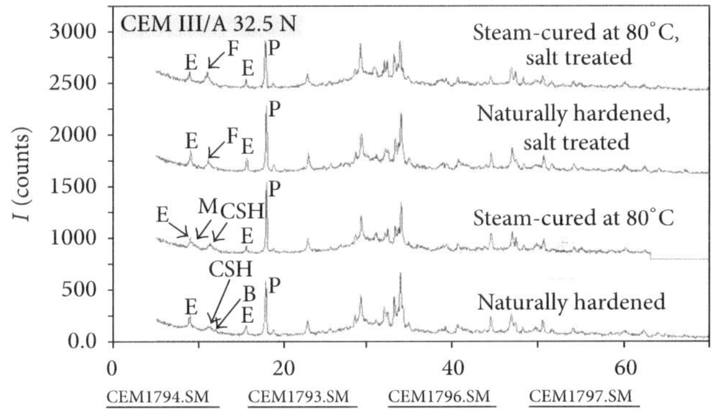

FIGURE 1: Influence of steam curing and salt-treatment on cement hydration, X-ray patterns of CEM III/A 32,5 N at the age of 180 days. *Note.* E: ettringite; F: Friedel's salt; M: monosulphate; P: portlandite; B: Brownmillerite; CSH: calcium-silicate-hydrate.

2.3. Results and Discussion

2.3.1. Results of XRD Analyses. X-ray patterns of salt-treated cement paste samples indicated the formation of Friedel's salt (calcium-aluminate-chloro-hydrate, $C_3A{\cdot}CaCl_2{\cdot}H_{10}$). The other Cl^--containing AFm phase, Kuzel's salt ($C_3A{\cdot}0{,}5CaSO_4 {\cdot}0{,}5CaCl_2{\cdot}H_{12}$), did not appear in the salt-treated cement paste samples [29, 30]. This statement is in agreement with the results of Balonis et al. [20]. Figure 1 shows the XRD patterns of CEM III/A 32,5 N at the age of 180 days. The absolute intensity of Friedel's salt in the steam-cured sample of CEM III/A 32,5 N was 35% higher (I_{abs} = 131 counts) comparing with the absolute intensity of non-steam-cured sample (I_{abs} = 97 counts). Monosulphate did not appear in the non-steam-cured samples. X-ray patterns indicated formation of monosulphate (AFm, $C_3A{\cdot}CaSO_4{\cdot}H_{12}$) parallel to ettringite formation (AFt, $C_3A{\cdot}3CaSO_4{\cdot}H_{32}$) in the steam-cured samples. The intensity of ettringite formation was higher in the salt-treated samples compared to the samples kept without salt-treatment. This observation can be explained by the reaction of sulphate ions released from the transformation *monosulphate → Friedel's salt*. The reaction of sulphate ions and the aluminate-ferrite mono- (AFm-) phases lead to the secondary ettringite formation observed here. The competitive reactions of AFm phase with chloride ion or with sulphate ion were observed. Higher absolute intensity of Friedel's salt was observed in all steam-cured samples than those of non-steam-cured samples of cements.

2.3.2. Results of TG/DTG/DTA Analyses. Figure 2 shows the first derivatives of thermogravimetric curves (DTG curves) in case of non-salt-treated (non-steam-cured, age: 90 days) and salt-treated samples steam-cured at 80°C at the age of 180 days. The following thermogravimetric changes were observed: (1) loss of humidity and dehydration of ettringite; (2) the second step of dehydration of Friedel's salt; (3) dehydration of portlandite; (4) decomposition of carbonates and CSH. The most remarkable difference between the results of thermal tests of salt-treated and non-salt-treated samples is the presence of a peak of DTG at approx. 360°C (salt-treated samples), which can be attributed to the formation of Friedel's salt (6 moles of H_2O of the total 10 mole H_2O). 40% of the water content (4 moles of H_2O) of Friedel's salt is lost below 200°C. The chemically bound chloride content was calculated from the second step of dehydration of Friedel's salt. Six moles of water represented 1 mole of Friedel's salt. The stoichiometric factor calculating the amount of Friedel's salt is f_{Fs} = 5,2. The chloride content in Friedel's salt is 12,6%.

Figure 3 shows the diagram of the chemically bound chloride content of salt-treated samples related to the mass of samples without the ignition loss. Test results indicated that steam-cured cements can bind higher amount of chloride ions than naturally hardened ones. Chloride ion binding capacity of tested cements in decreasing sequence at the age of tests (90 days or 180 days) was the following: (1) CEM III/B 32,5 (blast furnace cement); (2) CEM III/A 32,5 (blast furnace cement); (3) CEM II/A 32,5 (Portland-slag cement); (4) CEM

TABLE 2: Chemical composition of tested cements (m%) (data provided by Duna-Dráva Cement Ltd., HeidelbergCement Group).

	Chemical analysis									Free lime	Insoluble residue	Blaine surface
	LOI (%)	SiO_2 (%)	Al_2O_3 (%)	Fe_2O_3 (%)	CaO (%)	MgO (%)	SO_3 (%)	K_2O (%)	Cl (%)	(%)	(%)	(cm^2/g)
CEM I 42,5 N	2.64	20.11	5.15	3.34	62.82	2.48	2.99	0.05	0.015	0.66	0.33	3337
CEM II/B-M (V-L) 32,5 R	5.43	24.82	7.41	5.04	53.57	2.42	3.03	0.67	0.011	0.52	9.76	4152
CEM II/B-S 32,5 R	2.48	23.93	5.83	2.56	56.92	4.2	2.99	0.53	0.014	0.48	0.44	3594
CEM III/A 32,5 N	2.313	25.54	6.12	2.073	52.2	5.258	3.107	0.517	0.015	0.48	0.57	3505.4
CEM III/B 32,5 N-S	1.157	29.91	6.69	1.518	46.97	7.202	3.36	0.533	0.010	0.26	0.6	4142.8

Table 3: Experimental programme.

Series	Curing	Exposure
1	Natural hardening at 22°C and 100% r.h.	—
2	Natural hardening at 22°C and 100% r.h.	Salt-treatment
3	Steam curing at 80°C and 100% r.h.	—
4	Steam curing at 80°C and 100% r.h.	Salt-treatment

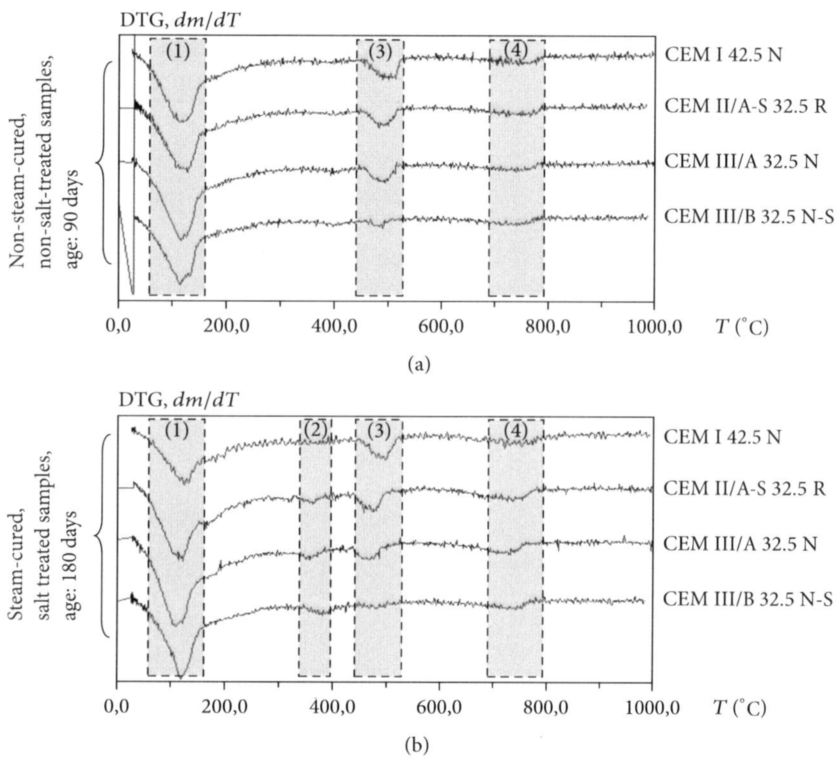

Figure 2: DTG curves of cement paste samples with increasing content of GGBS: (a) non-steam-cured and non-salt-treated samples at the age of 90 days; (b) steam-cured and salt-treated samples at the age of 180 days (salt-treatment: between the ages of 28–38 days). *Note.* (1) Loss of humidity and dehydration of ettringite; (2) the second step of dehydration of Friedel's salt; (3) dehydration of portlandite; (4) decomposition of carbonates and CSH.

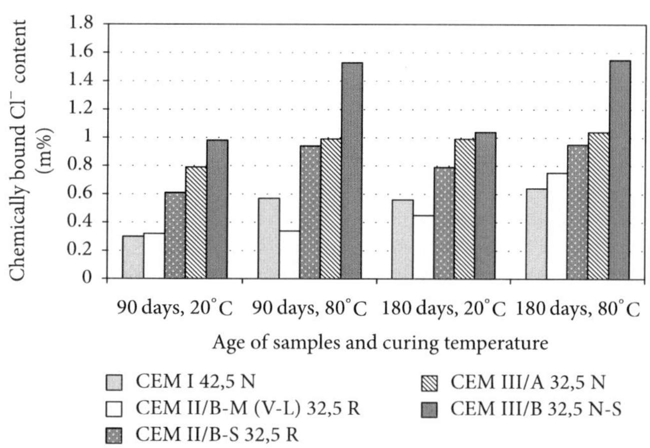

Figure 3: Amount of chemically bound Cl^- ions (m%) calculated from the second step of dehydration of Friedel's salt, related to the mass of samples without ignition loss as functions of ages and type of curing (cement paste samples were naturally hardened or steam-cured at 80°C).

Table 4: Standard composition of the tested cements according to EN 197-1:2000, m%.

Cement types	Constituents in standard composition of the tested cements, m%			
	Main			Minor
	Clinkers	GGBS	Fly ash	Gypsum
CEM I 42,5 N	95–100	0	0	5
CEM II/A-S 42,5 N	85–94	15 (6–20)	0	5
CEM II/B-S 42,5 R	65–79	20 (21–35)	0	5
CEM III/A 32,5 N	35–64	60 (36–65)	0	5
CEM III/B 32,5 N-S	20–34	75 (66–80)	0	5
CEM V/A (S-V) 32,5 N	40–64	30 (18–30)	30 (18–30)	5

II/B-M (V-L) 32,5 (Portland composite cement); (5) CEM I 42,5 (ordinary Portland cement).

In a previous study the chemically bound chloride content was calculated by indirect way from the total and water soluble chloride content. Both the total and water soluble chloride content were measured with analytical methods [29]. They found that blast furnace slag cement pastes have higher chloride ion binding capacity compared with the Portland cement control. This conclusion is in accordance with our test results obtained from the thermogravimetric mass loss of the second dehydration step of Friedel's salt using the thermogravimetric (TG) and derivative thermogravimetric (DTG) curves of thermal tests.

2.4. Summary of Observations for Chloride Binding Capacity. Purpose of this part of research was to study the effect of clinker substituting constituents (supplementary materials) and the influence of steam curing on chloride ion binding capacity of cements. Hydration as well as chloride ion binding mechanisms was investigated by TG/DTG/DTA and XRD. Five types of cements were selected with different amount of clinker substituting constituents. Salt-treatment of paste specimens was done between the ages of 28 and 38 days. Test results indicated that steam curing enhances the chloride binding capacity of cement paste samples comparing with the naturally hardened ones. Chloride ion binding capacity of tested cements in decreasing sequence at the age of 180 days was the following: (1) CEM III/B 32,5 N-S (with 62 m% GGBS); (2) CEM III/A 32,5 N (with 40 m% GGBS); (3) CEM II/B 32,5 R (with 26 m% GGBS); (4) CEM II/B-M (V-L) 32,5 R (with 18 m% V + 10 m% L); (5) CEM I 42,5 N (ordinary Portland cement, reference). The chemically bound chloride content increased with the increasing GGBS replacement level in case of slag cements. It was also experimentally shown that in the same type of cement (CEM II/B) the addition of ground granulated blast furnace slag (in Portland-slag cement, CEM II/B 32,5 R) is more effective in chloride ion binding as the fly ash and limestone in Portland composite cement (CEM II/B-M (V-L) 32,5 R). Limestone acts as an inert material for chloride binding.

3. Blended Cements for Enhanced Chloride Resistivity

3.1. Purpose of the Study. In this series of experiments different types of blended cements were used to study the influence of the clinker substituting constituents on chloride resistivity. The test variables were (i) types of cements, (ii) the water to cement ratios; (iii) presence or absence of air entraining agent. Another purpose was to study the influence of air entraining agent which can change the permeability of concrete. The test parameters were (i) the cement content in concrete mixtures (440 kg/m^3) and (ii) consistency of fresh concrete mixtures, while the studied characteristics were non-steady-state migration coefficients (D_{nssm}).

3.2. Experimental Studies

3.2.1. Tested Cement Types. In this study three Portland cements, two blast furnace cements, and one composite cement were selected with different amounts of GGBS (ground granulated blast furnace slag) (Table 4). Cements involved in our experiments are as follows: ordinary Portland cement (CEM I 42,5 N: reference cement), Portland-slag cements (CEM II/A-S 42,5 N, CEM II/B-S 42,5 R), blast furnace cements (CEM III/A-S 32,5 N, CEM III/B-S 32,5 N), and composite cement (CEM V/A (S-V) 32,5 N). In CEM V/A (S-V) 32,5 N composite cement the main constituents are GGBS and fly ash. According to the standard MSZ EN 197-1:2000 [35] main constituents are specially selected inorganic materials in a proportion exceeding 5% by mass related to the sum of all main and minor additional constituents. Amount of the main constituents different from clinkers in case of the cements involved in our study is summarized in Table 4. The composition of clinker minerals was the same in all types of cements as in CEM I 42,5 N ordinary Portland cement. All of the tested cements were standardised products of Duna-Dráva Cement Hungary Ltd. (HeidelbergCement Group). The chemical compositions of tested cements are given in the Table 5.

3.2.2. Concrete Mixtures. The amount of cement in the different concrete mixtures was the same (440 kg/m^3). This allowed us to make comparative study of chloride migration coefficients of concrete mixtures. Compositions of tested concrete mixes are given in Table 6. Amount of air entraining agent in the aerated mixtures was related to the mass of cement (0.35 m%). The consistency was kept at F5 [31] (according to its consistency, fresh concrete is subdivided into stiff (F1), plastic (F2), soft (F3), very soft (F4), free-flowing

TABLE 5: Chemical composition of tested cements (m%) (data provided by Duna-Dráva Cement Ltd., HeidelbergCement Group).

	Chemical analysis											
	LOI (%)	SiO_2 (%)	Al_2O_3 (%)	Fe_2O_3 (%)	CaO (%)	MgO (%)	SO_3 (%)	K_2O (%)	Cl (%)	Free lime (%)	Insoluble residue (%)	Blaine surface (cm^2/g)
CEM I 42,5 N	2.006	30.31	7,875	3.91	47,19	4.463	3.303	0.644	0.010	0.51	7.59	4052.7
CEM II/A-S 42,5 N	2.346	21,46	5.576	2.968	58.454	3.129	2.995	0.504	0.019	0,77	0.5	3650.1
CEM II/B-S 42,5 R	2,313	23.2	5,812	2.545	55,735	4.101	3.02	0.509	0.018	0.58	0,54	3520,4
CEM III/A 32,5 N	2.313	25.54	6.12	2.073	52,2	5,258	3.107	0.517	0.015	0.48	0.57	3505.4
CEM III/B 32,5 N-S	1,157	29.91	6.69	1.518	46.97	7.202	3.36	0.533	0.010	0.26	0.6	4142,8
CEM V/A (S-V) 32,5 N	2,006	30.31	7,875	3.91	47.19	4.463	3.303	0.644	0.010	0.51	7,59	4052.7

TABLE 6: Compositions of tested concrete mixes with water to cement ratios of 0.35, 0.45, and 0.55, with (AE) or without (NAE) air entraining agent.

	$w/c = 0.35$		$w/c = 0.45$		$w/c = 0.55$	
Mixture	NAE (kg/m^3)	AE (kg/m^3)	NAE (kg/m^3)	AE (kg/m^3)	NAE (kg/m^3)	AE (kg/m^3)
Cement	440	440	440	440	440	440
water ($w/c = 0.35$)	154	154	198	198	242	242
River sand 0-1 mm (25 m%)	451	451	425	424	408	396
River sand 0–4 mm (20 m%)	361	361	340	339	327	317
River gravel 4–8 mm (55 m%)	993	993	934	932	898	871
Superplasticizer (1.15 m%)	5.06	5.06	1.76	1.76	0	0
AE (0.35 m%)	—	1.54	—	1.54	—	1.54
Sum	2404.06	2405.60	2338.76	2336.30	2315.00	2267.54

(F5), and very free-flowing (F6) (EN 12350-5:2009 testing fresh concrete. Flow table test)).

3.2.3. Test Methods. The chloride resistivity was tested by the CTH, rapid electrical, test method described by Tang and Nilsson [32]. The non-steady-state migration coefficient (D_{nssm}) was determined on water treated concrete specimens. The specimens were kept in water bath until the age of testing. Then the specimens were cut into discs at the age of 27 days. The test started at the age of 28 days. Test duration was uniformly 8 hours and the applied DC was 30 V. The chloride penetration depths were determined by colorimetric method with application of 0.1 n $AgNO_3$ solution on the split specimens. D_{nssm} was calculated from the test parameters and variables [33, 34]. The results of the rapid electric migration test are also known as $D_{RCM,28}$. After the determination of the average chloride penetration depths, the discs were ground mm by mm with the use of profile grinder. Dry powder samples of the layers were collected to the further analysis. Wet analytical method (Mohr method) was used to determine the total chloride contents of the ground layers.

3.3. Results and Discussion. The non-steady-state migration coefficients (D_{nssm}) of the studied concrete mixtures are shown in Figures 4–9.

In Figure 10 typical test results of chloride profiles are shown (example obtained from concrete samples made of CEM II/A 42,5 N Portland-slag cement) with different water to cement ratios, with or without air entraining agents. Test results indicated that chloride concentration in concrete is higher by using air entraining agents considering the same water to cement ratio and the same depth. Another important observation is the higher the water to cement ratio of the concrete specimen, the higher the total chloride concentration in the same depth.

Based on the test results the following observations can be done:

(i) Increase in water to cement ratio results in increase in depth of chloride penetration considering the same cement.

(ii) Chloride penetration depths decrease by increase in slag content of cements considering the same water to cement ratio.

(iii) Observed depths of chloride penetration and those of chloride migration coefficients (D_{nssm}) resulted in the following sequence of efficiency for our tested cements (from the best): CEM III/B 32,5 N > CEM V/A (S-V) 32,5 N > CEM III/A 32,5 N > CEM II/B-S 42,5 R > CEM II/A-S 42,5 N > CEM I 42,5 N.

(iv) Chloride penetration depths decrease by increasing the slag content of cement.

(v) Chloride migration coefficients are lower for blended cements compared to CEM I 42,5 N ordinary Portland cement.

(vi) Chloride concentration in concrete is higher by using air entraining agents considering the same water to cement ratio and at the same depth.

(vii) Our experimental results indicated the lowest chloride diffusion coefficient, hence the highest chloride resistivity for concrete of cement CEM III/B 32,5 N among our tested cements.

4. Conclusions

In order to support durability design present study concentrated on control of chlorides in concrete. Results of two series of extensive experimental studies are presented concerning (i) chloride binding and (ii) chloride penetration influenced

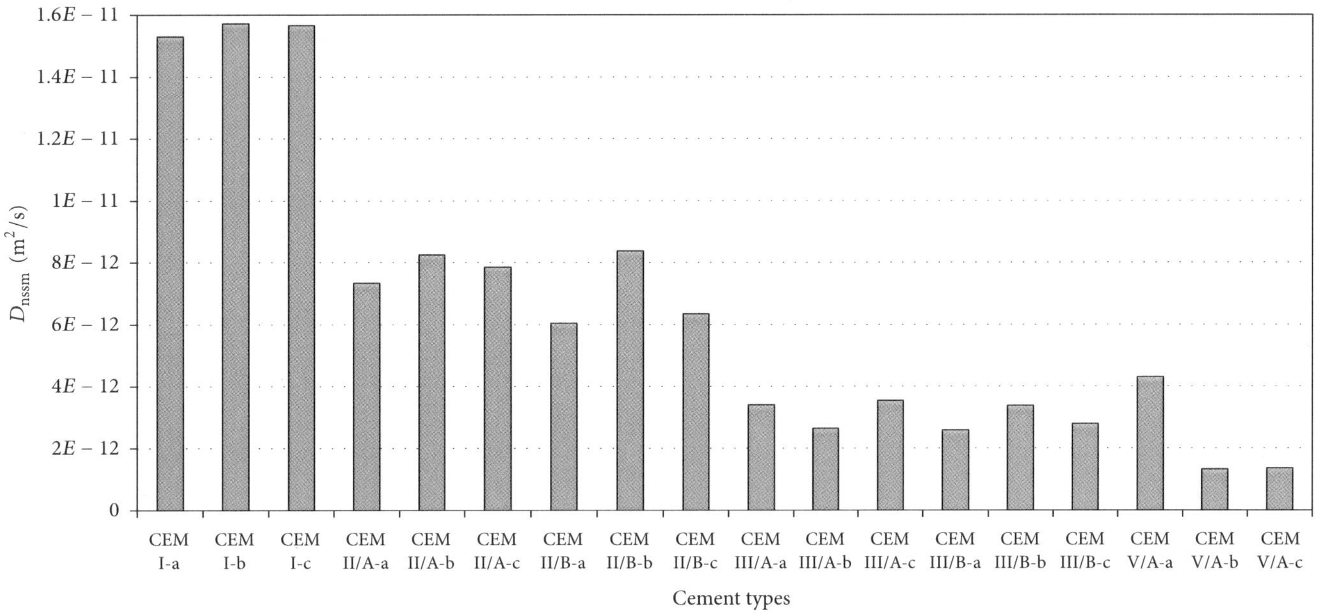

FIGURE 4: Non-steady-state migration coefficients (D_{nssm}) of the studied concrete mixtures as a function of cement types, *water to cement ratio of 0.35, and concrete mixtures made with air entraining agents.*

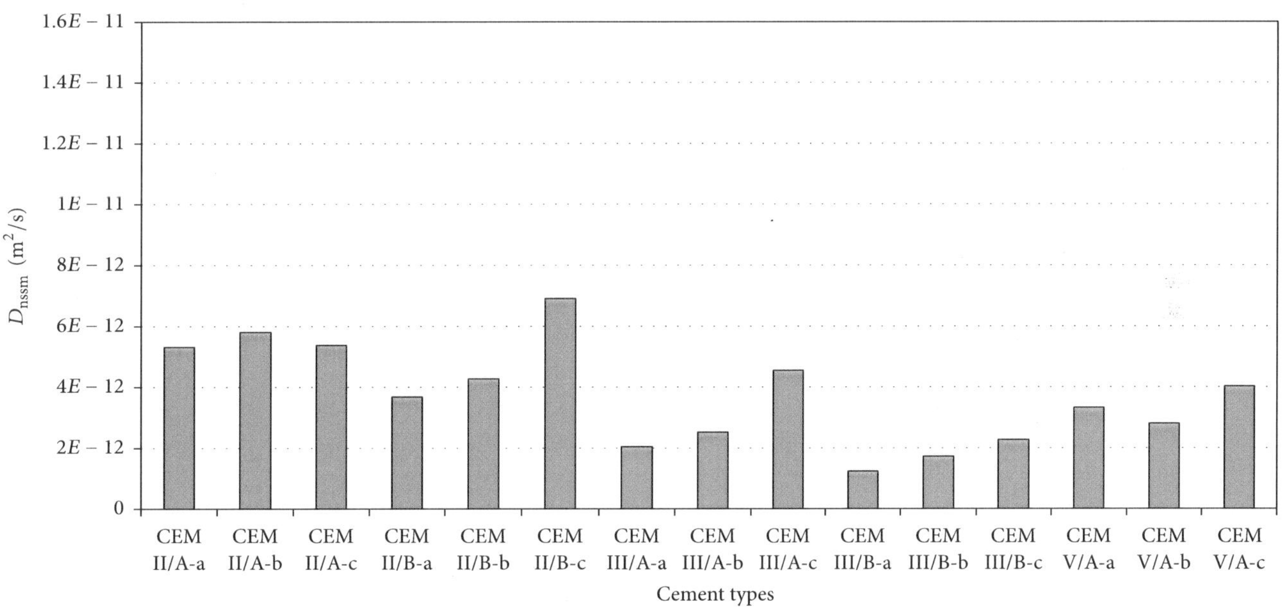

FIGURE 5: Non-steady-state migration coefficients (D_{nssm}) of the studied concrete mixtures as a function of cement types, *water to cement ratio of 0.35, and concrete mixtures made without air entraining agents.*

by type of binary or ternary blended cements. The chloride binding capacities of the different cements were studied on cement paste specimens while the non-steady-state migration coefficients were tested on high-strength concrete, with or without application of air entraining agent for different water to cement ratios. Test methods to determine chloride binding capacities included X-ray powder diffraction (XRD) and thermal analyses (TG/DTG/DTA). Results of chloride binding is obtained from the thermogravimetric (TG) mass loss of the second dehydration step of Friedel's salt (TG). Test method to determine the chloride resistivity of concrete was rapid chloride migration test (CTH-test).

(i) Chloride Binding

(1) Cements with increasing slag contents have improving capacity for chloride binding. In our experiments the chemically bound chloride content increased with the increasing GGBS replacement level in case of slag cements.

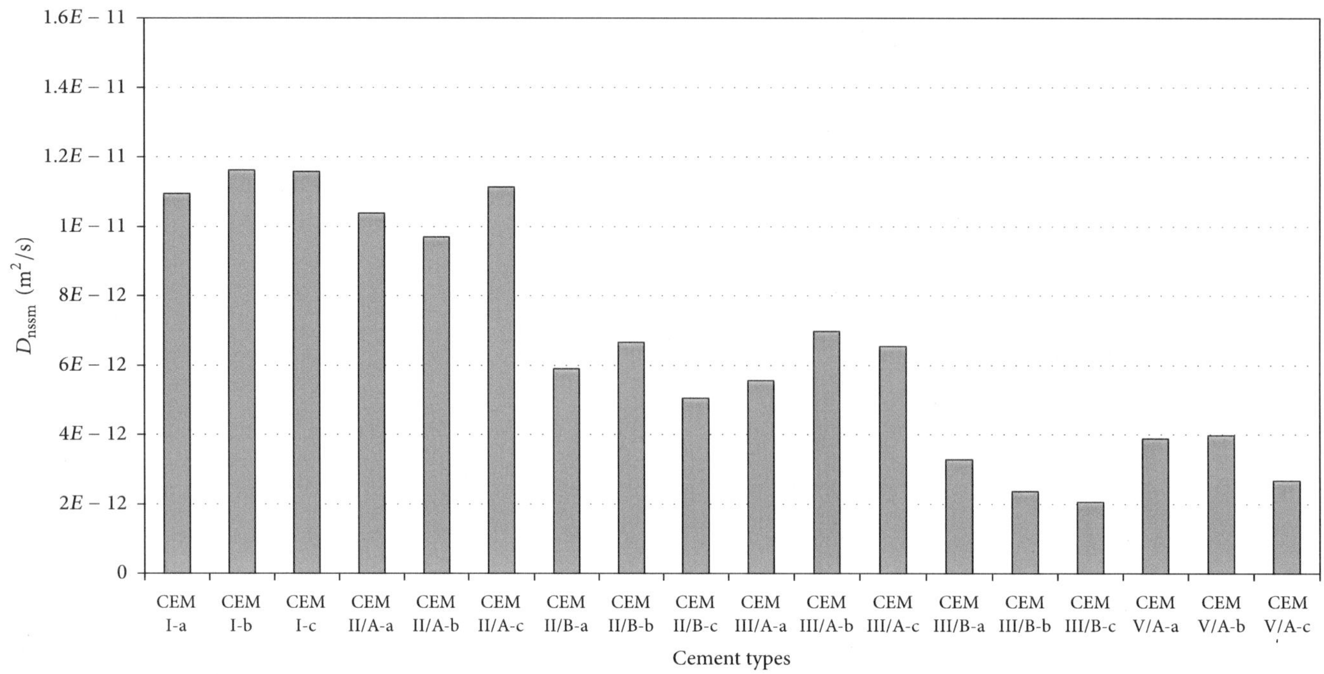

FIGURE 6: Non-steady-state migration coefficients (D_{nssm}) of the studied concrete mixtures as a function of cement types, *water to cement ratio of 0.45, and concrete mixtures made with air entraining agents.*

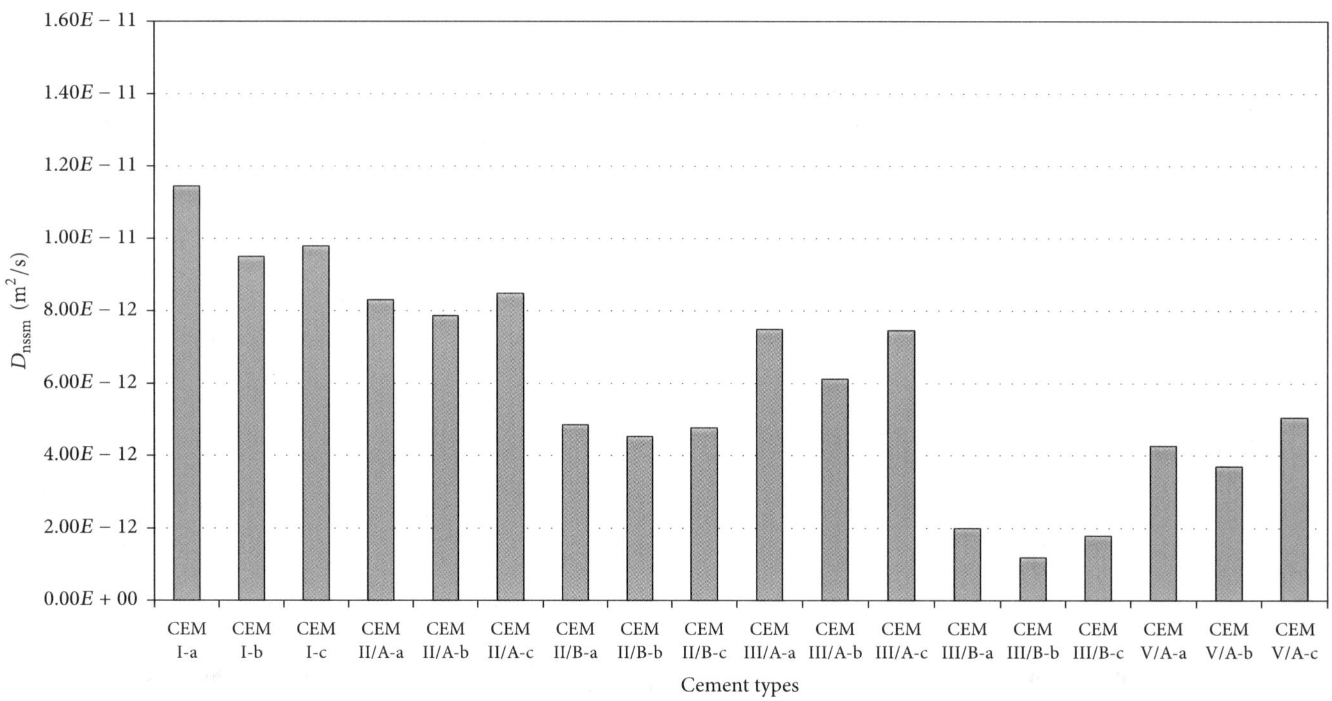

FIGURE 7: Non-steady-state migration coefficients (D_{nssm}) of the studied concrete mixtures as a function of cement types, *water to cement ratio of 0.45, concrete mixtures made without air entraining agents.*

(2) It was also experimentally shown that in the same type of cement (CEM II/B) the addition of ground granulated blast furnace slag (in Portland-slag cement, CEM II/B 32,5 R) is more effective in chloride ion binding than in Portland composite cement with fly ash and limestone (CEM II/B-M (V-L) 32,5 R). Limestone acts as an inert material for chloride binding.

(3) During steam curing AFm aluminate hydrates also formed in addition to AFt phase (ettringite). Formation of monosulphate is accelerated by the higher temperature of steam curing. Higher amounts of

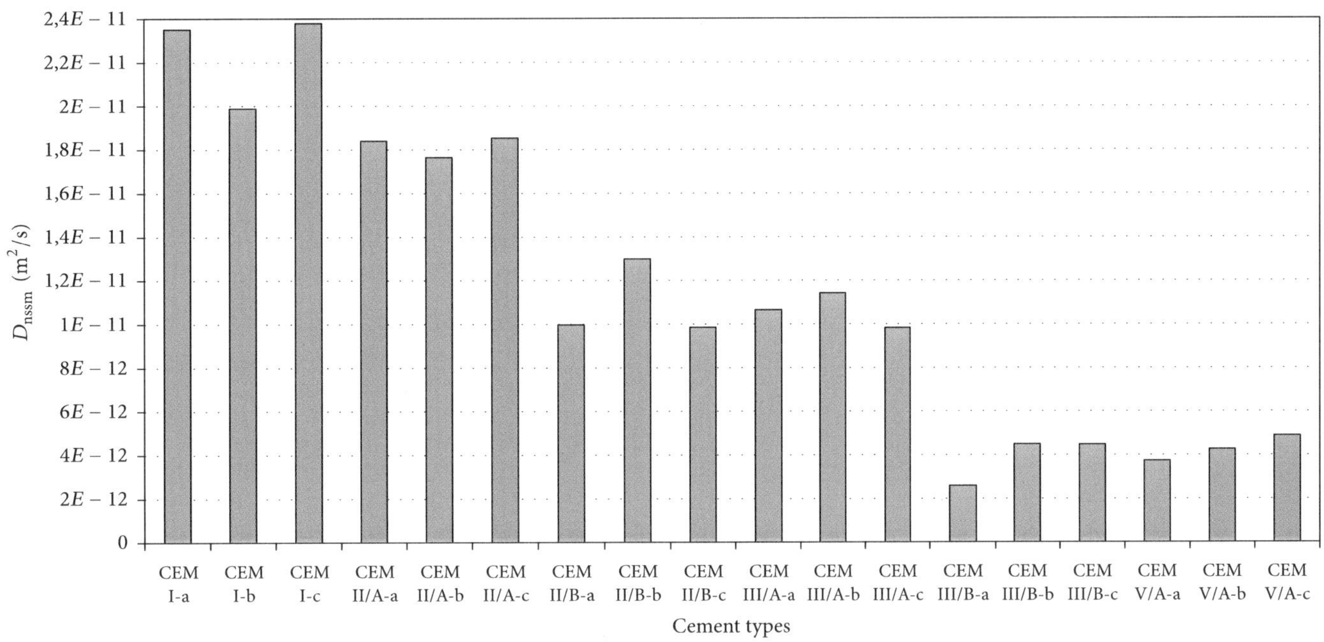

FIGURE 8: Non-steady-state migration coefficients (D_{nssm}) of the studied concrete mixtures as a function of cement types, *water to cement ratio of 0.55, and concrete mixtures made with air entraining agents.*

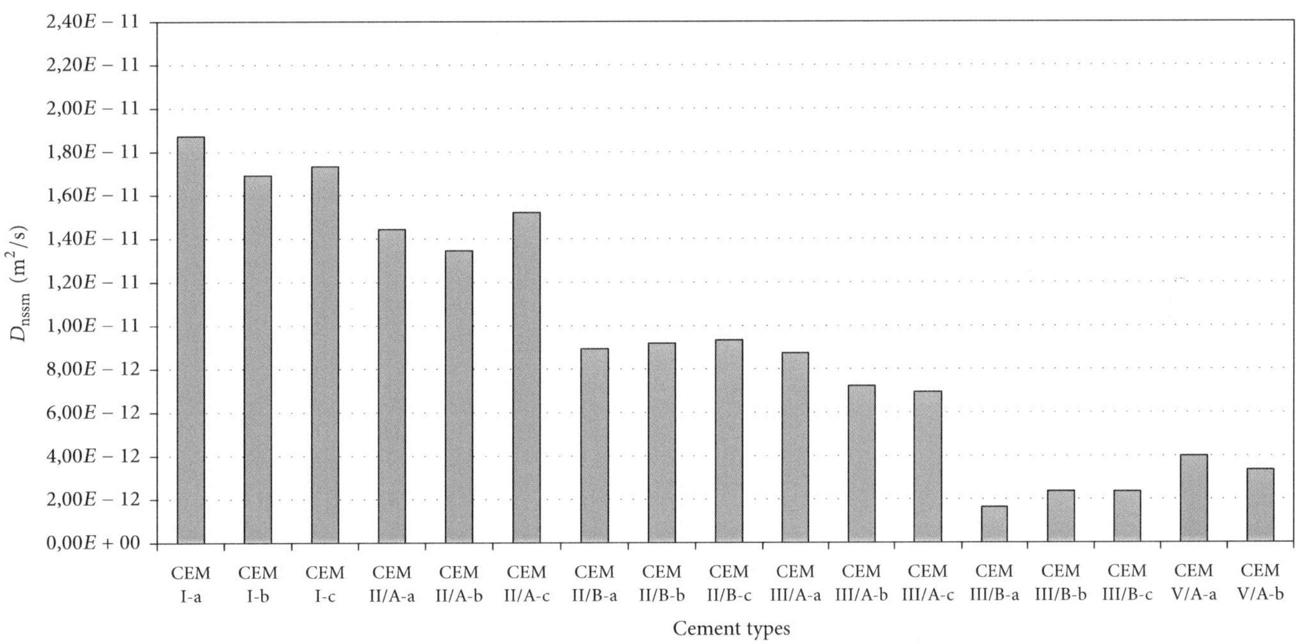

FIGURE 9: Non-steady-state migration coefficients (D_{nssm}) of the studied concrete mixtures as a function of cement types, *water to cement ratio of 0.55, and concrete mixtures made without air entraining agents.*

monosulphate are available in steam-cured cement paste samples compared to non-steam-cured samples. In presence of chloride ions, sulphate ions released by the chloride binding promote the formation of the secondary ettringite. Steam curing (used for precast elements) contributes to improved chloride binding of cements.

(4) The chloride ion binding capacities of cements improved by the age of samples.

(ii) Chloride Resistivity. Test results indicated that the increasing substitution of clinkers by GGBS improves the chloride resistivity in concrete made with the same water to cement ratio. The application of air entraining agent increases the values of D_{nssm} (reducing the chloride resistivity), because the permeability increases by the increased porosity.

(5) Chloride migration coefficients decrease with increase in slag content of cements; chloride penetration depths decrease as well by increase in slag content

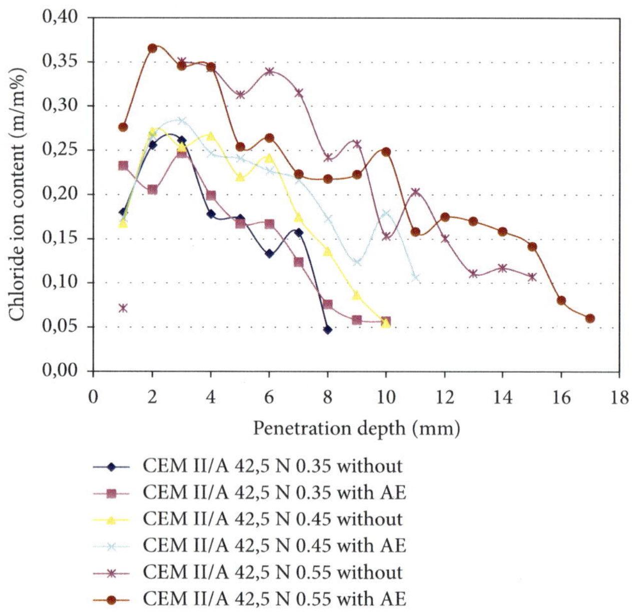

FIGURE 10: Chloride profiles with Portland-slag cement CEM II/A-S 42,5 N with different water to cement ratios, with or without air entraining agents.

of cements considering the same water to cement ratio.

(6) Application of air entrainment negatively influences the chloride migration coefficients; that is, chloride penetration depths are increasing in case of air entrained concrete.

(7) Chloride concentration in concrete is higher by using air entraining agents considering the same water to cement ratio and the same depth of specimens.

(8) Observed depths of chloride penetration and those of chloride migration coefficients resulted in the following sequence of efficiency for the tested cements (from the best): CEM III/B 32,5 N > CEM V/A (S-V) 32,5 N > CEM III/A 32,5 N > CEM II/B-S 42,5 R > CEM II/A-S 42,5 N > CEM I 42,5 N.

(9) Our experimental results indicated the lowest chloride migration coefficient, hence the highest chloride resistivity for cement CEM III/B 32,5 N among our tested cements.

Our observations give guidance to durability evaluations as well as mix design of concrete.

Conflicts of Interest

The authors declare that there are no conflicts of interest regarding the publication of this article.

Acknowledgments

The authors acknowledge the support by the Hungarian Research Grant NVKP_16-1-0019 "Development of Concrete Products with Improved Resistance to Chemical Corrosion, Fire or Freeze-Thaw."

References

[1] L-O. Nilsson, *Prediction Models for Chloride Ingress And Corrosion Initiation in Concrete Structures*, Chalmers University of Technology, 2001.

[2] K. Tuutti, "Corrosion of steel in concrete," CBI Research Report 4:82, Swedish Cement and Concrete Research Institute, Stockholm, Sweden, 1982.

[3] O. A. Kayyali and M. N. Haque, "Cl^-/OH^- ratio in chloride-contaminated concrete - a most important criterion," *Magazine of Concrete Research*, vol. 47, no. 172, pp. 235–242, 1995.

[4] J. M. Frederiksen, L-O. Nilsson, E. Poulsen, P. Tang L, and A. Andersen, "HETEK, a system for estimation of chloride ingress into concrete, theoretical background. the danish road directorate," The Danish Road Directorate 83, HETEK, 1997.

[5] C. Alonso, C. Andrade, M. Castellote, and P. Castro, "Chloride threshold values to depassivate reinforcing bars embedded in a standardized OPC mortar," *Cement and Concrete Research*, vol. 30, no. 7, pp. 1047–1055, 2000.

[6] A. Neville, "Chloride attack of reinforced concrete: an overview," *Materials and Structures*, vol. 28, no. 2, pp. 63–70, 1995.

[7] S. Matthews, *Design of Durable Concrete Structures*, vol. ISBN-13:978-1848061750, IHS BRE Press, 2014.

[8] fib, "Model Code for Concrete Structures," Wiley, 2010.

[9] fib, "Benchmarking of deemed-to-satisfy provisions in standards," fib Bulletin 76 ISBN 978-2-88394-116-8, Lausanne, Switzerland, 2016.

[10] P. M. Friedel, "Sur un chloro-aluminate de calcium hydraté se maclant par compression," *Bulletin de la Société Française de Minéralogie*, vol. 19, pp. 122–136, 1897.

[11] M. V. A. Florea and H. J. H. Brouwers, "Chloride binding related to hydration products: Part I: Ordinary Portland Cement," *Cement and Concrete Research*, vol. 42, no. 2, pp. 282–290, 2012.

[12] K. Larsen, *Chloride binding in concrete-effect of surrounding environment and concrete composition [Ph.D. thesis]*, The Norwegian University of Science and Technology, 1998.

[13] J. J. Beaudoin, V. S. Ramachandran, and R. F. Feldman, "Interaction of chloride and CSH," *Cement and Concrete Research*, vol. 20, no. 6, pp. 875–883, 1990.

[14] S. Nagataki, N. Otsuki, T.-H. Wee, and K. Nakashita, "Condensation of chloride ion in hardened cement matrix materials and on embedded steel bars," *ACI Materials Journal*, vol. 90, no. 4, pp. 323–332, 1993.

[15] O. Wowra and M. J. Setzer, "About the interaction of chloride and hardened cement paste," in *Proceedings of the Proceedings of 2nd International RILEM Workshop "Testing and Modelling Chloride Ingress into Concrete"*, C. Andrade and J. Kropp, Eds., pp. 3–12, Paris, France, 2000.

[16] A. K. Suryavanshi, J. D. Scantlebury, and S. B. Lyon, "Mechanism of Friedel's salt formation in cements rich in tri-calcium aluminate," *Cement and Concrete Research*, vol. 26, no. 5, pp. 717–727, 1996.

[17] H. Zibara, *Binding of external chloride by cement pastes [Ph.D. thesis]*, University of Toronto, Department of Building Materials, 2001.

[18] F. P. Glasser, A. Kindness, and S. A. Stronach, "Stability and solubility relationships in AFm phases: part I. chloride, sulfate

and hydroxide," *Cement and Concrete Research*, vol. 29, no. 6, pp. 861–866, 1999.

[19] H. Justnes, "A review of chloride binding in cementitious systemes," *Nordic Concrete Research*, vol. 21, pp. 1–6, 1996.

[20] M. Balonis, B. Lothenbach, G. Le Saout, and F. P. Glasser, "Impact of chloride on the mineralogy of hydrated portland cement systems," *Cement and Concrete Research*, vol. 40, no. 7, pp. 1009–1022, 2010.

[21] K. Kopecskó, *Chloride ion binding capacity of clinker minerals and cements influenced by steam curing [Ph.D. thesis]*, in Hungarian, Budapest University of Technology and Economics, 2006.

[22] J. Hill and J. H. Sharp, "The mineralogy and microstructure of three composite cements with high replacement levels," *Cement and Concrete Composites*, vol. 24, no. 2, pp. 191–199, 2002.

[23] R. D. Hooton and M. P. Titherington, "Chloride resistance of high-performance concretes subjected to accelerated curing," *Cement and Concrete Research*, vol. 34, no. 9, pp. 1561–1567, 2004.

[24] R. K. Dhir, M. A. K. El-Mohr, and T. D. Dyer, "Chloride binding in GGBS concrete," *Cement and Concrete Research*, vol. 26, no. 12, pp. 1767–1773, 1996.

[25] R. Luo, Y. Cai, C. Wang, and X. Huang, "Study of chloride binding and diffusion in GGBS concrete," *Cement and Concrete Research*, vol. 33, no. 1, pp. 1–7, 2003.

[26] K. Kopecskó, "Presence of chloride ions in concrete. MŰSZAKI SZEMLE (Technical Review)," Hungarian Technical Scientific Society of Transylvania, 2009.

[27] "Cement composition, specifications and conformity criteria for common cements," MSZ EN 197-1:2000.

[28] "Methods of testing cement. Determination of setting times and soundness," MSZ EN 196-3:2005.

[29] H. J. Kuzel, "Röntgenuntersuchung im System $3CaO{\cdot}Al_2O_3{\cdot}CaSO_4{\cdot}nH_2O$ - $3CaO{\cdot}Al_2O_3{\cdot}$ $CaCl_2{\cdot}nH_2O$ – H_2O"," *Neues Jahrbuch Mineralog. Monatsh*, pp. 193–200, 1966.

[30] H. F. W. Taylor, *Cement Chemistry*, Academic Press, New York, NY, USA, 1990.

[31] "Testing fresh concrete. Flow table test," MSZ EN 12350-5:2009.

[32] L. Tang and L. O. Nilsson, "Chloride diffusivity in high strength concrete," *Nordic Concrete Research*, vol. 11, pp. 162–170, 1992.

[33] T. Luping, L. Nilsson, and P. Basheer, *Resistance of Concrete to Chloride Ingress: Testing and Modelling*, ISBN: 978-0-203-88241-2, CRC Press, 2011.

[34] "NT Build 492," Concrete, mortar and cement-based repair materials: chloride migration coefficient from non-steady-state migration experiments.

3

Flexural and Thermal Properties of Novel Energy Conservation Slotted Reinforced Concrete Beams

Gao Ma,[1] Liang Huang,[1] Libo Yan,[2,3] Hui Wang,[1] and Peng Yin[1]

[1]*College of Civil Engineering, Hunan University, Changsha 410082, China*
[2]*Department of Organic and Wood-Based Construction Materials, Technical University of Braunschweig, Hopfengarten 20, 38102 Braunschweig, Germany*
[3]*Centre for Light and Environmentally-Friendly Structures, Fraunhofer Wilhelm-Klauditz-Institut WKI, Bienroder Weg 54E, 38108 Braunschweig, Germany*

Correspondence should be addressed to Liang Huang; lianghuanghnu@gmail.com

Academic Editor: João M. P. Q. Delgado

Conventional solid reinforced concrete (RC) beams were modified to slotted beams for consideration as thermal insulation structural components. The slotted beam consisted of an outer and an inner beam, respectively, with a slot located near the middle of the beam along its width direction for filling thermal insulation material. Flexural and thermal behavior of the slotted beams were investigated. Three RC reference solid beams and six slotted beams were fabricated and tested under four-point bending tests. The test results indicated that the failure mode of both slotted beams and the solid beams was flexural failure. However, the damage process of the slotted beams was different from that of the solid beams at the final loading stage. The moment curvature analysis indicated that the tensile reinforcement ratio of the outer and inner beams had an important effect on the flexural behavior, especially the ductility of the slotted beams. Thermal study indicated that the heat transfer coefficient of the slotted beam was greatly reduced and the thermal inertia factor increased a lot, compared with the solid beam. In addition, FE simulation results showed that a new frame structure using slotted beams exhibited obvious and attractive thermal insulation property.

1. Introduction

Building energy conservation is one of the most important energy-saving policies in developing and developed countries [1]. For a building, thermal bridge refers to a component with a relatively higher heart transfer than the surrounding parts, which leads to a significant reduction in the thermal insulation of the building. Thermal bridges generally exist in building structures [2, 3], such as frame structures, shear wall structures, and masonry structures. For a reinforced concrete (RC) frame structure, the beams and columns all contribute to thermal bridges. Generation and characteristics of thermal bridge are related to the material properties of the structure and the layout of the structure. In a building, its walls, windows, beams, and columns are the main components to generate thermal bridges. If these components can be designed to have an enhanced thermal insulation and reduced heat conduction performance, the issue of thermal bridges in building structures can be solved.

In general, the popular external wall thermal insulation can be classified into exterior insulation and inner insulation of the external wall. Inner insulation of external wall is a technique which uses thermal insulating layer on the inner surface of the external wall of a structure. Because the insulating layer hides inside the walls, this technique reduces the usable area of the structure, affects the redecoration, and generates mildew and moisture condensation in the walls. To date, the technique of exterior insulation of the external wall is still not so mature; for instance, the construction is difficult and the energy-saving investment is high. Meanwhile, the outside insulation layer is under the risk of fire disaster, is easy to drop off and gets seepage in rainy weather; it is difficult to paste ceramic tile or hang marble on the thermal insulating layer and the durability of the exterior insulation system is reduced because it is directly exposed to the outdoor atmosphere. To solve these problems, self-thermal insulation building blocks such as thermal insulation concrete block [4],

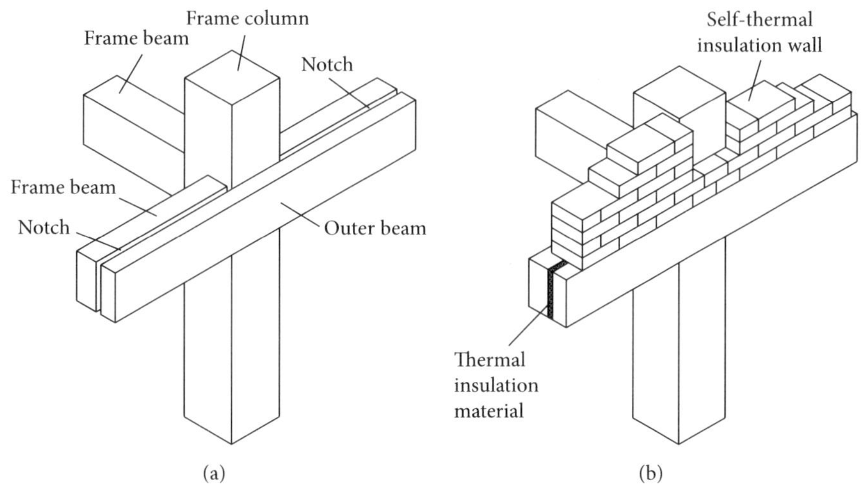

FIGURE 1: Sketch of energy conservation slotted beam-wall system: (a) constructional details of the slotted beam and (b) the view after self-thermal insulation wall constructed.

hollow block [5], and vacuum insulation window [6–8] have been developed and adopted in practical application.

For RC frame structures, the use of self-thermal insulation building blocks and vacuum insulation windows prevents the walls and the windows from generation of thermal bridges, and hence thermal bridges mainly exist in beams and columns. This paper presents a new type of energy conservation slotted beam-wall system, as shown in Figure 1. Conventional reinforced concrete (RC) beams are modified into slotted beams, and the gap between the outer and inner beams is used to fill with thermal insulation materials. Self-thermal insulation wall is constructed on the slotted beams. It is obvious that this type of slotted beam-wall structure possesses a favourable thermal insulation property, because the thermal bridges at the beams and columns can be reduced effectively. Meanwhile, this new energy conservation wall system is convenient in construction.

In this paper, the feasibility of slotted RC beams as thermal insulation structural components was investigated. Three reference solid beams and six slotted beams were fabricated and tested under four-point bending test. The flexural behavior of the slotted beams was investigated in detail, including failure mode, load-deflection response, and moment curvature analysis. The effects of reinforcement ratio and width ratio of the inner beam to the outer beam on the flexural behavior of slotted beams were investigated, and then a design method for slotted beam was proposed. The thermal properties of the slotted and solid RC beams were analysed and compared. In addition, the thermal performance of the slotted beam-wall system was investigated through finite element method (FEM) analysis.

2. Experimental

2.1. Fabrication of Beams. In order to study the flexural behavior of the slotted beams, a total of 9 beams were designed and constructed, including 6 slotted beams and 3 reference solid beams. As displayed in Figure 2(a), the width of the slot is 30 mm and the distance from the slot to the left/right end is 150 mm. The total width of the outer and inner beam of a slotted beam is 250 mm, which is equal to the width of the solid beam. The depth and the length of the slotted beams are identical to those of the reference solid beams, with a depth of 250 mm and a length of 3000 mm. The inner and the outer beam were designated as "beam A" and "beam B", respectively (Figure 2(a)).

The reinforcement details of the tested beams are also shown in Figure 2. Detailed descriptions of the specimens are given in Table 1. Constructional reinforcement was used for the compressive region of the beam. The following aspects were considered in the tests: the reinforcement ratio of beam A and beam B and the width ratio (b_A/b_B). The amount of the tensile reinforcement, the compressive reinforcement, and the shear reinforcement of the slotted beam were designed to be close to those of the corresponding reference solid beam, in order to facilitate comparison of the results. In Table 1, SB denotes the slotted beams and RB denotes the reference solid beams. The number behind SB denotes the section type, reflecting the width ratio (b_A/b_B). The diameter of the longitudinal reinforcement rebar in beam B is given at the last part of the specimen denomination. For example, SB1-14 represents a slotted beam with section type 1, and the diameter of the rebar in beam B is 14 mm. The reference solid beam for SB1-14 and SB2-14 is RB-14.

The mix proportion of concrete is listed in Table 2. The concrete compressive strength of each specimen was determined from the compression test on 150 × 150 × 300 mm standard concrete prism. Deformed steel was used for longitudinal reinforcement and plain steel was used for hoop reinforcement. Mechanical properties of the rebars are given in Table 3.

2.2. Experimental Program. Figure 3(a) shows a slotted beam under four-point bending test. The locations of the two concentrated loads and the supports are given in the schematic view in Figure 3(b). Dial indicators with an accuracy of 0.01 mm were fixed at the bottom of the beams to record the deflection responses, as shown in Figure 3(a). For slotted

Table 1: Detailed descriptions of the specimens.

Specimen identifier	f_c (MPa)	Section $b \times h$ (mm × mm)	Reinforcement details (unit: mm)				Tensile reinforcement ratio ρ (%)		Shear reinforcement ratio ρ_v (%)		
			①	②	③	④					
RB-10	31.0	250 × 250	2B16	2B10	2B8	A8@100	1.0		0.4		
RB-12	28.8	250 × 250	2B16	2B12	2B8	A8@100	1.1		0.4		
RB-14	29.3	250 × 250	2B16	2B14	2B8	A8@100	1.3		0.4		
		$(b_A + b_B) \times 250$					Beam A ρ_A	Beam B ρ_B	Beam A $\rho_{v,A}$	Beam B $\rho_{v,B}$	Width ratio b_A/b_B [(1)]
SB1-14	29.0	(170 + 80) × 250	2B16	2B14	2B6	A6@100	1.1	1.8	0.3	0.7	2.1
SB2-10	29.7	(150 + 100) × 250	2B16	2B10	2B6	A6@100	1.2	0.7	0.4	0.6	1.5
SB3-12	31.3	(130 + 120) × 250	2B16	2B12	2B6	A6@100	1.4	0.9	0.4	0.5	1.1
SB1-12	29.7	(170 + 80) × 250	2B16	2B12	2B6	A6@100	1.1	1.3	0.3	0.7	2.1
SB2-14	29.0	(150 + 100) × 250	2B16	2B14	2B6	A6@100	1.2	1.4	0.4	0.6	1.5
SB3-10	28.9	(130 + 120) × 250	2B16	2B10	2B6	A6@100	1.4	0.6	0.4	0.5	1.1

Note: [(1)]the width ratio for slotted beam is the ratio of the width of beam A to that of beam B.

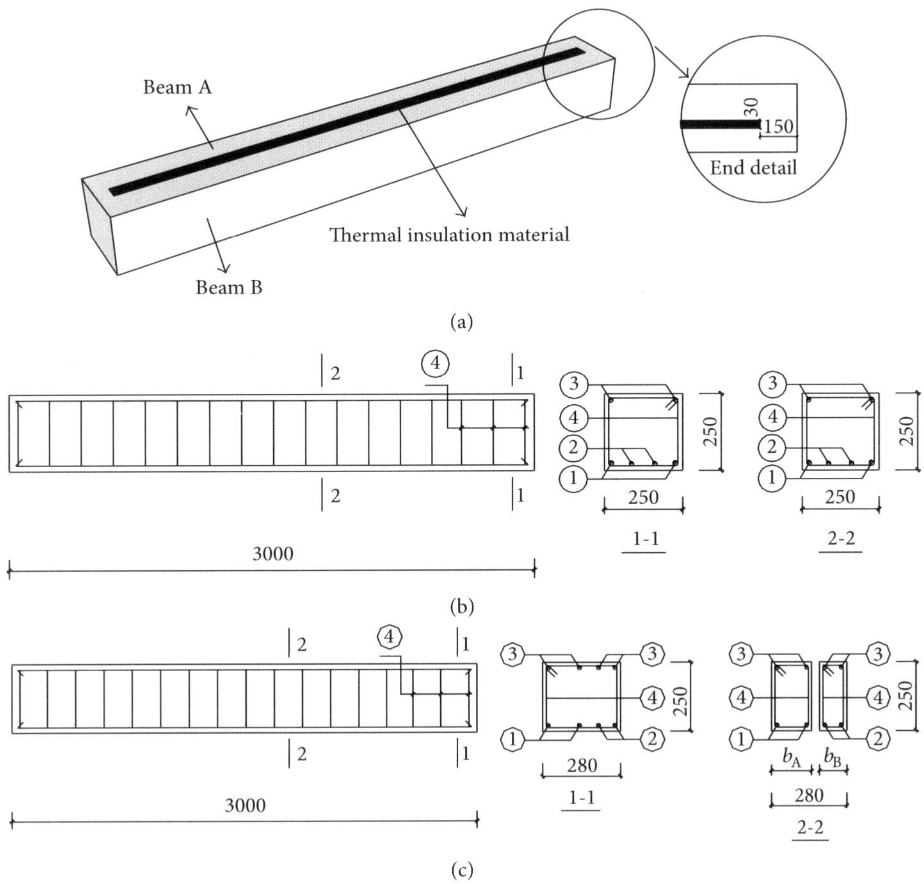

FIGURE 2: Specimen dimensions and reinforcement: (a) sketch of a slotted beam; (b) solid RC beam; and (c) slotted beam (units: mm).

TABLE 2: Concrete mix proportion.

Component	Water	Ordinary Portland cement	Sand	Coarse aggregate
Quantity (kg/m^3)	173	392	487	1249

TABLE 3: Mechanical properties of the reinforcement rebars.

Material type	Yield strength (MPa)	Ultimate strength (MPa)
B10	382	536
B12	376	556
B14	379	543
B16	385	549
A6	346	464
A8	352	478

beams, two dial indicators were used for each position to record the deflection of beam A and beam B, respectively. Dial indicators were also fixed at the left and right ends of each beam to record the displacement of the beam ends. Crack width observation instrument with an accuracy of 0.02 mm was carefully used to measure the maximum crack width of the beams. The spacing between the cracks and the crack number were also measured and marked at load intervals. Strain gauges on the tensile rebars and concrete surface at the middle span of the beams were recorded. During the test, the loads corresponding to the appearance of the first tensile crack and tensile rebar yielding were also recorded. The four-point bending test was conducted according to the standard for test method of concrete structures in China [9].

3. Four-Point Bending Test Results and Analysis

3.1. Crack and Damage Patterns of the Solid and Slotted Beams. The damage patterns of the solid beams and slotted beams are shown in Figure 4. For the solid beams, the first crack was tensile crack and appeared in the flexural zone near the middle span of the beam. With an increase of the load, new tensile cracks appeared and the former tensile cracks developed into the upper part of the beam. Shear cracks were observed near the ends of the beams at the late loading stage, because the shear force and moment were both significant in this region. Concrete crush was observed on the compressive region of the middle span of the beam at the final loading stage. Therefore, it can be concluded that the final failure mode of the solid beams was the flexural failure.

For the slotted beams, the first tensile crack appeared on the surface of the middle span of beam A and beam B, respectively. Then, the cracks in both beam A and beam B developed on the surfaces of the beams with the increase

(a)

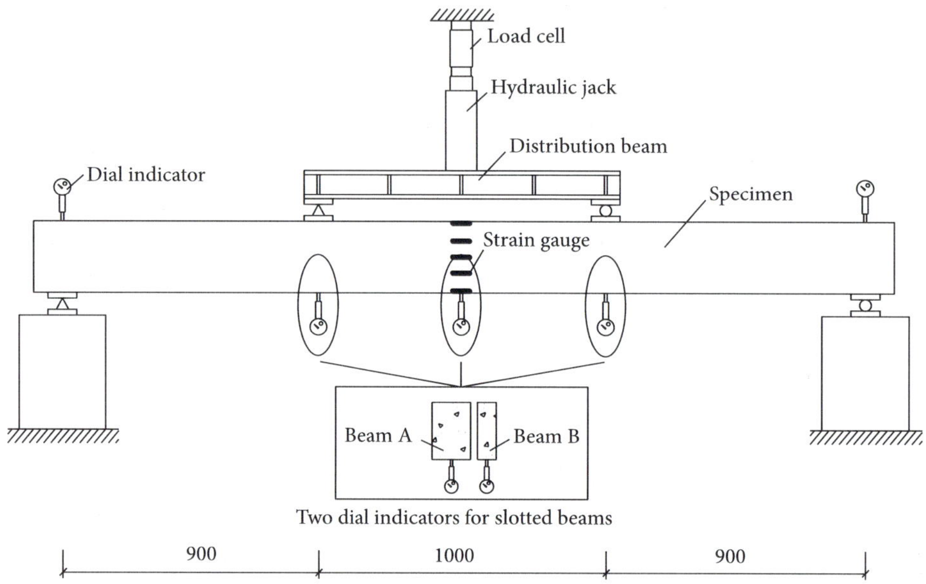

(b)

FIGURE 3: Four-point bending test system of the RC beams: (a) photo of the test devices; (b) sketch of the test devices (units: mm).

(a)

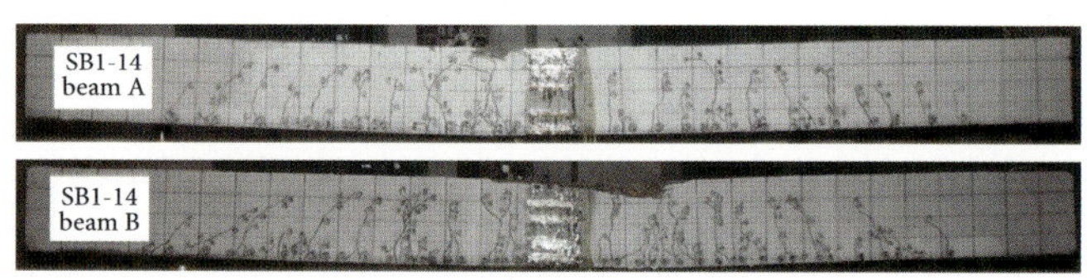

(b)

FIGURE 4: Damage patterns of the beams: (a) RB-14 and (b) SB1-14.

of the load. At the final loading stage, concrete crushing began to appear in the compression region at the middle span of beam A and beam B. The mechanical behavior of beam A and mechanical behavior of beam B of a slotted beam were different from each other, due to the difference in the section dimension and reinforcement ratio. Since the deflections of beam A and beam B at the middle span were almost similar to each other caused by the loading scheme used in the test, the concrete crushing at the middle span of beam A and beam B did not appear simultaneously, resulting in the different ductility behavior of beam A and beam B. The failure process of the two parts of the slotted beams observed during the final loading stage demonstrated this inference. More in-depth analysis concerning the ductility behavior of the two parts of the slotted beams will be given in Section 3.3. Though the mechanical behavior of beam A and beam B was different, severe concrete crushing was observed at the middle span of both beam A and beam B at the failure stage, indicating that the final failure model of the slotted beams was flexural failure. Therefore, it can be concluded that

TABLE 4: Test results of the specimens.

Specimen identifier	Yielding deflection Δ_y (mm)	Yielding load (kN)	Peak load capacity (kN)	Ultimate deflection Δ_u (mm)	Ductility ratio μ_Δ [(1)]
RB-10	11.0	98.6	113.0	69.6	6.3
RB-12	11.9	108.0	120.6	67.1	5.6
RB-14	12.3	128.6	141.0	60.7	5.0
SB1-14	12.3	129.5	132.9	51.2	4.2
SB2-10	14.9	106.7	109.9	56.1	3.8
SB3-12	12.2	108.6	120.0	52.6	4.3
SB1-12	12.6	113.8	118.9	64.4	5.1
SB2-14	12.4	129.5	135.0	61.2	4.9
SB3-10	14.8	96.6	103.8	46.8	3.2

Note: [(1)] the ductility of the beam was calculated by Δ_u/Δ_y.

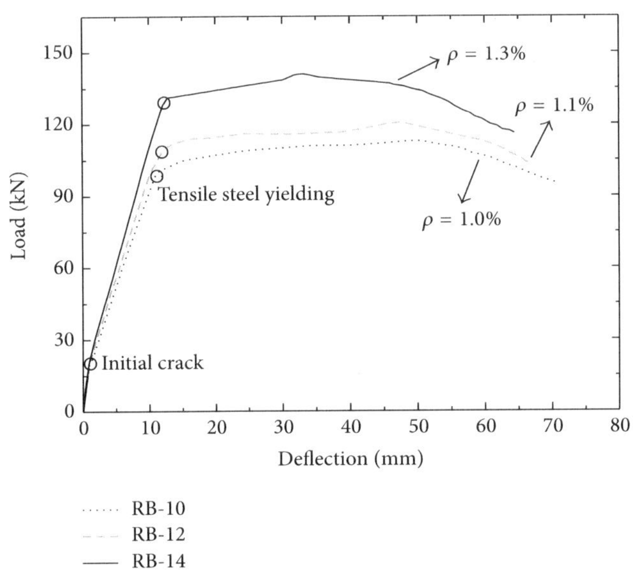

FIGURE 5: Deflection curves of the solid beams.

the damage process of the slotted beams exhibited a certain difference from that of the solid beams.

3.2. Load-Deflection Curve Analysis. The flexural load-deflection curves of the solid RC beams are given in Figure 5. These curves can be divided into three stages: precrack stage, preyield stage, and postyield stage. The initial linear portion of the curves shows that the stiffness of the beams remained constant prior to the cracking load. The initial cracking point of the three solid beams were almost identical, because the section dimension and the concrete strength of the three beams were almost the same. After the cracking load, the stiffness of the solid beams decreased gradually, as displayed in Figure 5. The yield strength and load capacity increased with the increase in the tensile reinforcement ratio. Because all the three beams were under-reinforced beams, with tensile reinforcement ratio of 1.0%, 1.1%, and 1.3%, respectively, obvious ductility behavior can be observed from the load-deflection curves. The mechanical properties of the solid beam are given in Table 4, where the yielding load was determined when the strain gauges pasted on tensile steel reached yielding strain; the ultimate deflection corresponded to the displacement when vertical load decreased to 85% of the peak load capacity. In Table 4, it can be found that the ductility ratio of the solid beams decreased with the increase in the tensile reinforcement ratio.

It was difficult to fix load cells at the two fulcrums of the steel distribution beam during the four-point bending test, so the concentrated force applied on the two parts of the slotted beams was not measured during the tests. Because both beam A and beam B entered into nonlinear stage after cracking point, it was improper to calculate the actual load applied on each part of the two beams based on the initial elastic cross-section stiffness of each beam. For a slotted beam, the dial indicators showed that the deflections of beam A and beam B were almost the same. Therefore, the total load applied on the top of the distribution beam and the average readings of the dial indicators were used to depict the load-deflection curves of the slotted beams.

Figure 6 shows the deflection curves of the slotted beams and the corresponding solid beams. Before the deflection curves entered into the descending branch, the load-deflection curves of the slotted RC beams were close to that of the corresponding solid RC beam, indicating that the cracking load and yielding load of the slotted beams were almost similar to those of the corresponding solid beam. For SB2-10, the ductility of the load-deflection curves of the slotted beams was lower than that of the corresponding solid RC beam RB-10, and the ductility of SB3-10 became obviously lower than that of SB2-10. Obvious lower ductility was also observed for the slotted beams SB3-12 and SB1-14, comparing with their corresponding solid RC beam. However, the load-deflection curves of SB1-12 and SB2-14 were close to that of the corresponding solid beam and it can be found that the difference between the reinforcement ratio of beam A and beam B of SB1-12/SB2-14 was obviously smaller than that of the other slotted beams. This experimental phenomena indicated that the reinforcement ratio of beam A and beam B had an obvious effect on the flexural behavior of the slotted beams. Detailed analysis for this phenomena will be given in Section 3.3.

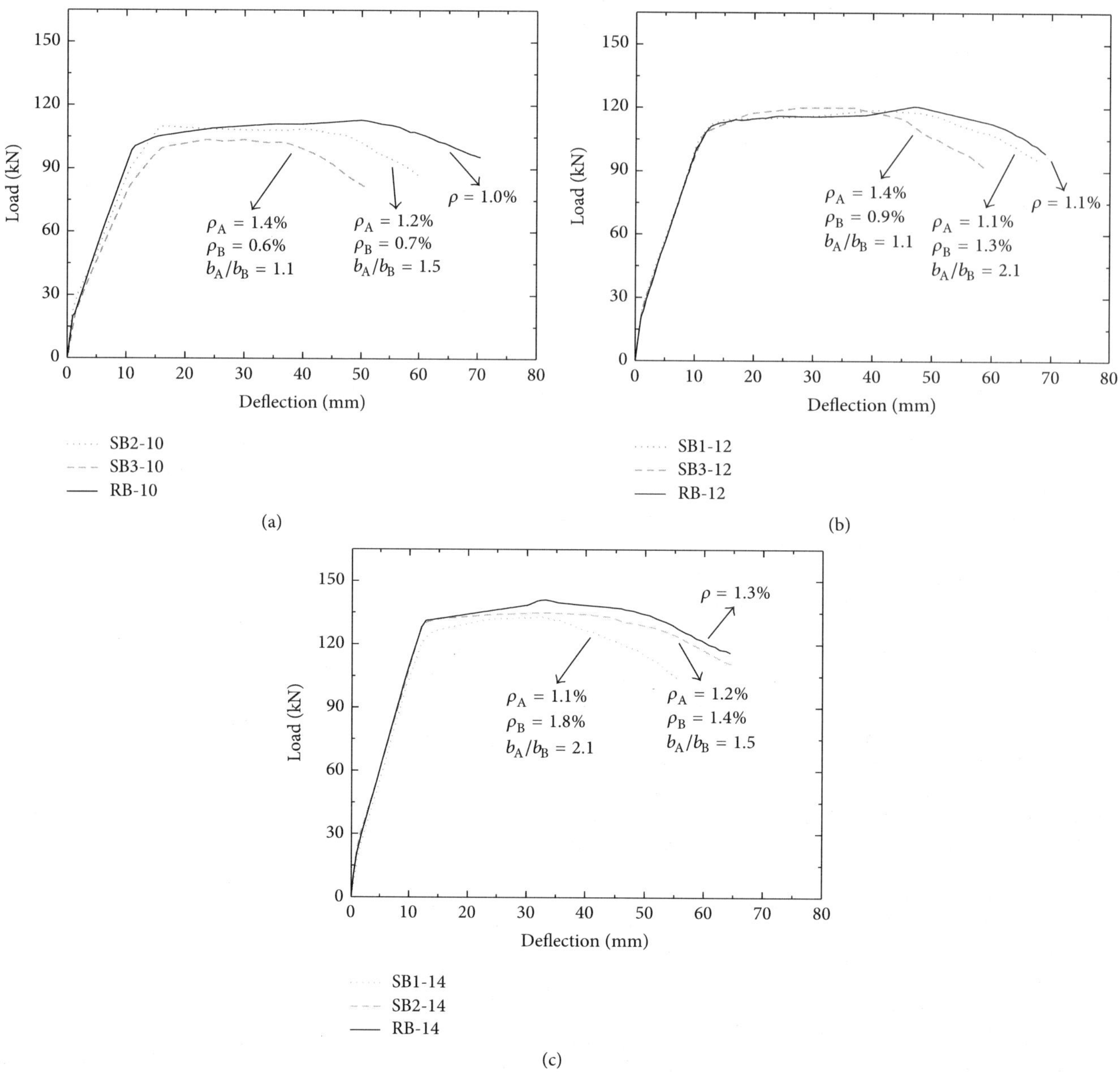

FIGURE 6: Load-deflection curves of the slotted beams: (a) SB2-10 and SB3-10; (b) SB1-12 and SB3-12; and (c) SB1-14 and SB2-14.

3.3. Moment Curvature Analysis. The moment curvature relationship of a beam section can be calculated by using the information of the beam section size, the reinforcement, and the stress-strain curves of the material. In the calculation, the stress-strain curves of the steel rebars obtained from the axial tensile tests were used, and the stress-strain curves of the concrete followed the modified Kent-Park model [10, 11]. The calculation process of moment curvature relationship of a beam section was described in detail in relevant reference [12] and hence is omitted here. The moment curvature curve of all the solid beams is shown in Figure 7, and the values corresponding to the yield point, peak point, and failure point are given in Table 5. The results show that the moment capacity of the beam section increased with the increase in the reinforcement ratio; however, the curvature ductility exhibited a decrease trend, because increased tensile reinforcement ratio can lead to an increase of the height of the compressive region (resulting in decreased curvature).

The deflection curves of beam A and beam B for a slotted beam during the tests were almost the same, based on the readings of the dial indicators fixed on the bottom surfaces of beam A and beam B. Therefore, it can be inferred that the curvatures of the section at the middle span of the two beams are almost identical during the tests, based on the deflection theory of beam. Hence, the moment curvature curve of the entire slotted beam can be derived through superimposing the moment of beam A and beam B at a certain curvature. The moment curvature curves of the two parts of the slotted beams are shown in Figure 8, where the moment curvature

TABLE 5: Characteristic points on the calculated moment curvature curves of the specimens.

Specimen identifier	Yielding curvature ϕ_y (rad)	Yielding moment (kN·m)	Peak moment capacity (kN·m)	Ultimate curvature ϕ_u (rad)	Ductility ratio μ_ϕ (1)
RB-10	0.012	41.6	48.1	0.158	12.7
RB-12	0.013	45.8	52.6	0.139	11.0
RB-14	0.014	52.0	58.0	0.122	8.9
SB1-14	0.014	51.5	57.3	0.112	8.0
SB2-10	0.013	41.1	47.5	0.136	10.6
SB3-12	0.014	46.3	52.0	0.122	8.8
SB1-12	0.013	46.0	52.9	0.141	11.2
SB2-14	0.014	52.1	58.1	0.123	8.9
SB3-10	0.013	41.3	46.7	0.120	9.0

Note: (1) the curvature ductility of the beam section was calculated by ϕ_u/ϕ_y.

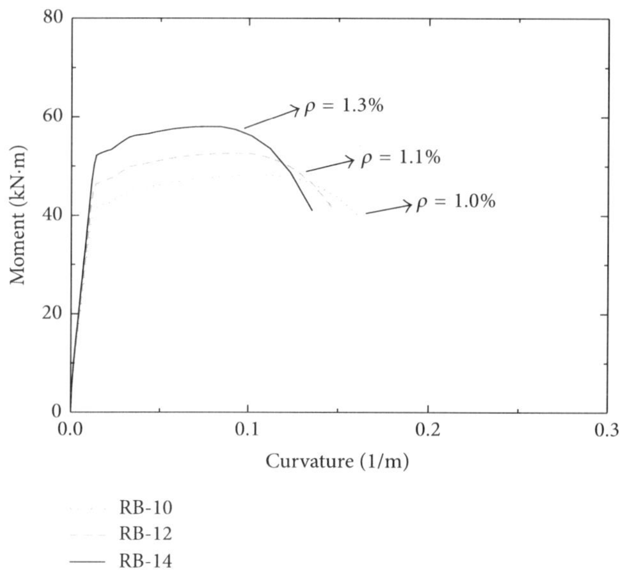

FIGURE 7: Moment curvature curves of the solid beams.

curves of the corresponding entire slotted beam are also given for the purpose of comparison.

For SB1-12, the ductility of the moment curvature curves of beam A and beam B was almost the same, and the tensile reinforcement ratio of beam A and beam B of SB2-14 was 1.1% and 1.3%, respectively, similar to each other. Similar phenomenon was also found for SB2-14. The similar reinforcement ratio resulted in the fact that the heights of the compression region of beam A and beam B at the concrete crush point were almost equivalent to each other [12]. Hence, the ultimate curvature of beam A and beam B of SB1-12/SB2-14 was nearly the same, because the height of the section was identical for all beams.

For SB2-10, SB3-10, and SB3-12, the ductility of the moment curvature curves of beam B was obviously greater than that of the corresponding beam A, due to the fact that the reinforcement ratio of beam B was much less than that of the corresponding beam A. For SB1-14, the ductility of beam B for SB1-14 was smaller than that of beam A, because the reinforcement ratio of beam B was larger than that of beam A. In Figures 8(a), 8(b), 8(d), and 8(e), it can be found that the ductility of the entire slotted beam was determined by the beam, of which the ductility was worse than the other part. Therefore, the ductility of both beam A and beam B should be considered and designed carefully in the application of slotted beams in practical engineering.

It should be noted that the width ratio b_A/b_B of SB1-12 was 2.1, which was the largest width ratio in the tests. However, the ductility of SB1-12 and RB-12 was nearly the same. For SB3-12, the width ratio of beam A and beam B was 1.1, which was the smallest width ratio in the tests. However, the ductility of SB3-12 and RB-12 was obviously different. Therefore, it can be concluded that the reinforcement ratio rather than the width ratio of beam A to beam B is the key factor affecting the ductility behavior of the slotted beams.

The moment curvature curves of the slotted beams and corresponding solid beams are shown in Figure 9. The moment curvature curves of SB1-12 and RB-12 were almost coincident, because the reinforcement ratio of RB-12 and the two parted beams of SB1-12 were almost the same. Similar phenomenon was also found for SB2-14 and RB-14. In Figure 9(a), it can be clearly found that the larger the difference between ρ_A and ρ_B, the worse the ductility of the slotted beam. Therefore, the two parted beams of the slotted beam should be designed to pose similar tensile reinforcement ratio, in order to achieve a similar ductility capacity of the corresponding solid beam.

3.4. Load Capacity Analysis. The flexural capacities of RC beams can be determined when the geometry, the reinforcement, and the strength of the material are known. The flexural capacities of the solid and slotted beams, calculated according to the concrete code of China [13], are given in Table 6, where the experimental results are also given. In Table 6, M_{cap}^{exp} is yield capacity of the beams obtained from experiments and M_{cap}^{cal} is theoretically calculated flexural capacity of the beams. Because the yield strength of the steel rebar is used to calculate the flexural capacity of the beam in the concrete code, the yield capacities of the beams obtained from the tests were used for the comparison with the theoretically calculated values.

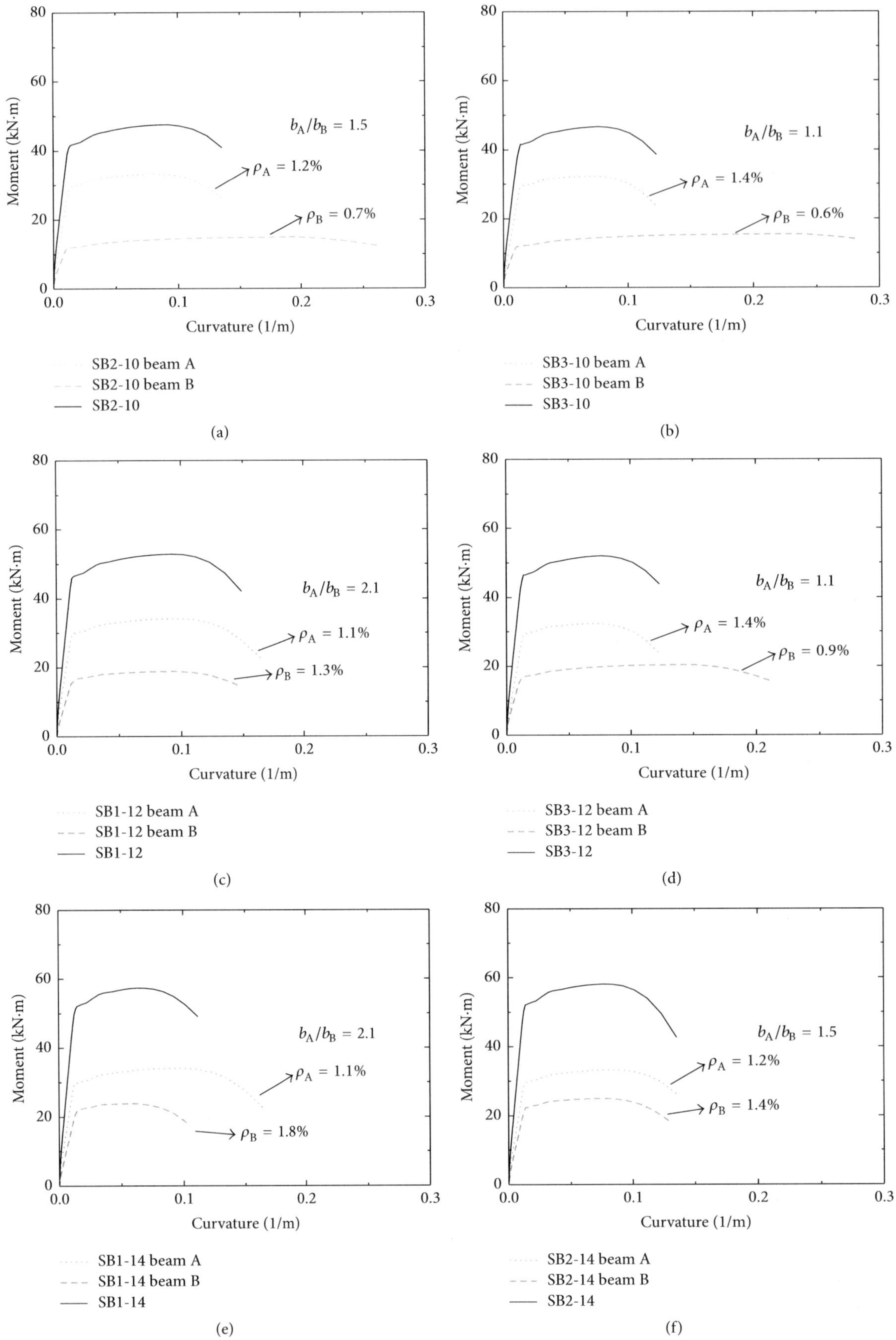

FIGURE 8: Moment curvature curves of the slotted beams: (a) SB2-10; (b) SB3-10; (c) SB1-12; (d) SB3-12; (e) SB1-14; and (f) SB2-14.

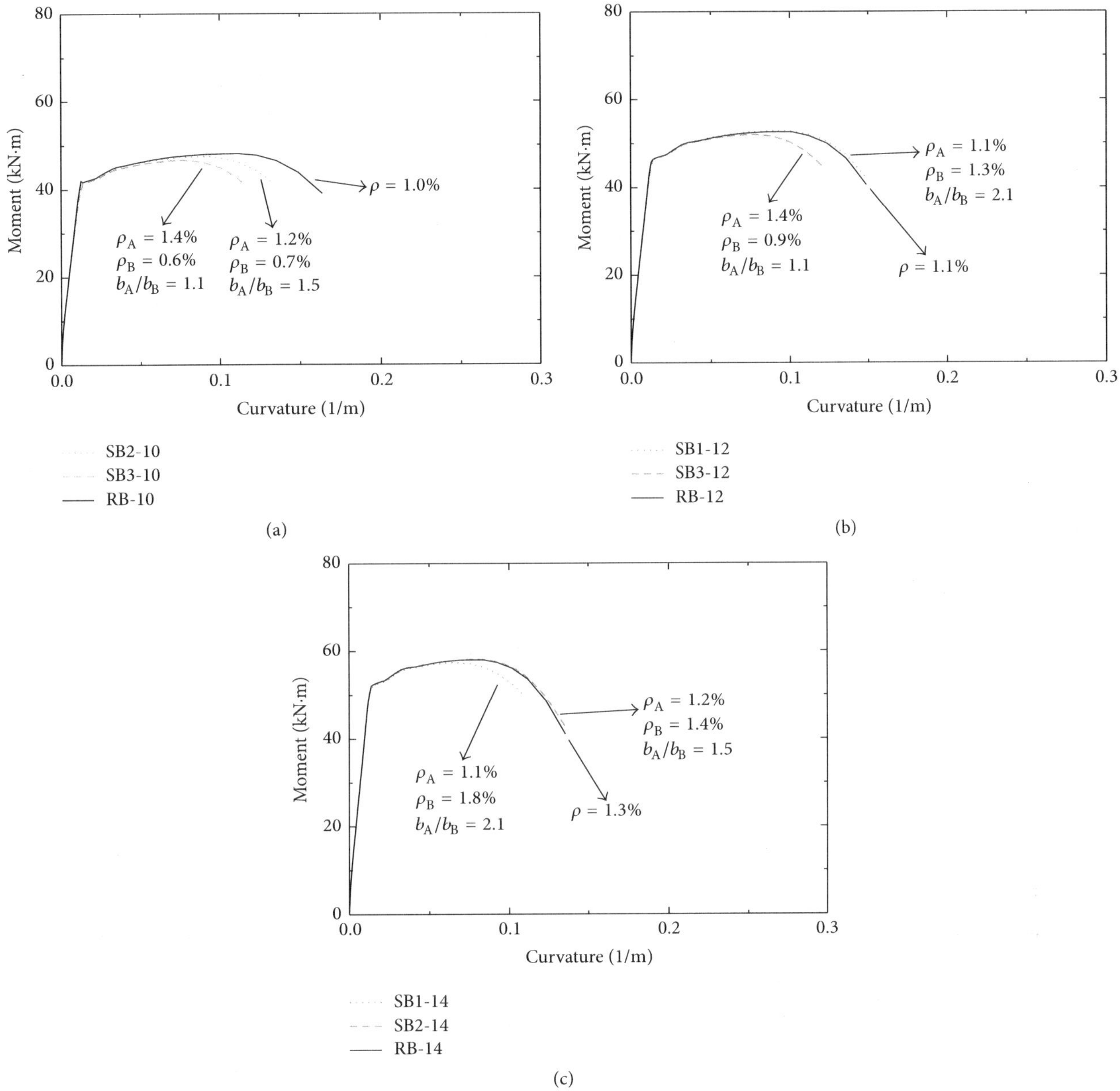

FIGURE 9: Comparison of the moment curvature curves of the slotted beams and corresponding solid beams: (a) SB2-10, SB3-10, and RB-10; (b) SB1-12, SB3-12, and RB-12; and (c) SB1-14, SB2-14, and RB-14.

Table 6 shows that the calculated flexural capacity M_{cap}^{cal} was close to the value obtained from the tests for both the solid and slotted beams, indicating that the method used in the code was effective in predicting the flexural capacity of these two types of beams. Meanwhile, the calculated flexural capacities were conservative compared with the test results. It was found that the sum of the flexural capacities of beam A and beam B of the slotted beams was equal to that of the corresponding solid beam. This observation was coherent with the phenomena shown in Figure 6, where it was found that the load-deflection curves of the solid beam and slotted beam were close to each other before the descending branch. However, it should be noted that the slotted beam exhibited decreased ductility capacity compared with the corresponding solid beam, as discussed in Section 3.2.

4. Structural Design Advices for the Slotted RC Beams

Based on the above experimental results and analysis, several design advices can be made for the slotted RC beams, in order to obtain favourable mechanical behavior:

(1) The target failure mode of the slotted RC beam should be typical flexural failure mode, in order to ensure sufficient ductility. Therefore, sufficient tensile

TABLE 6: Yield capacities of the specimens.

Specimen identifier	M_{cap}^{exp} (kN·m)		M_{cap}^{cal} (kN)		$M_{cap}^{exp}/M_{cap}^{cal}$
RB-10	44.4		40.8		1.09
RB-12	48.6		45.6		1.07
RB-14	57.9		51.6		1.12
		Beam A	Beam B	Beam A + beam B	
SB1-14	58.3	29.4	22.2	51.6	1.13
SB2-10	49.5	29.4	11.4	40.8	1.21
SB3-12	48.5	29.4	16.2	45.6	1.06
SB1-12	51.2	29.4	16.2	45.6	1.12
SB2-14	55.9	29.4	22.2	51.6	1.08
SB3-10	44.6	29.4	11.4	40.8	1.09

reinforcement is needed at the bottom of the beam. Meanwhile, sufficient stirrup reinforcement for shear is also needed in order to prevent shear failure.

(2) The ductility of beam A and ductility of beam B of a slotted beam should be close to each other in order to insure that these two beams underwent similar damage process under flexural load and reach the final failure nearly at the same time. In order to achieve the above flexural behavior of a slotted beam, the reinforcement ratios of beam A and beam B should be close to each other.

(3) In addition, the concrete strength, the span limitation, and the ratio of beam width to beam height should meet the current specification requirement for reinforced beam in the code.

5. Thermal Insulation Behavior of the Slotted Beams

Thermal performance is an important parameter to evaluate the feasibility of the slotted beam as a thermal insulation structural component. The heat transfer coefficient is a frequently used index to evaluate the thermal properties of materials. This index is influenced by the thermal conductivity and thickness of the materials through which heat is transferred. The heat transfer coefficient K of the slotted beam can be calculated using the following equations based on the thermal deign code for civil building in China [14]:

$$R = \frac{\delta_A}{\lambda_c} + \frac{\delta_{insul}}{\lambda_{insul}} + \frac{\delta_B}{\lambda_c},$$
$$R_0 = R_{in} + R + R_{out}, \quad (1)$$
$$K = \frac{1}{R_0},$$

where R is sum of thermal resistance values of all the material layers for the slotted beam; δ_A, δ_{insul}, and δ_B are thickness of beam A, the thermal insulation layer, and beam B, respectively; λ_c and λ_{insul} is thermal conductivity of the concrete and the thermal insulation layer, respectively; R_0 is thermal resistance value of the slotted beam; R_{in} and R_{out} is thermal resistance of the inner and outer surface against heat convection, respectively, and are set to be $0.11\,m^2 \cdot K/W$ and $0.04\,m^2 \cdot K/W$, respectively, based on the thermal design code for civil building in China [14]; and K is heat transfer coefficient of the beam. The larger the coefficient is, the easier the heat is transferred through the beam. For the solid beam, (1) is changed to $R = \delta_c/\lambda_c$ and the calculation process becomes simpler.

Thermal inertia index, as a dimensionless quantity, represents the resistance of the material to variations in temperature. The thermal inertia index D of the slotted beam can be computed using the following equation [14]:

$$D = \frac{\delta_A}{\lambda_c} S_c + \frac{\delta_{insul}}{\lambda_{insul}} S_{insul} + \frac{\delta_B}{\lambda_c} S_c, \quad (2)$$

where S_c and S_{insul} are thermal storage coefficient of the concrete and the thermal insulation layer. Materials with high thermal inertia show small changes in temperature under the effect of outer environmental temperature, showing good heat insulation effect. From (1), it is easy to find that, for the slotted beam with a certain sum width of beam A and beam B, the width ratio of beam A to beam B has no effect on the thermal resistance value R and only δ_{insul} and λ_{insul} of the thermal insulation layer have effect on R. Similarly, only δ_{insul}, λ_{insul}, and S_{insul} have effect on the thermal inertia index of the slotted beam when the sum width of beam A and beam B is determined.

Rigid extruded polystyrene foam board, also called XPS, can provide excellent insulation against both cold and heat and has the following advantages: high corrosion resistance, moisture proof, light weight, and antiaging capacity. Hence, XPS has been widely used as thermal insulation material in practical engineering. In this paper, XPS was used to fill the gap in the slotted beams. In addition to XPS, it should be noted that other insulation materials can also be used as the thermal insulation layer in the slotted beam. The values of λ and S for reinforced concrete and XPS recommended in the Chinese code [14, 15] are given in Table 7. The thermal calculation result of the beams is given in Table 8. It can be seen that the heat transfer coefficient of the slotted beam decreases by 77.4% compared with the solid RC beam. The thermal inertia factor of the slotted beam is 2.83, higher than that of the solid beam, 2.47. These data indicate that the

Table 7: Thermophysical parameters of the materials.

Material type	Thermal conductivity λ (W/(m·K))	Thermal storage coefficient S (W/(m·K))
Reinforced concrete	1.74	17.2
XPS	0.03	0.36

Table 8: Thermal calculation results of the beams.

Beam type	Thermal resistance R (m^2·K/W)	Heat transfer coefficient K (W/(m^2·K))	Thermal inertia factor D
Solid beam	0.29	3.45	2.47
Slotted beam	1.29	0.78	2.83

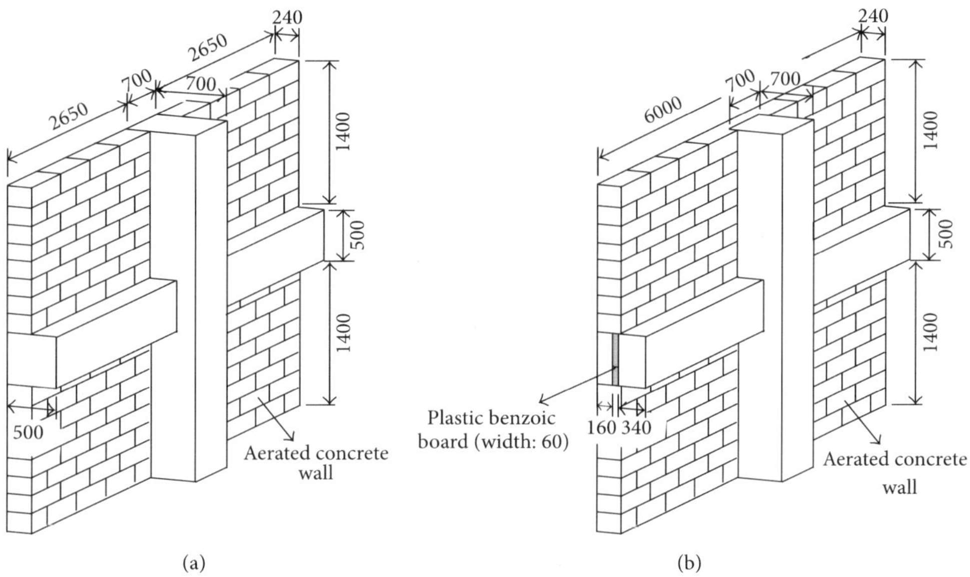

Figure 10: Geometry of the unit frame built in COMSOL Multiphysics: (a) normal frame and (b) frame using the slotted beam-wall system (units: mm).

thermal bridge of the slotted beam is effectively mitigated and the thermal performance of the slotted beam becomes much better than that of the conventional solid RC beam.

6. Thermal Insulation Behavior of the Slotted Beam-Wall System

6.1. FEM Model Building. A FEM software, COMSOL Multiphysics [16], which has passed the model tests required by the European standard EN 10211: 2007 for thermal bridges in building constructions [17] and was found to be suitable for simulating the heat transport through buildings [18], was used to simulate the thermal insulation behavior of the frame structure using the slotted beam-wall system. The thermal insulation behavior of this new type of structure in winter was considered in the FEM model and a normal frame structure FEM model was also built for comparison purpose. The temperatures inside and outside the room are set to 18°C and 0°C, respectively, according to the design standard for energy efficiency of residential buildings in hot summer and cold winter zone in China [19]. These temperatures are typical for the buildings using heating in the south area of China in winter. The plan and elevation layout are always regular for residential frame buildings; therefore only one unit of the frame was modeled in COMSOL due to the symmetrical structure of the frame structure, for the purpose of conciseness. The dimension information of the frame FEM model is shown in Figure 10, where the beam span, the floor height, and the size of columns and beams are typical for 10-story height residential frame buildings in China. The exterior vacuum insulation windows, beam perpendicular to the exterior wall, and the floor slabs were not modeled, in order to make the comparison of the thermal properties between the normal frame and frame using the slotted beam-wall system more specific. The boundary condition of the four sides of the model was thermal insulation, due to the symmetry of the model. 3D steady state thermal bridge was considered in COMSOL analysis, meaning the field variables are independent of time. More detailed modeling and analysis process can be referred to in COMSOL user manual. Thermophysical parameters of the materials used in the FEM model are given in Table 9 and it is assumed that these thermal properties are independent of the temperature during the FEM analysis.

TABLE 9: Thermophysical parameters of the materials.

Material type	Density (kg/m^3)	Thermal conductivity (W/(m·K))	Thermal storage coefficient ($W/(m^2 \cdot K)$)	Heat capacity (J/(kg·K))
Reinforced concrete	2500	1.74	17.20	920
XPS	30	0.03	0.36	1380
Aerated concrete wall	500	0.19	2.81	1050

TABLE 10: Comparison of the FEM simulation results.

Structure type	Heat transfer area (m^2)	Total heat flux (W)	Heat flux on walls (W)	Heat flux on columns (W)	Heat flux on beams (W)
Normal RC frame	19.8	460.3	224.1	102.2	134.0
RC frame using slotted beams	19.8	287.1	201.7	49.4	36.1

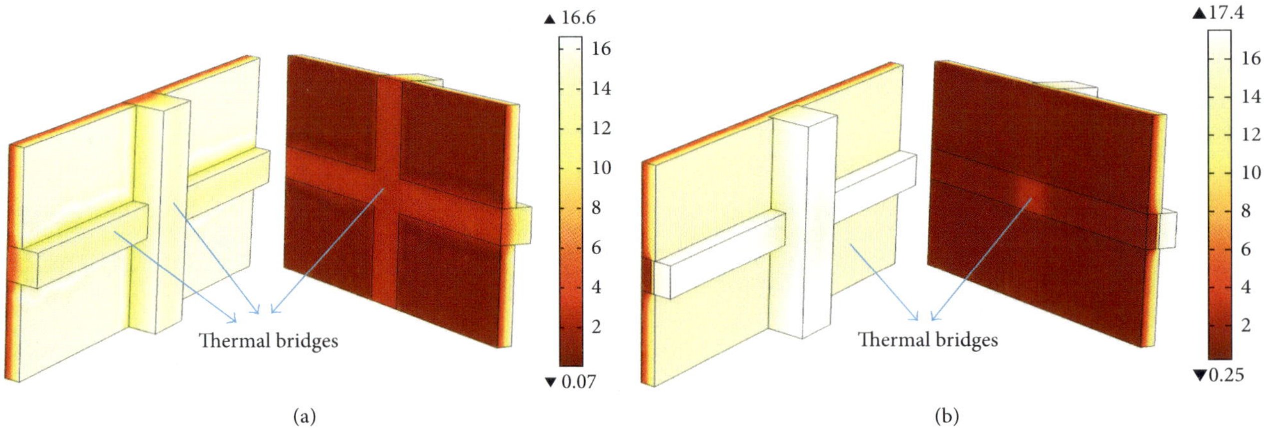

FIGURE 11: Surface temperature distribution of the FEM models (units: °C): (a) normal frame and (b) frame using the slotted beam-wall system.

6.2. Simulation Result and Discussion. The temperature distribution on the surfaces of the FEM models is shown in Figure 11. It is obvious that the beams and columns constitute the thermal bridges for the normal frame in Figure 11(a). However, the thermal bridges have been controlled after the slotted beam-wall system have been utilized, as shown in Figure 11(b), where the location of beam-column joint becomes the thermal bridge. In fact, this thermal bridge can be reduced in practical engineering because it is easy to carry out heat insulation treatment on this small area.

The results of the calculated heat flux are given in Table 10. The heat flux in the normal frame and frame using the slotted beam-wall system was 460.3 W and 287.1 W, respectively. The heat flux through the beams and columns in the normal frame structure was 236.2 W (51.3% of the total heat flux of the normal structure), indicating that the thermal bridge effect was obvious in the normal frame structure. However, the heat flux through the beams and columns in the frame using the slotted beam-wall system was 85.5 W (29.8% of the total heat flux of this new type structure). The thermal bridge effect has been controlled in the new frame structure through using the slotted beam-wall system. In addition, though the area of the walls of this new type structure was larger than that of the normal frame, the heat flux on the walls of this new type structure was 22.4 W less than that of the normal frame.

7. Conclusions

In this study, a novel slotted beam-wall system was proposed for thermal insulation of RC frame buildings. Four-point bending test was carried out to investigate the flexural behavior of the slotted beams. The flexural behavior of the slotted beams was investigated in detail, including failure mode, load-deflection curves, and moment curvature analysis. Thermal insulation behavior of the slotted beam was analyzed through theoretical calculation. Finally, the thermal behavior of the frame using the slotted beam-wall system was simulated through FEM method. The following conclusions can be drawn:

(1) Though the final failure mode of the slotted beams was flexural failure mode, the failure process of the slotted beams was different from that of the solid beams at the final loading stage.

(2) Before the final failure stage, the load-deflection curve of the slotted beam was close to that of its corresponding solid beam. The tensile reinforcement ratio rather

than the width ratio of the two parted beams had an important effect on the flexural behavior of the slotted beams. The tensile reinforcement ratios of the two parted beams of a slotted beam should be close to each other, in order to obtain similar ductility behavior of their corresponding solid beam.

(3) The slotted beams exhibited attractive thermal insulation properties. The heat transfer coefficient of the slotted beam was greatly reduced and the thermal inertia factor increased, compared with the corresponding solid beams. Hence, the slotted beam has a great potential for the practical thermal insulation application of RC frame buildings.

(4) The FEM simulation results showed that the frame structure using the slotted beam-wall system exhibited obvious and attractive thermal insulation property, compared with the normal frame structure. It is believed that this new type of structure can become an effective measure to solve the thermal insulation problem.

Competing Interests

The authors declare that they have no competing interests.

Acknowledgments

This research was sponsored by the Science and Technology Plan of Changsha, China (Grant no. CSCG-HNTF-GK20130576), the National Natural Science Foundation of China (Grant no. 51408211), the Natural Science Foundation of Hunan Province, China (Grant no. 2015JJ3032), the Fundamental Research Funds for the Central Universities through the project of Young Teacher Growth of Hunan University (Grant no. 531107040799), and the Open Fund of Hunan Province Engineering Laboratory of Bridge Structure (Changsha University of Science & Technology, Grant no. 14KD02).

References

[1] J. Iwaro and A. Mwasha, "A review of building energy regulation and policy for energy conservation in developing countries," *Energy Policy*, vol. 38, no. 12, pp. 7744–7755, 2010.

[2] B. P. Jelle, "Traditional, state-of-the-art and future thermal building insulation materials and solutions—properties, requirements and possibilities," *Energy and Buildings*, vol. 43, no. 10, pp. 2549–2563, 2011.

[3] T. G. Theodosiou and A. M. Papadopoulos, "The impact of thermal bridges on the energy demand of buildings with double brick wall constructions," *Energy and Buildings*, vol. 40, no. 11, pp. 2083–2089, 2008.

[4] K. S. Al-Jabri, A. W. Hago, A. S. Al-Nuaimi, and A. H. Al-Saidy, "Concrete blocks for thermal insulation in hot climate," *Cement and Concrete Research*, vol. 35, no. 8, pp. 1472–1479, 2005.

[5] S. Kumar, "Fly ash-lime-phosphogypsum hollow blocks for walls and partitions," *Building and Environment*, vol. 38, no. 2, pp. 291–295, 2003.

[6] J. Wang, P. C. Eames, J. F. Zhao, T. Hyde, and Y. Fang, "Stresses in vacuum glazing fabricated at low temperature," *Solar Energy Materials and Solar Cells*, vol. 91, no. 4, pp. 290–303, 2007.

[7] N. Ng, R. E. Collins, and L. So, "Thermal conductance measurement on vacuum glazing," *International Journal of Heat and Mass Transfer*, vol. 49, no. 25-26, pp. 4877–4885, 2006.

[8] N. Ng, R. E. Collins, and L. So, "Characterization of the thermal insulating properties of vacuum glazing," *Materials Science and Engineering B*, vol. 138, no. 2, pp. 128–134, 2007.

[9] Code of China, *Standard for Test Method of Concrete Structures (GB/T 50152-2012)*, China Architecture & Building Press, Beijing, China, 2012 (Chinese).

[10] D. C. Kent and R. Park, "Inelastic behaviour of reinforced concrete members with cyclic loading," *Bulletin of the New Zealand Society for Earthquake Engineering*, vol. 4, no. 1, pp. 108–125, 1971.

[11] B. D. Scott, R. Park, and M. J. N. Priestley, "STRESS-strain behavior of concrete confined by overlapping hoops at low and high strain rates," *Journal of the American Concrete Institute*, vol. 79, no. 1, pp. 13–27, 1982.

[12] J. K. Wight and J. G. MacGregor, *Reinforced Concrete: Mechanics and Design*, Pearson Education, 6th edition, 2011.

[13] Code of China, *Code for Design of Concrete Structures (GB 50010-2010)*, China Architecture & Building Press, Beijing, China, 2010 (Chinese).

[14] Code of China, *Thermal Design Code for Civil Building (GB50176-93)*, China Planning Press, Beijing, China, 1993 (Chinese).

[15] Code of China, "Rigid extruded polystyrene foam board for thermal insulation (XPS)," GB/T 10801.2-2002, Standards Press of China, Beijing, China, 2002 (Chinese).

[16] COMSOL, *Multiphysics User's Guide*, Comsol AB, Stockholm, Sweden, 2010.

[17] European Committee for Standardization. EN ISO 10211, "Thermal bridges in building construction—heat flows and surface temperatures—detailed calculations," ISO 10211:2007, 2007.

[18] A. W. M. van Schijndel, "Multiphysics modeling of building physical constructions," *Building Simulation*, vol. 4, no. 1, pp. 49–60, 2011.

[19] Code of China, "Design standard for energy efficiency of residential buildings in hot summer and cold winter zone," JGJ 134-2010, China Architecture & Building Press, Beijing, China, 2010 (Chinese).

4

Microwave Dielectric Properties of CCTO/PVA Composites

M. P. F. Graça,[1] K. D. A. Sabóia,[2,3] F. Amaral,[1,4] and L. C. Costa[1]

[1]*I3N and Physics Department, University of Aveiro, 3810-193 Aveiro, Portugal*
[2]*Universidade Estadual do Ceará, Fortaleza, Brazil*
[3]*Laboratório de Telecomunicações e Ciência e Engenharia de Materiais–LOCEM, Departamento de Física, Universidade Federal do Ceará, Fortaleza, Brazil*
[4]*Coimbra Health School, Polytechnic Institute of Coimbra, Coimbra, Portugal*

Correspondence should be addressed to F. Amaral; filipe.amaral79@hotmail.com

Academic Editor: Fernando Rubio-Marcos

The $CaCu_3Ti_4O_{12}$ (CCTO) ceramic powder was inserted in the polyvinyl alcohol (PVA) polymeric matrix, with an increasing weight fraction of the filler, to form a flexible and high dielectric constant composite at the GHz region. The structural characterization of the samples was performed using X-ray diffraction and scanning electron microscopy (SEM). The complex permittivity was calculated by the small perturbation theory using two resonant cavities (2.7 GHz and 5.0 GHz). Several classical models (Maxwell Garnett, Lichtenecker, effective medium theory (EMT), and Yamada) were used to fit the real part of the complex permittivity of the composite as a function of the weight fraction of CCTO powder inserted in the PVA matrix. The best predictions for the dielectric behavior of these samples were obtained with the EMT and Yamada models.

1. Introduction

Materials with excellent dielectric properties, such as a high-dielectric constant, low dissipation factor, and good stability over a broad frequency and temperature ranges, are required for some of modern electronic applications. In the last few years, it has emerged an increasing interest in size reduction and performance enhancement of embedded passive components, especially for embedded capacitors [1–3]. Materials for such kind of applications must present (beyond the already mentioned properties) good dielectric and thermal stability over a broad frequency and temperature ranges. Furthermore, the final piece should be manufactured in a simple way, with low production costs. To achieve this goal, many trials were performed with ceramic-polymer composites with high-dielectric constant [4–7].

Generally, ceramics with high-dielectric constant values are very promising for capacitive applications in microelectronics and microwave devices (like dielectric resonators for cell mobile phones, for example) where miniaturization is crucial. However, it is well known that, for some applications, they are too brittle which becomes a critical factor. On the contrary, polymers that have normally lower dielectric constant, but present a much lower dielectric loss, are flexible, easy to process, and traditionally used in low-leakage capacitors. The association of high-dielectric constant ceramics with a polymer results in composites with acceptable dielectric constant and loss factor values, with lower dependence on frequency and temperature and, simultaneously, more flexible than ceramics [5–7].

In this work, the polyvinyl alcohol (PVA) was used as a polymeric matrix, and $CaCu_3Ti_4O_{12}$ (CCTO) ceramic particles were used as a filler. PVA is a water-soluble polymer widely applied in adhesives, sealants, coatings, paints, plastics, textiles, etc. It is also a polymer of great interest because of its characteristics, for various pharmaceutical and biomedical applications. PVA is produced by polymerization of vinyl acetate to polyvinyl acetate (PVAc), followed by hydrolysis of PVAc to PVA. Because the hydrolysis reaction does not go to completion, the final result of the polymerization process is a copolymer of PVA and PVAc. The PVA is normally supplied in powder form, and several grades are available with varying viscosity and solubility characteristics [8]. Besides the simplicity associated with the

PVA synthesis process, we have chosen this polymer matrix because of its high thermal and chemical stabilities and nontoxicity. PVA also presents a high-dielectric strength and good charge storage capacity, which makes it usable as the polymer matrix in composite-based electronic components [9, 10].

On the contrary, $CaCu_3Ti_4O_{12}$ (CCTO) is a known oxide of the $ACu_3M_4O_{12}$ family. Subramanian et al. have been the first to report the exceptional dielectric properties of CCTO [11], and since then, a significant effort has been devoted to understand the polarization mechanisms that justify the colossal dielectric constant ($\varepsilon' > 10^5$) observed for poly- [12, 13] and single-crystalline ceramics [14]. The CCTO is a perovskite-like compound that, depending on the synthesis and sintering process, can present a dielectric constant (at 1 kHz and room temperature) as high as 10^5 and such dielectric properties are almost temperature- and frequency-independent for large ranges [15–17]. The CCTO has a high potential for technological applications, like memory devices based on capacitive components [18].

Many CCTO/polymer composites have been investigated in the last years [6, 19, 20]. Some of them, CCTO/epoxy and CCTO/polyvinylidene fluoride, for instance, are very promising to be used as new embedded capacitors [19, 21–23]. The dielectric properties of such composites are, however, very dependent on temperature, and the dissipation factor is high. Besides, their preparation process is complex and harsh thermal environments are necessary to improve their thermal properties.

Rarely the dielectric response of CCTO/polymers at the microwave frequency region has been reported. Considering this lack, this paper is focused on the development of a new kind of CCTO/polymer composite prepared by a simple method and with desirable dielectric properties, particularly at the microwave frequency region, such as a high-dielectric constant and low dissipation factor. The complex permittivity of the composites was calculated by the small perturbation theory using two resonant cavities, at 2.7 GHz and 5.0 GHz. Several classical models (Maxwell Garnett, Lichtenecker, effective medium theory (EMT), and Yamada) were used to fit the real part of the complex permittivity of the composite as a function of the weight of CCTO powder inserted in the PVA matrix.

2. Experimental Methods

2.1. Sample Preparation. The CCTO powder was prepared by the solid-state method using stoichiometric quantities of calcium carbonate ($CaCO_3$) (Aldrich, 99%), titanium oxide (TiO_2) (Aldrich, 99%), and copper (II) oxide (CuO) (Merck). The powders were mixed in a Fritsch 7.0 planetary mill for 30 min at 370 rpm. Milling was performed in agate vials and balls, under air, with a powder/balls weight fraction of 1 : 9. Therefore, the milling was used to guarantee the homogeneity of the powder. Then, the mixed powder was submitted to calcination in air at about 1,320 K for 24 h [24, 25]. The PVA (polyvinyl alcohol [-CH_2CHOH-]$_n$, from Merck) flakes were milled in a Fritsch 7.0 planetary mill to obtain a powder with granulometry lower than 200 μm. The X-ray analysis confirmed the presence of CCTO as the major phase. The CCTO powder was added to PVA, to obtain composites with weight fraction between 0 and 70% to study the dielectric response in the microwave region. The mix of CCTO and PVA was milled for 100 min, and the resulting material was uniaxially pressed into small cylinders, with 15 mm in height and 2 mm in diameter, applying 4.0 MPa during 20 min. These cylindrical samples were submitted to a heat treatment of about 420 K for two hours.

Composites with a weight fraction higher than 70% produced brittle samples; thereafter, their use as an electronic component would be compromised. We also point out that the preparation method used in this work to produce the CCTO/PVA composites presents some advantages compared to the wet techniques. In particular, the maximum weight fraction that is admitted for CCTO/PVA composites prepared by the wet techniques is about 30%, which, along with a more complex procedure, makes our method suitable for future industrial applications [26, 27].

The adopted nomenclature to point the weight fraction of the composites is CCTO(x), where x stands for the percentage of CCTO in the sample. For instance, CCTO(30) represents a sample with 70% (weight) of PVA and 30% (weight) of CCTO.

2.2. Structural and Morphological Characterization. The structure of the obtained samples was examined by X-ray diffraction using a X'Pert MPD Philips diffractometer (CuKα radiation, $\lambda = 1.54056$ Å) at 40 kV, and 30 mA, with a curved graphite monochromator, an automatic divergence slit, a progressive receiving slit, and a flat plane sample holder in a Bragg–Brentano parafocusing configuration with a step of 0.04° in 1 s, in the 2θ angle range of 10–60°. The samples morphology was characterized by scanning electron microscopy (SEM) using a Hitachi S4100. To perform this analysis, all the samples were covered with carbon.

2.3. Dielectric Measurements. To study the dielectric response of the composites in the high-frequency region, we used the resonant cavity method. The measurement of the complex permittivity was performed using two resonant cavities (2.7 GHz and 5.0 GHz) [28]. When the sample is introduced inside the cavity, a perturbation of the electromagnetic field is induced. As a consequence, there is a change in the transmission in the cavity, shifting the resonance frequency, and degrading the quality factor. The shift in the cavity resonant frequency, Δf (Figure 1), is related to the real part of the complex permittivity, ε', while the change on the inverse of the quality factor of the cavity, $\Delta(1/Q)$, is related to the imaginary part of the complex permittivity, ε''. It is possible to obtain a simple expression when we consider only the first-order perturbation in the electric field caused by the sample [29]:

$$\frac{\Delta f}{f_0} + i\Delta\left(\frac{1}{2Q}\right) = \varepsilon_0\left(\varepsilon^* - 1\right)\frac{\int_{V_s} E_i^* E_0 dV}{\int_V \varepsilon_0 E_0^2 dV}, \quad (1)$$

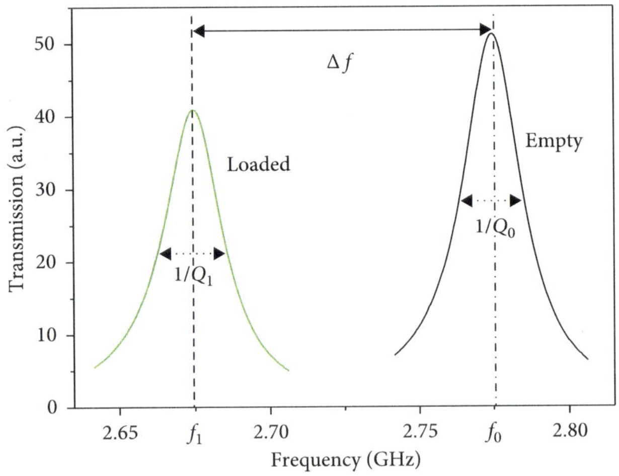

FIGURE 1: Transmission of the cavity with and without PTFE samples, showing the two resonances separated by Δf.

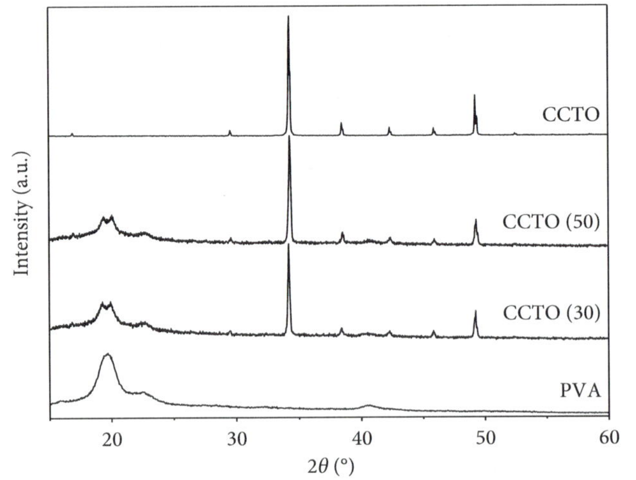

FIGURE 2: XRD patterns of the CCTO, PVA, and CCTO(x) for x = 50 and 30.

where f_0 is the resonance frequency of the cavity, $\varepsilon* = \varepsilon' - i\varepsilon''$ is the complex permittivity of the material, and E_i and E_0 are the electric fields inside and outside the material, respectively. The integration is performed in the volume of the sample, V_S, and in the volume of the cavity, V. Splitting the real and imaginary parts, we can obtain the expressions for ε' and ε'':

$$\varepsilon' = K\frac{\Delta f}{f_0}\frac{V}{V_S} + 1, \quad (2)$$

$$\varepsilon'' = \frac{K}{2}\Delta\left(\frac{1}{Q}\right)\frac{V}{V_S}, \quad (3)$$

where K is a constant related to the depolarization factor, which depends upon the geometric parameters. This parameter can be determined using a sample of known dielectric constant. In our case, we used a polytetrafluorethylene (PTFE) sample with the same size and shape of the studied samples. In these measurements, were used rectangular cavities operating in the $TE_{1,0,5}$ and $TE_{1,0,11}$ modes with the resonant frequencies of 2.7 GHz and 5.0 GHz, respectively. The transmission of the cavities was measured using a HP 8753D Network Analyzer, at room temperature.

3. Results and Discussion

3.1. X-Ray Analysis. The XRD patterns were obtained at room temperature (300 K) by step scanning using powdered samples. Figure 2 shows the XRD pattern of the base samples (CCTO and PVA) and of the CCTO(x) for $x = 30$ and $x = 50$. The pattern obtained from the database of Inorganic Crystal Structures Database (ICSD) confirms the presence of CCTO as a major phase, presenting a cubic structure, with space group 1 m–3 m and space group number 229. Using Celref program [30], the lattice parameters of the CCTO powders were obtained: $a = b = c = 7.3998 \pm 0.0027$ Å and $\alpha = \beta = \gamma = 90.00°$, which is in accordance with the ICSD data of CCTO ($a = 7.3930$ Å and $\alpha = \beta = \gamma = 90.0000°$). From the patterns of the base samples, we could see in Figure 2 that the patterns of the samples with $x = 30$ and $x = 50$, for example, are formed by peaks that can only be associated with PVA and CCTO. Although it is not presented here, the samples with different values of x show the same pattern. This result was already expected, once the temperature that we used for the heat treatment of the cylindrical samples is too low to degrade the polymer or to initiate any chemical reaction between the CCTO and the polymer.

3.2. SEM Analysis. Figure 3 shows the micrographs of some of the prepared composites. In Figure 3(a), we present the micrograph of the PVA sample. From the micrograph of the surface of PVA polymer sample, we can identify large polymer platelets. Between these structures, it is possible to identify many holes uniformly distributed throughout the sample. Figure 3(b) shows the micrograph of the CCTO particles. From this micrograph, it is possible to assume that the CCTO grains have dimensions ranging between 3 μm and 15 μm, presenting a spherical-like shape.

Figure 3(c) shows the micrograph of the CCTO(20) sample, and Figure 3(d) shows the micrograph of the CCTO(50) sample, both with an amplification of 100 times. From the analysis of Figures 3(c) and 3(d), we can point that the porosity is low, particularly for the higher CCTO percentage (CCTO(50)). Combining these micrographs with those with higher amplification (x1k), we can state that, in both cases, the polymeric phase completely enfolds the ceramic particles, which means that a good affinity between these two phases is expected. The polymer is responsible for the aggregation of the ceramic powder and for the stiffness of the final composite, keeping the piece with enough mechanical strength for technological applications.

3.3. Dielectric Analysis. Figure 4 presents the signal transmission along the 2.7 GHz resonant cavity for the following samples: PVA, CCTO(x) (for x = 10, 20, 30, 40), PTFE, and empty cavity. The observed curves of the transmission for

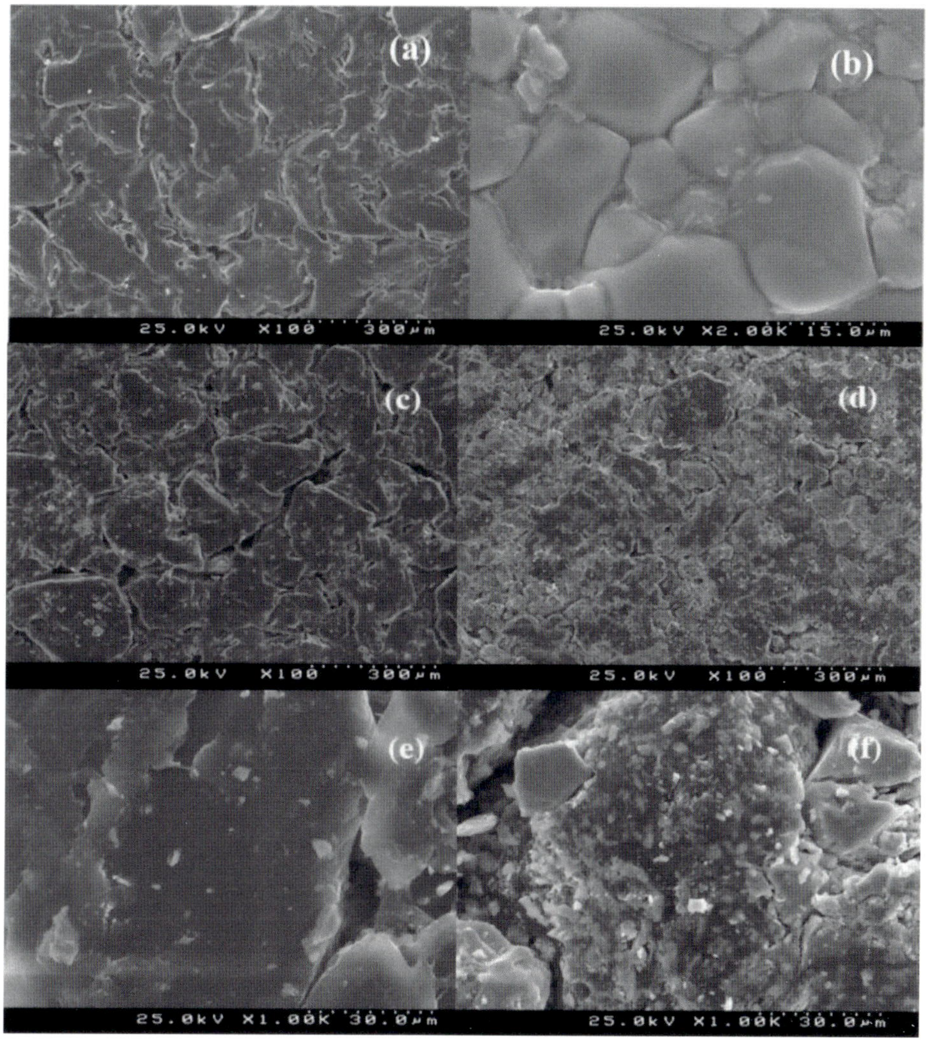

FIGURE 3: SEM micrographs of (a) PVA sample (100x), (b) CCTO powders (2000x), (c) CCTO(20) (100x), (d) CCTO(50) (100x), (e) CCTO(20) (1000x), and (f) CCTO(50) (1000x).

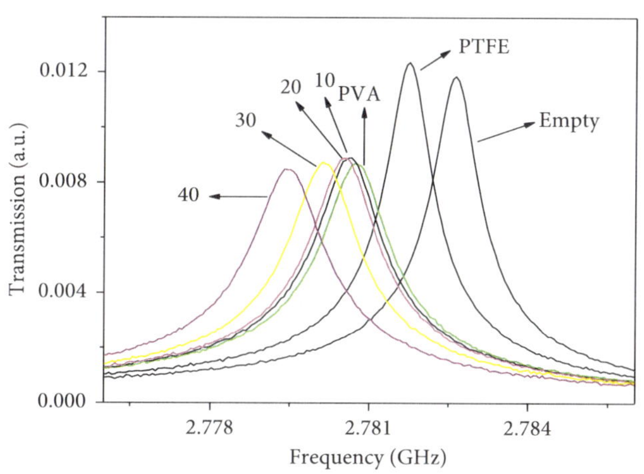

FIGURE 4: Transmission of the 2.7 GHz resonant cavity for PVA, CCTO(x) (for $x = 10, 20, 30, 40$), PTFE, and empty cavity.

the samples with $x = 50$, 60, and 70 are very close to the values found for $x = 40$. These are not presented to avoid overlapping of the curves. Increasing the weight fraction of CCTO powder of the composite implies a larger perturbation in the electromagnetic field configuration, leading to a frequency shift to the lower frequencies (red shift) and a pronounced attenuation of the intensity of the transmission signal.

Figure 5 shows the dielectric constant, as a function of the weight fraction of CCTO in the composites, for 2.7 GHz and 5.0 GHz cavities, measured at room temperature. It also shows the imaginary part of the complex permittivity (ε'') as a function of the weight fraction of CCTO for the 5.0 GHz cavity. The values of ε'' in the cavity of 2.7 GHz were the same for all the samples which means that ε'' is too low to be measured by this method. As expected, the dielectric constant (ε') increases with the weight fraction of CCTO, for each of the analyzed frequencies, and a similar behavior is observed for ε''. The real part of the complex permittivity for the pure PVA samples was 5.5 at 2.7 GHz and 5.3 at 5.0 GHz (and $\varepsilon'' = 0.11$). For the sample CCTO(10), it was measured 7.4 at 2.7 GHz cavity and 6.7 at 5.0 GHZ ($\varepsilon'' = 0.11$). The dielectric constants for the sample CCTO(70) were 17.4 at 2.7 GHz and 16.2 at 5.0 GHZ ($\varepsilon'' = 0.31$).

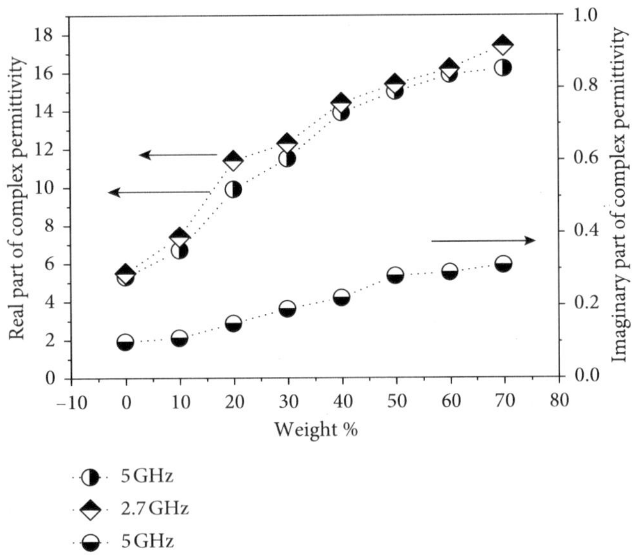

FIGURE 5: Real and imaginary parts of complex permittivity measured in 2.7 GHz and 5 GHz resonant cavity, for different weight fractions of CCTO.

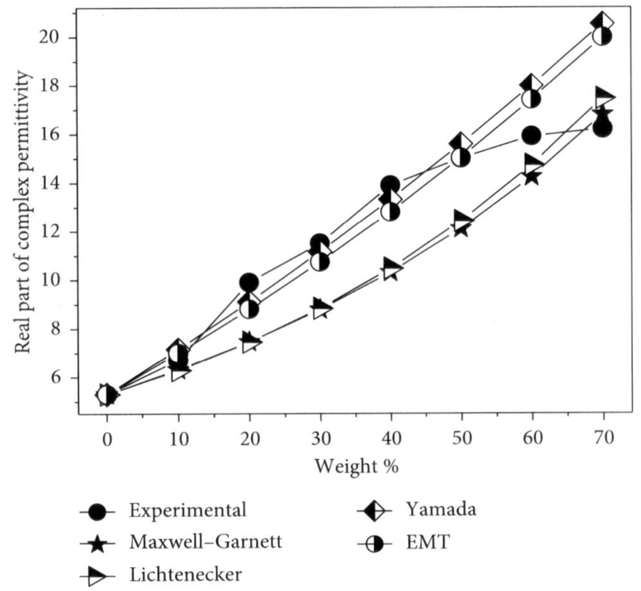

FIGURE 6: Real part of complex permittivity of the composite samples, as a function of the weight fraction of CCTO, obtained for the 5 GHz cavity, and respective fitting using four classic mixing models.

Several models were used in order to predict the dielectric constant of the composite material. Figure 6 shows the experimental dielectric constant at 5.0 GHz, and, for comparison, the dielectric constants predicted by the Maxwell Garnett, Lichtenecker, EMT, and Yamada models are also included. The Maxwell Garnett [31] and the Lichtenecker [32, 33] models describe the dielectric property of a biphasic dielectric mixture comprising spherical crystallites, with high-dielectric constant, dispersed in a matrix of low-dielectric constant. According to the Maxwell Garnett model, the effective dielectric constant of the composite is given by

$$\varepsilon' = \varepsilon_1\left(1 + \frac{3V_f\beta}{1 - V_f\beta}\right),$$
$$\beta = \frac{\varepsilon_2 - \varepsilon_1}{\varepsilon_2 + 2\varepsilon_1}, \quad (4)$$

where ε_1 and ε_2 are the permittivities of the polymer matrix and of the CCTO, respectively. V_f is the volume fraction of the ceramic fillers. The Lichtenecker model is expressed by

$$\ln\varepsilon' = V_f \ln\varepsilon_2 + (1 - V_f)\ln\varepsilon_1. \quad (5)$$

From Figure 6, one can see that the predicted values for ε', for both Maxwell Garnett and Lichtenecker models, deviate too much from that of the measured experimental values, only correctly fitting for the samples with lower (10%) and higher (70%) weight fractions of CCTO. The micrograph images (Figure 3(b)) show that the CCTO grains do not present a perfect spherical geometry, and the discrepancy between experimental and predicted values could be directly related to this fact.

Since CCTO particles dispersed in the matrix have no spherical shape, as could be evidenced by SEM, the models that take into account different morphologies for the particles are more suitable. One example of this is the effective medium theory (EMT) model. The real part of the dielectric constant for the composite in the EMT model is given as follows [34]:

$$\varepsilon' = \varepsilon_1\left(1 + \frac{V_f(\varepsilon_2 - \varepsilon_1)}{\varepsilon_1 - n(1 - V_f)(\varepsilon_2 - \varepsilon_1)}\right), \quad (6)$$

where n is a ceramic morphology fitting factor. Small values of n indicate the filler particles have a shape nearly spherical, while high values of n indicate that the shape of the particles is very irregular. The better value for the shape parameter obtained for the experimental results was $n = 0.1$. Although this model is suitable for particles with dimensions less than 1 μm, Figure 6 shows a good approximation between the values predicted by the EMT model and the experimental data obtained for the composites with a weight fraction up to 50%. However, for higher weight fractions, the model begins to deviate, which should be related to an increasing aggregation of the CCTO particles as the polymer matrix becomes less significant.

From Figure 3(b), it can be seen that the CCTO grain morphology and granulometry are not homogenous for all the particle distribution, presenting differences in size and shape. In addition, from Figure 3(d), one can note that some irregular dispersion of the CCTO particles in the polymeric matrix is present. Thereby, another model was used to try a better agreement with the experimental curve: the Yamada model. The Yamada model [35] predicts the following values for real part of the dielectric constant of a biphasic composite:

$$\varepsilon' = \varepsilon_1\left(1 + \frac{\gamma V_f(\varepsilon_2 - \varepsilon_1)}{\gamma\varepsilon_1 - (1 - V_f)(\varepsilon_2 - \varepsilon_1)}\right), \quad (7)$$

where γ is also a fitting factor associated with the ceramic grain morphology. The best fitting was obtained for $\gamma = 14.309$. Globally, the fitting performed with the EMT model presents a better approximation to the experimental data. However, both Yamada and EMT models fail to predict the dielectric behavior of these CCTO/PVA composites for weight fractions higher than 50%. Once again, this discrepancy should be related to a deficient dispersion of the minor phase.

4. Conclusions

Two resonant cavities with 2.7 GHz and 5.0 GHz resonant frequencies were used to measure the complex permittivity of polymer matrix composites, prepared by the dispersion of CCTO ($CaCu_3Ti_4O_{12}$) powders, prepared by the solid-state method, in a polyvinyl alcohol (PVA) matrix. Different weight fractions of the filler were used to form a flexible and high-dielectric constant composite at GHz region. PVA was chosen as the polymeric matrix because of its high thermal and chemical stability, nontoxicity, and for presenting a high-dielectric strength and good charge storage capacity, which makes it usable as the polymer matrix in composite-based electronic components. Within this study, we were able to prepare, by a simple method, a ceramic/polymer composite with desirable dielectric properties, particularly at the microwave frequency region, such as a high-dielectric constant and low dissipation factor. As expected, the dielectric constant increased with the weight fraction of CCTO for both analyzed frequencies, and a similar behavior is observed for ε''. In an attempt to predict the dielectric constant of the CCTO/PVA composites as a function of the weight fraction of the filler, several classical mixing models were used. This way, the EMT (with $n = 0.1$) and Yamada (with $\gamma = 14.309$) models were identified as the most suitable to fit the experimental data. Best fitting results, using EMT and Yamada models, were obtained for weight fractions lower than 50%. Both EMT and Yamada models are more associated with nonspherical shape particles dispersed in a matrix, a fact that could be evidenced by the SEM.

Disclosure

An earlier version has been presented as a poster at the First International Symposium on Dielectric Materials and Applications (ISyDMA'2016).

Conflicts of Interest

The authors declare that they have no conflicts of interest.

Acknowledgments

We thank CAPES Brazilian agency (Coordenação de Aperfeiçoamento de Pessoal de Nível Superior) for the scholarship of K. D. A. Sabóia. This work was funded by FEDER funds through the COMPETE 2020 Programme and National Funds through FCT-Portuguese Foundation for Science and Technology under the project UID/CTM/50025/2013.

Supplementary Materials

(1a) Porosity of the sample PVA (resume); (1b) porosity of the sample PVA (full data); (2a) porosity of the sample CCTO(20) (resume); (2b) porosity of the sample CCTO(20) (full data); (3a) porosity of the sample CCTO(50) (resume); (3b) porosity of the sample CCTO(50) (full data); (4) micrograph of the PVA sample with representation of pores (red); (5) micrograph of CCTO(20) with representation of pores (red); (6) micrograph of CCTO(50) with representation of pores (red). (*Supplementary Materials*)

References

[1] P. Hu, Y. Shen, Y. Guan et al., "Topological-structure modulated polymer nanocomposites exhibiting highly enhanced dielectric strength and energy density," *Advanced Functional Materials*, vol. 24, no. 21, pp. 3172–3178, 2014.

[2] Z. B. Pan, L. M. Yao, J. W. Zhai, D. Z. Fu, B. Shen, and H. T. Wang, "High-energy-density polymer nanocomposites composed of newly structured one-dimensional $BaTiO_3$@Al_2O_3 nanofibers," *ACS Applied Materials & Interfaces*, vol. 9, no. 4, pp. 4024–4033, 2017.

[3] B. Xie, H. B. Zhang, Q. Zhang et al., "Enhanced energy density of polymer nanocomposites at a low electric field through aligned $BaTiO_3$ nanowires," *Journal of Materials Chemistry A*, vol. 5, no. 13, pp. 6070–6078, 2017.

[4] Y. Mobarak, M. Bassyouni, and M. Almutawa, "Materials selection, synthesis, and dielectrical properties of PVC nanocomposites," *Advances in Materials Science and Engineering*, vol. 2013, Article ID 149672, 6 pages, 2013.

[5] Y. H. Jin, N. Xia, and R. A. Gerhardt, "Enhanced dielectric properties of polymer matrix composites with $BaTiO_3$ and MWCNT hybrid fillers using simple phase separation," *Nano Energy*, vol. 30, pp. 407–416, 2016.

[6] F. Amaral, C. P. L. Rubinger, F. Henry, L. C. Costa, M. A. Valente, and A. Barros-Timmons, "Dielectric properties of polystyrene–CCTO composite," *Journal of Non-Crystalline Solids*, vol. 354, pp. 5321-5322, 2008.

[7] Y. Shen, A. Gu, G. Liang, and L. Yuan, "High performance $CaCu_3Ti_4O_{12}$/cyanate ester composites with excellent dielectric properties and thermal resistance," *Composites Part A: Applied Science and Manufacturing*, vol. 41, no. 11, pp. 1668–1676, 2010.

[8] H. N. Friedlander, H. E. Harris, and J. G. Pritchard, "Structure–property relationships of poly(vinyl alcohol). I. Influence of polymerization solvents and temperature on the structure and properties of poly(vinyl alcohol) derived from poly(vinyl acetate)," *Journal of Polymer Science Part A-1: Polymer Chemistry*, vol. 4, no. 3, pp. 649–664, 1966.

[9] S. K. Pasha, K. Deshmukh, M. B. Ahamed, K. Chidambaram, M. K. Mohanapriya, and N. A. Raj, "Investigation of microstructure, morphology, mechanical, and dielectric properties of PVA/PbO nanocomposites," *Advances in Polymer Technology*, vol. 36, no. 3, pp. 352–361, 2017.

[10] K. Deshmukh, M. B. Ahamed, R. R. Deshmukh et al., "Synergistic effect of vanadium pentoxide and graphene oxide

in polyvinyl alcohol for energy storage application," *European Polymer Journal*, vol. 76, pp. 14–27, 2016.

[11] M. A. Subramanian, D. Li, N. Duran, B. A. Reiner, and A. W. Sleight, "High dielectric constant in $ACu_3Ti_4O_{12}$ and $ACu_3Ti_3FeO_{12}$ phases," *Journal of Solid State Chemistry*, vol. 151, no. 2, pp. 323–325, 2000.

[12] M. S. Ivanov, F. Amaral, V. A. Khomchenko, J. A. Paixão, and L. C. Costa, "Investigation of micro- and nanoscale barrier layer capacitance mechanisms of conductivity in $CaCu_3Ti_4O_{12}$ via scanning probe microscopy technique," *RSC Advances*, vol. 7, no. 65, pp. 40695–40704, 2017.

[13] S. Jesurani, S. Kanagesan, S. Hashim, I. Ismail, and I. Ibrahim, "Microstructural and dielectric properties of Zr doped microwave sintered $CaCu_3Ti_4O_{12}$ synthesized by sol-gel route," *Advances in Materials Science and Engineering*, vol. 2014, Article ID 187420, 6 pages, 2014.

[14] F. Amaral, L. C. Costa, M. A. Valente et al., "Colossal dielectric constant of poly-and single-crystalline $CaCu_3Ti_4O_{12}$ fibers grown by the laser floating zone technique," *Acta Materialia*, vol. 59, no. 1, pp. 102–111, 2011.

[15] M. A. Subramanian, W. J. Marshall, T. G. Calvarese, and A. W. Sleight, "Valence degeneracy in $CaCu_3Cr_4O_{12}$," *Journal of Physics and Chemistry of Solids*, vol. 64, no. 9-10, pp. 1569–1571, 2003.

[16] S. V. Kalinin, J. Shin, G. M. Veith et al., "Real space imaging of the microscopic origins of the ultrahigh dielectric constant in polycrystalline $CaCu_3Ti_4O_{12}$," *Applied Physics Letters*, vol. 86, no. 10, pp. 102902–102905, 2005.

[17] F. Amaral, C. P. L. Rubinger, M. A. Valente, L. C. Costa, and R. L. Moreira, "Enhanced dielectric response of GeO_2-doped $CaCu_3Ti_4O_{12}$ ceramics," *Journal of Applied Physics*, vol. 105, no. 3, p. 034109, 2009.

[18] A. P. Ramirez, M. A. Subramanian, M. Gardel et al., "Giant dielectric constant response in a copper-titanate," *Solid State Comunications*, vol. 115, no. 5, pp. 217–220, 2000.

[19] X. You, N. Chen, and G. Du, "Constructing three-dimensionally interwoven structures for ceramic/polymer composites to exhibit colossal dielectric constant and high mechanical strength: $CaCu_3Ti_4O_{12}$/epoxy as an example," *Composites Part A: Applied Science and Manufacturing*, vol. 105, pp. 214–222, 2018.

[20] Z. M. Dang, T. Zhou, S. H. Yao et al., "Advanced calcium copper titanate/polyimide functional hybrid films with high dielectric permittivity," *Advanced Materials*, vol. 21, no. 20, pp. 2077–2082, 2009.

[21] P. Mishra and P. Kumar, "Enhancement of dielectric properties of 0.2[BZT-BCT]–0.8[(1 $-$ x)epoxy- x CCTO] (x = 0.02, 0.04, 0.06, 0.08 & 0.1) composites for embedded capacitor and energy harvesting applications," *Journal of Alloys and Compounds*, vol. 617, pp. 899–904, 2014.

[22] X. L. Chao, P. Wu, Y. Zhao, P. F. Liang, and Z. P. Yang, "Effect of $CaCu_3Ti_4O_{12}$ powders prepared by the different synthetic methods on dielectric properties of $CaCu_3Ti_4O_{12}$/polyvinylidene fluoride composites," *Journal of Materials Science: Materials in Electronics*, vol. 26, no. 5, pp. 3044–3051, 2015.

[23] B. Shri Prakash and K. B. R. Varma, "Dieletric behavior of CCTO/epoxy and Al-CCTO/epoxy composites," *Composites Science and Technology*, vol. 67, no. 11-12, pp. 2363–2368, 2007.

[24] A. G. Pinheiro, F. M. Pereira, M. R. P. Santos, H. H. B. Rocha, and A. S. B. Sombra, "Electric properties of $Bi_4Ti_3O_{12}$(BIT)-$CaCu_3Ti_4O_{12}$ (CCTO) composite substrates for high dielectric constant devices," *Journal of Materials Science*, vol. 42, no. 6, pp. 2112–2120, 2007.

[25] A. F. L. Almeida, P. B. A. Fechine, J. C. Góes, M. A. Valente, M. A. R. Miranda, and A. S. B. Sombra, "Dielectric properties of $BaTiO_3$ (BTO) – $CaCu_3Ti_4O_{12}$ (CCTO) composite screen-printed thick films for high dielectric constant devices in the medium frequency (MF) range," *Materials Science and Engineering B*, vol. 111, no. 2-3, pp. 113–123, 2004.

[26] L. A. Ramajo, M. A. Ramírez, P. B. Bueno, M. M. Reboredo, and M. S. Castro, "Dielectric behaviour of $CaCu_3Ti_4O_{12}$-epoxy composites," *Materials Research*, vol. 11, no. 1, pp. 85–88, 2008.

[27] M. Abdelaziz and E. M. Abdelrazek, "Effect of dopant mixture on structural, optical and electron spin resonance properties of polyvinyl alcohol," *Physica B: Condensed Matter*, vol. 390, no. 1-2, pp. 1–9, 2007.

[28] K. D. Sabóia, H. Girão, A. Sombra et al., "Microwaves dielectric properties of $Y_3Fe_5O_{12}$-$CaCu_3Ti_4O_{12}$ composites," in *Proceedings of IEEE International Conference in Microwave and Optoelectronics (IMOC'11)*, pp. 537–540, Natal, Brazil, October 2011.

[29] L. C. Costa, S. Devesa, and F. Henry, "A 5 GHz resonant cavity for complex permittivity measurements: design, test performances and application," *Materials Science Forum*, vol. 514–516, pp. 1561–1565, 2006.

[30] J. Laugier and A. Filhol, *CELREF Program*, ILL, Grenoble, France, 1978.

[31] J. C. Maxwell, *A Treatise on Electricity and Magnetism*, Dover Corporation, Downers Grove, IL, USA, 1954.

[32] S. George and M. T. Sebastian, "Three-phase polymer–ceramic–metal composite for embedded capacitor applications," *Composites Science and Technology*, vol. 69, no. 7-8, pp. 1298–1302, 2009.

[33] H. S. Nalwa, *Ferroelectric Polymers*, Marcel Dekker, New York, NY, USA, 1995.

[34] Y. Rao, J. Qu, T. Marinis, and C. P. Wong, "A precise numerical prediction of effective dielectric constant for polymer-ceramic composite based on effective-medium theory," *IEEE Transactions on Components and Packaging Technologies*, vol. 23, no. 4, pp. 680–686, 2000.

[35] T. Yamada, T. Ueda, and T. Kitayama, "Piezoelectricity of a high-contne lead zirconate titanate/polymer composite," *Journal of Applied Physics*, vol. 53, no. 6, pp. 4328–4332, 1982.

5

Experimental Study of Aged and Seriously Damaged RC Beams Strengthened Using CFRP Composites

Ning Zhuang,[1,2] Honghan Dong,[1,2] Da Chen,[1,2] and Yeming Ma[1,2]

[1]*Jiangsu Key Laboratory of Coast Ocean Resources Development and Environment Security, Hohai University, Nanjing, China*
[2]*College of Harbor Coastal and Offshore Engineering, Hohai University, Nanjing, China*

Correspondence should be addressed to Da Chen; chenda@hhu.edu.cn

Academic Editor: Alain Portavoce

This paper presents results from experiments on aged and seriously damaged reinforced concrete (RC) beams strengthened with different arrangements of external carbon fiber-reinforced polymer (CFRP) laminates and end anchorages. Seven RC beams from an old bridge, measuring 250 × 200 × 2300 mm, were tested. All specimens were loaded to yield load to evaluate initial mechanical properties. Then, these seriously damaged specimens were repaired using different CFRP-reinforcing schemes and reloaded to failure. The yield load growth due to CFRP reinforcement ranged from 5% to 36%. Different parameters including CFRP dimension and position, bonding length, and end anchorage were investigated and facilitated conclusions on beam ductility, load-midspan deflection response, and failure mode. This research contributes to knowledge about the CFRP repair of aged and seriously damaged beams to ensure better performance in overloaded conditions.

1. Introduction

Reinforced concrete (RC) is one of the most widely used building materials in civil engineering applications. Aging of the material and overloading caused by misuse or unusual events may damage RC structures. Thus, RC components must be maintained and sometimes strengthened to improve their service life. Given the complex nature of RC construction methods, which require the use of heavy equipment, conventional reinforcement techniques are not appropriate. Carbon fiber-reinforced polymer (CFRP) products, which have a number of advantages (such as corrosion resistance, favorable strength-to-weight ratio, durability in adverse environments, and ease of application), have been widely accepted as new effective technologies for restoring or improving the performance of RC structures [1–6].

Bonding a CFRP sheet transverse to the axis of the member, or perpendicular to potential shear cracks, will increase the strength of the member [7–9]. Experiments have also shown that beams strengthened with CFRP can increase ultimate strength by 83–125% [10]. However, premature debonding failure due to separation of laminates has been shown to be the main problem of this method due to excessive stress concentration at the end of laminates [11]. Such failure can be avoided using different anchorage systems [12, 13].

On the contrary, the performance of aged RC elements of existing structures is different from that of newly constructed specimens in the laboratory because aged samples have been deteriorated by reinforcement corrosion, layer cracks, and concrete cover spalling for many years [14]. Very limited work has been conducted on aged RC elements strengthened by CFRP composites. Pantelides and Fitzsimmons [15] investigated the seismic rehabilitation with CFRP of RC bridge bents. Zanardo [16] studied the mechanical properties of a 40-year-old, four-span RC slab bridge that was retrofitted using the CFRP laminate. An in-service 1950s vintage RC deck girder bridge repaired with externally bonded carbon fiber laminates for shear strengthening was inspected by Williams and Higgins [17].

In addition to above studies, some laboratory experiments have been conducted on newly cast specimens to investigate the mechanical properties of damaged beams [18, 19]. For example, some experimental tests have investigated members strengthened with CFRP under preloading or precracking

[20, 21]. However, these studies focused on the mechanical performance of RC members that were slightly damaged. Actually, structures may be overloaded and damaged seriously, but little information is found in the literature regarding aged and seriously damaged structural members. Therefore, the main objective of this study was to examine the effectiveness of different CFRP sheet arrangements and end anchorages for improving the performance of aged and seriously damaged RC beams.

2. Experimental Programme

2.1. Test Programme. In the test programme, beam specimens from an old bridge were first tested in four-point bending with an effective span of 2,300 mm and a shear span of 900 mm to evaluate their initial strength, as illustrated in Figure 1. Load was applied at a displacement rate of 1.5 mm/min. The beam specimens were instrumented with linear variable differential transformers to measure vertical displacement at the midspan of the beam. Beam specimens were unloaded immediately when the load reached the yield load. Then, these seriously damaged beams were removed from the testing frame and repaired using different CFRP-reinforcing schemes. After maintenance for seven days, these repaired beams were reloaded to failure.

2.2. Specimen Configuration. Beam specimens had cross sections of 200 mm width by 250 mm height and length of 2300 mm (Figure 2). The test specimens were divided into two groups. Group one included three beams that had two steel bars of 16 mm nominal diameter for tension reinforcement with a clear cover of 40 mm. Group two consisted of four beams that had three steel bars of 12 mm nominal diameter for tension reinforcement with a clear cover of 35 mm. Table 1 and Figures 3-4 show the different CFRP-reinforcing schemes of the beam specimens.

2.3. Material Properties. The concrete compressive strength achieved from the core test was 50 MPa. The steel reinforcement was extracted from beam specimens after the test. A mass loss analysis was performed on the reinforcing steel bars in accordance with the standard ASTM GI-90 [22] to obtain the degree of corrosion. The average steel mass loss was 6.8% and 8.5% for Group one and Group two. The flexural reinforcement in test specimens was 16 and 12 mm nominal diameter steel with a yield strength of 360 MPa. The mechanical properties of the carbon fibers used are shown in Table 2.

2.4. CFRP Repairing. Special consideration was given to surface preparation before strengthening the specimens with the CFRP sheet. The weakened surficial layer of concrete was removed from the beam by sandblasting, and then cracks caused in the first loading were sealed by applying an epoxy gel adhesive. Finally, the surface was grounded and cleaned using a high-pressure air jet. The CFRP sheet was bonded to the tension face of the beams with the fibers oriented parallel to the longitudinal axis of the beams. When bonding the CFRP sheet, a primary layer of epoxy was applied to the bonding face first before the CFRP sheet was applied. Then, a secondary layer of epoxy was applied to the surface of the CFRP sheet and forced into the sheet using a paint roller. The properties of the epoxy gel adhesive used for sealing the cracks are shown in Table 3.

3. Test Results and Discussion

3.1. Yield Load Growth

3.1.1. Effect of CFRP Sheet Dimension and Position. The load capacity growth was greatly influenced by the dimension and the position of the strengthened CFRP sheet [23]. The percent increase of yield load could be evaluated as follows:

$$\alpha_{\text{yield}} = \frac{f_{\text{str}} - f_{\text{bef}}}{f_{\text{bef}}} \times 100\%, \quad (1)$$

where α_{yield} = the percent increase of the yield load, f_{str} = the yield load after strengthening (MPa), and f_{bef} = the yield load before strengthening (MPa).

The yield load grew by 5 to 36 percent as beams were strengthened with different CFRP schemes, as shown in Figure 5. In Group one, the yield load of specimens WA, LEU, and WE increased by 5%, 32%, and 36%, respectively. Beam WE showed a higher yield load growth compared with beam LEU. The main reason was that beam LEU was locally strengthened in the middle flexure zone (800 mm), while beam WE was strengthened in full span. Seriously damaged beams tend to crack more easily with increasing loads especially as RC materials have already aged. Locally strengthening flexural span could not prevent or restrain the formation of shear cracks, which could lead to excessive distress. In Group two, the yield load of beams LEA-S, WEA-S, LE-S, and WPA-S increased by 22%, 26.6%, 8.6%, and 12%, respectively. The yield load growth of beam WEA-S was approximately 13.2% higher than that of beam WPA-S because beam WPA-S was also locally strengthened in the middle flexure zone. Thus, when flexural strengthening was applied to a seriously damaged beam, its shear strength often required enhancement.

For aged and seriously damaged beams, flexural/shear cracks would also easily propagate in the beam side faces. A beam strengthened in the side face by the CFRP sheet was expected to exhibit superior performance compared to beams not strengthened in the side face [24]. So, in testing Group two beams, CFRP sheets were bonded onto the beam side faces to prevent premature debonding failure causing by side flexural/shear cracks.

3.1.2. Effect of Bond Length. As the CFRP sheet was used in conjunction with epoxy-laminating resin, the system was representative of the “dry lay-up” application technique for composite strengthening. The tension in the CFRP sheet was transferred to the concrete mainly via shear stress in the adhesive within a short distance nearest to the load application point. As the cracking of the concrete near the loading

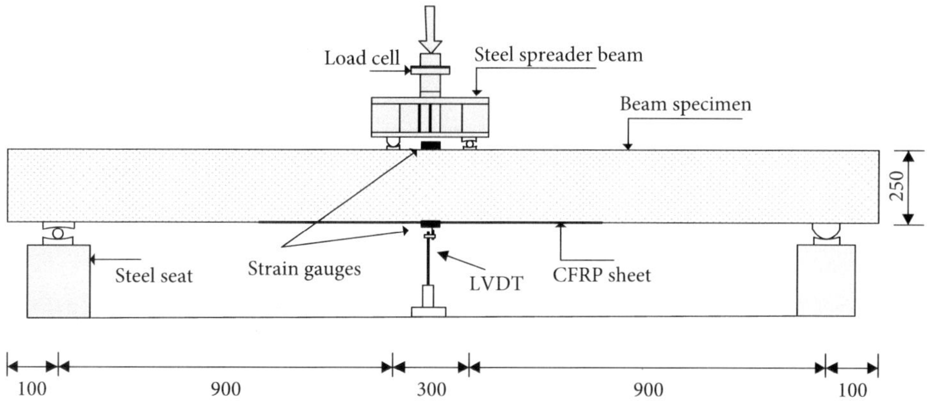

Figure 1: Schematic diagram of the test beam (dimensions in mm).

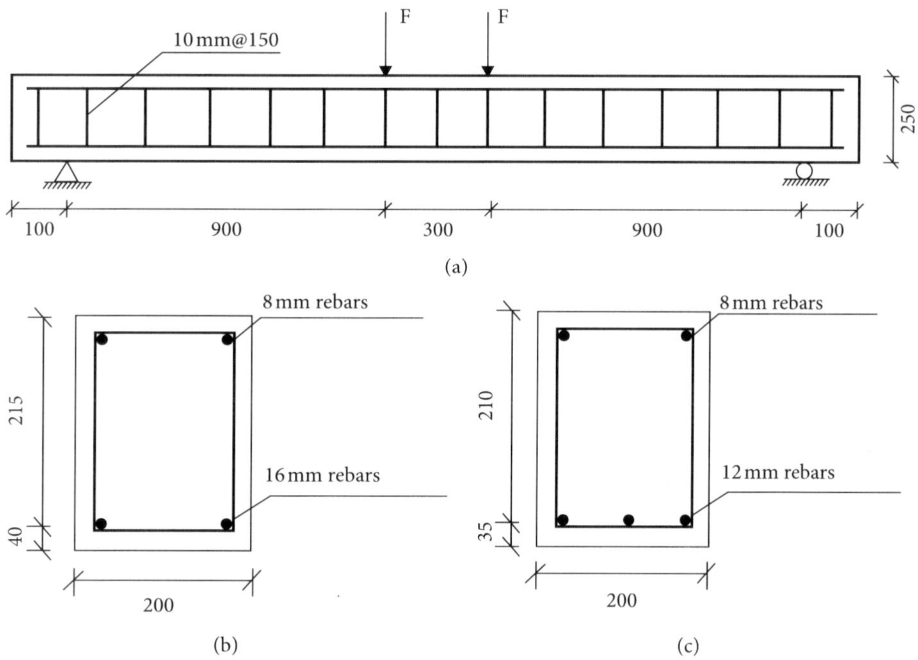

Figure 2: Configuration and reinforcement details of beams: (a) stirrup reinforcement; (b) beam cross sections of Group one specimens; (c) beam cross sections of Group two specimens (dimensions in mm; F is the loading force).

Table 1: Test matrix.

Group	Specimen	Dimension and position of the CFRP sheet	Anchoring schemes	Greater than effective bond length	Beam side face strengthening
One	WE	Full span	No anchorage	Yes	No
	LEU	800 mm in the middle	U-shaped wraps	Yes	No
	WA	Full span (only CF anchors bonded with epoxy resin)	CF anchors	No	No
Two	WEA-S	Full span	CF anchors	Yes	Yes
	LEA-S	800 mm in the middle	CF anchors	Yes	Yes
	WPA-S	Full span (only 800 mm bonded with epoxy resin)	CF anchors	Yes	Yes
	LE-S	800 mm in the middle	No anchorage	Yes	Yes

L: local span strengthened with the CFRP sheet; W: whole span strengthened with the CFRP sheet; A: CFRP sheet fixed by CF anchors; U: U-shaped CFRP wraps as end anchorages; E: CFRP sheet bonded in the full span with epoxy; P: part of the CFRP sheet bonded in the beam span with epoxy resin; S: CFRP sheets were bonded onto the beam side faces.

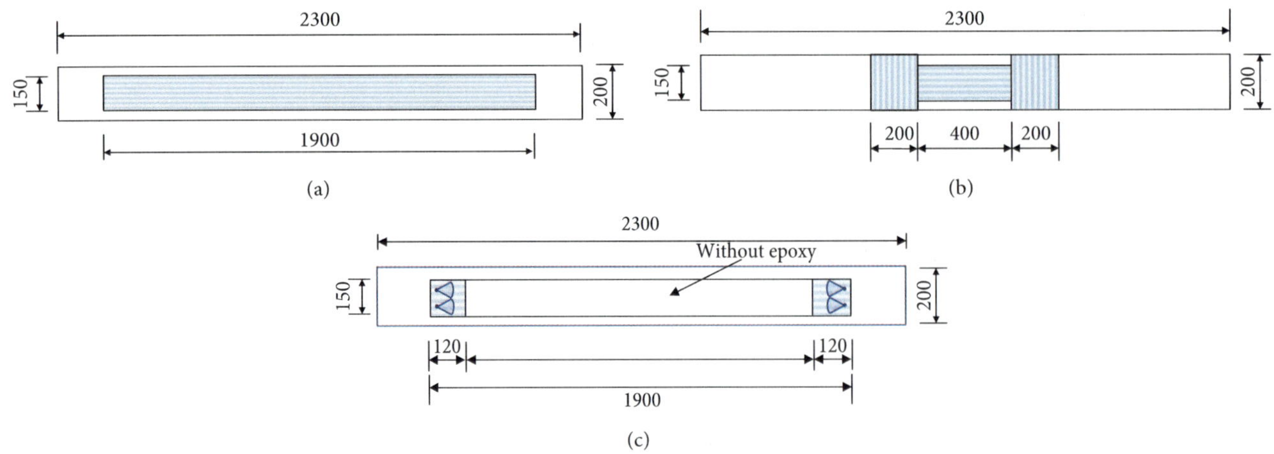

FIGURE 3: Layout of the CFRP sheet on Group one beams: (a) beam WE; (b) beam LEU; (c) beam WA (dimensions in mm).

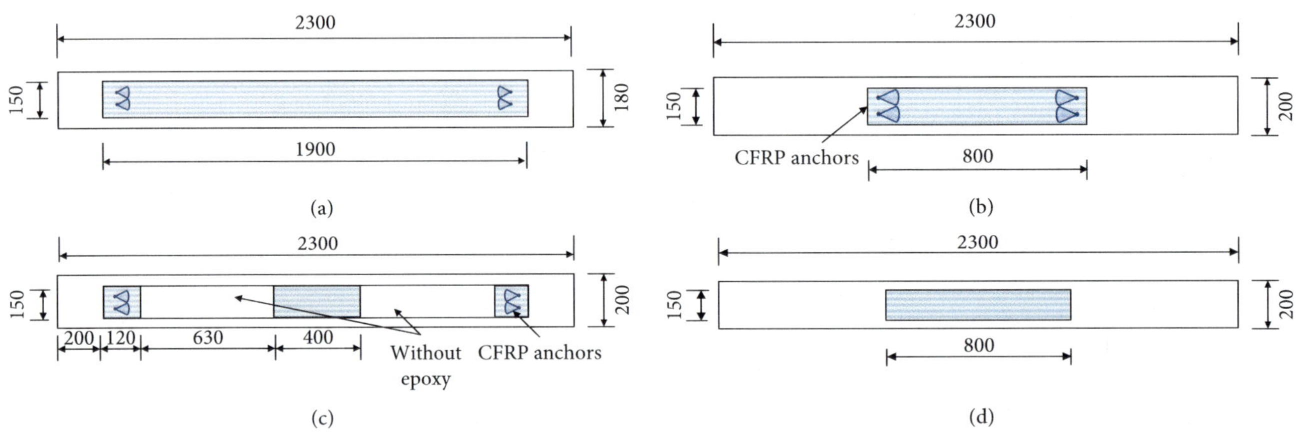

FIGURE 4: Layout of the CFRP sheet on Group two beams: (a) beam WEA-S; (b) beam LEA-S; (c) beam WPA-S; (d) beam LE-S (dimensions in mm).

TABLE 2: Mechanical properties of carbon fibers.

Property	Value
Fiber density	1.8 g/cm^3
Ply thickness	0.381 mm
Ply weight	230 g/m^2
Elongation	1.5%
Tensile strength	3.45 GPa
Tensile modulus	230 GPa

TABLE 3: Mechanical properties of the epoxy gel adhesive.

Property	Value
Modulus of elasticity	1980 MPa
Tensile strength	37.6 MPa
Compressive strength	80.5 MPa
Elongation	1.8%
Bending strength	60 MPa

point propagated, bond resistance was gradually lost in the zone nearest to the load but simultaneously activated a new bond zone further away from the load. This action was repeated until delamination propagated completely through the whole length of the CFRP. The length of CFRP that included the active bonded area was termed the "effective bond length." A bond model of the effective bond length L_e (m) was given by Chen and Teng [25]:

$$L_e = \sqrt{\frac{nE_p t_p}{\sqrt{f'_c}}}, \tag{2}$$

where E_p and t_p are the elastic modulus (MPa) and thickness (mm) of the bonded plate, respectively, and f'_c and n are the cylinder compressive strength of the concrete block (MPa) and the number of CFRP reinforcements, respectively.

The load capacity that a CFRP strip carries can increase when the bond length is less than the effective bond length. When the bonded length exceeds the effective bond length, the load capacity is not dependent on the actual bonded length. Therefore, the bonding length of aged and seriously damaged beams must be greater than the effective bond length. For beam WA, whose bond length was shorter than the effective bond length, its yield load growth was obviously much lower than that of other beams.

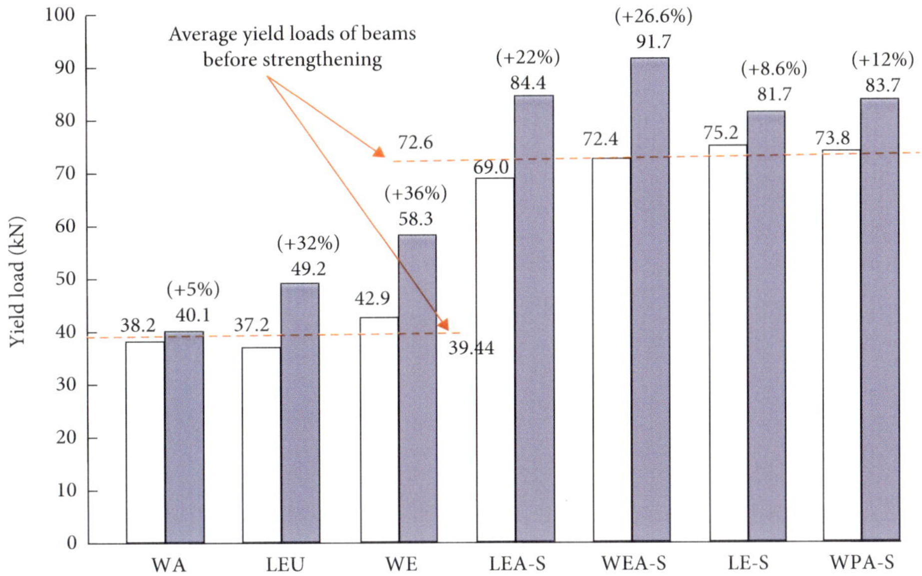

Figure 5: Increases in yield load of nonstrengthened beams and beams strengthened using CFRP.

3.1.3. Contribution of Anchoring Schemes. As illustrated in Figure 6 [26], the shear stress and normal stress were high at the ends of the CFRP laminate and decreased rapidly towards the midspan. Therefore, the separation between the CFRP laminate and the concrete face was very easy at beam ends because of stress concentrations.

Two types of anchoring schemes were used to minimize this kind of debonding tendency: U wraps (U-shaped CFRP wraps) and CF anchors (carbon fiber anchors). U wraps are a type of traditional anchorage scheme that has been studied by many scholars [27, 28]. CF anchors originally developed by the Shimizu Corporation in Japan offered a new way to anchor FRP sheets [29].

CF anchors were cut from the same CFRP sheet material that was bonded onto the beam span. Then, the CF anchors were saturated with epoxy and installed immediately into the holes. The CF anchor depth was 110 mm. The ends of the CF anchors were fanned out over the CFRP sheet. The details of CF anchors and U wraps are shown in Figure 7.

As shown in Figure 5, beams LEA-S (CF anchors) and LEU (U-shaped CFRP wraps) both exhibited higher yield load growth compared with beam LES (not anchored) as other conditions were the same. This indicated that CF anchors and U-shaped CFRP wraps were both effective methods for mounting CFRP sheets to improve beam strength.

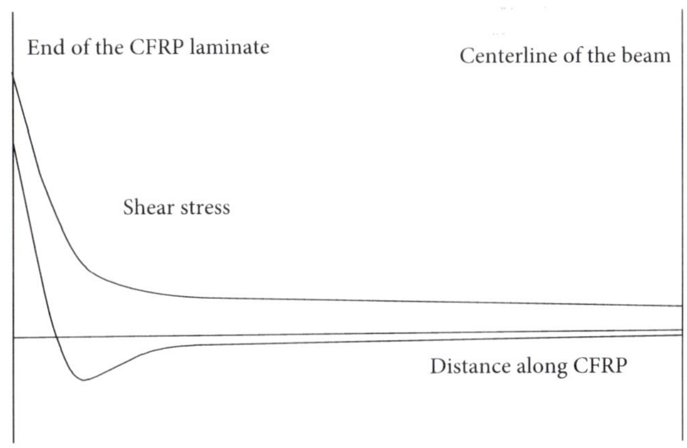

Figure 6: Stress variation with distance from the CFRP sheet end.

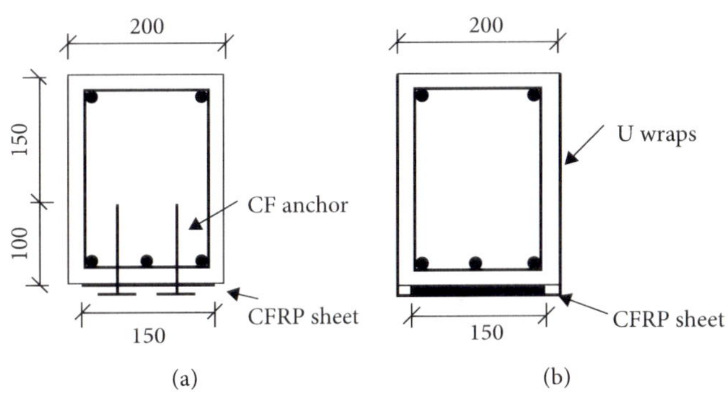

Figure 7: Detailed schemes of CF anchors (a) and U wraps (b) (dimensions in mm).

3.2. Load-Midspan Deflection Response. In Group one, the failure deflections of beams WE, LEU, and WA were 33 mm, 48.9 mm, and 70.1 mm, respectively (Figure 8). The deflections of beams WA and LEU were higher than those of the unanchored beam WE. During the postyielding stage, the load-deflection curves of beams LEU and WA changed the slope and then remained linear until the beams failed. In contrast, there were several sudden decreases in the postyielding stage curve of beam WE because of the sequential rupturing of the CFRP sheet.

In Group two, the failure deflections of beams LEA-S, WEA-S, WPA-S, and LE-S were 26 mm, 51 mm, 71 mm, and 30 mm, respectively (Figure 9). Due to the CF anchors, the beam WEA-S did not have sudden decreases in the load-deflection curve compared with beam WE. The deflection of

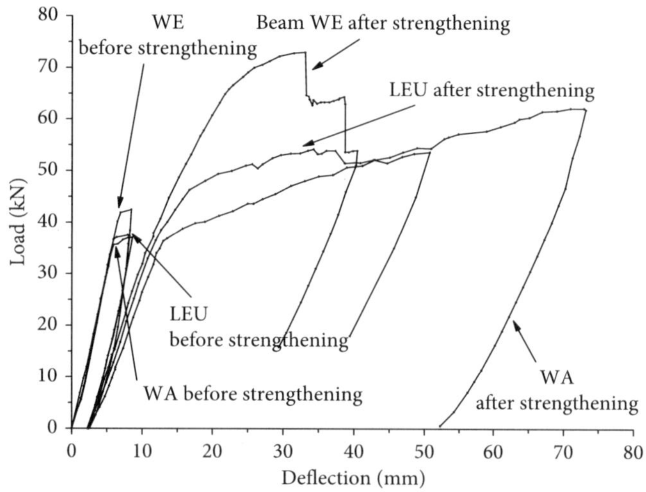

FIGURE 8: Load-deflection curves of Group one beams.

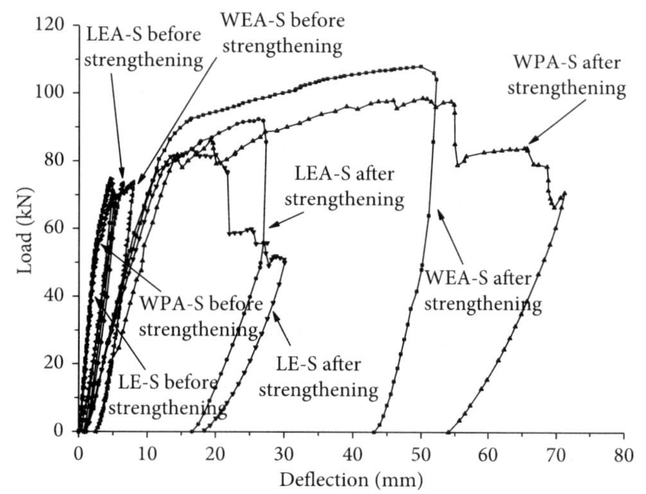

FIGURE 9: Load-deflection curves of Group two beams.

beam LEA-S was slightly smaller, but the postyield stiffness was greater than that of beam LE-S, which indicated that CF anchors could also effectively increase beam stiffness after yield load.

Beam WPA-S was strengthened in full span, but its CFRP sheet in the shear span was not bonded to the concrete face and therefore could not fully increase shear strength, in contrast to beam WEA-S. So, the yield load growth of WPA-S was 15% smaller, while the failure deflection was 34% larger, than that of WEA-S. And the CFRP sheet in the middle flexural span of beam WPA-S would debond with the increase of load, which also caused a sudden decrease in the load-deflection curve.

During the postyielding stage, the load-deflection curve of beam LEU (U-shaped CFRP wraps) was smoother than that of beam LEA-S (CF anchors). This indicated that the strength of both increased slightly during the yield stage, but the stiffness of beam LEA-S degenerated slowly compared with beam LEU. CF anchors and U-shaped CFRP wraps could both prevent premature end-peeling of the CFRP sheet, while CF anchors could improve beam stiffness more effectively than U wraps, especially in the postyield stage.

3.3. Beam Ductility. Structural ductility can be defined as a material's ability to sustain applied loads beyond the elastic limit without significant loss of load-carrying capacity until failure. Strengthening schemes using externally bonded CFRP are an important technique to influence structural ductility. Ductility can be evaluated based on a deformation ratio (or index) employing the ultimate and failure deflections. The ductility index at ultimate load μ_{ult} and failure μ_{fail} can be computed as follows:

$$\mu_{\text{ult}} = \frac{\delta_{\text{ult}}}{\delta_{\text{yield}}}, \qquad \mu_{\text{fail}} = \frac{\delta_{\text{fail}}}{\delta_{\text{yield}}}, \tag{3}$$

where δ_{ult} = the beam's midspan deflection at ultimate load (mm), δ_{yield} = the beam's midspan deflection at yield load (mm), and δ_{fail} = the beam's midspan deflection at failure. The calculated ductility indices of the beams are shown in Table 4.

The ductility index decreased after beams were strengthened with the CFRP sheet. These responses indicated that the increase of load capacity would cause a reduction of beam ductility. In Group two, the yield load growth of beam LEA-S (strengthened with CF anchors) was higher than that of beam LE-S (without CF anchors). Table 4 also shows that the ductility index at failure μ_{fail} of beam LEA-S was lower, while the ductility index at ultimate load μ_{ult} was higher than that of beam LE-S. This observation indicated that beams strengthened with CF anchors would increase ductility before ultimate load.

3.4. Failure Mode. Beams strengthened with CFRP failed in a variety of modes of flexural cracks, shear cracks alone, shear cracks and debonding, flexural cracks and debonding, and bar bond splitting, as shown in Table 4 and Figure 10.

Beam WEA-S was initiated by vertical cracks in the middle flexural span and failed by concrete crushing at the top compression zone of the midspan (Figure 10(a)). In beams WPA-S and LE-S which were not strengthened in shear spans, the shear cracks initiated first and failed due to CFRP debonding in the middle span (Figure 10(b)). The shear cracks first initiated in the shear span of beam LEU that was "U"-wrapped; then, these cracks propagated with an inclination passing through the CFRP sheets, but without CFRP debonding (Figure 10(c)).

Beam LEA-S (strengthened by CF anchors) failed in the bar bond splitting failure mode (Figure 10(d)). The bar bond splitting failure occurred suddenly by the slip between reinforcing bar and concrete interface, and the beam had a smaller deflection compared with other beams when failed. As the beam failed, the sliding between reinforcing bar and

TABLE 4: The test results of beam specimens.

Group	Beams	Yield load growth (%)	Yield deflection (mm)	Ultimate load deflection (mm)	Failure deflection (mm)	μ_{ult}	μ_{fail}	Failure mode
One	WE	36	16.6	75.8	33	1.93	2.41	Flexural cracks and debonding
	LEU	32	13.5	55.6	48.9	2.74	3.62	Shear cracks
	WA	5	10.2	62.4	70.1	6.87	6.87	Flexural cracks
Two	LEA-S	22	12.6	91.7	26	2.02	2.06	Bar bond splitting
	WEA-S	26.6	12.5	109.2	51	3.91	4.08	Flexural cracks
	WPA-S	12	13.7	98.9	71	4.01	5.18	Shear cracks and debonding
	LE-S	8.6	13.2	91.7	30	1.52	2.40	Shear cracks and debonding

(a) (b) (c) (d) (e) (f)

FIGURE 10: Failure modes of beams: (a) flexural cracks; (b) shear cracks and debonding; (c) shear cracks; (d) bar bond splitting; (e) flexural cracks and debonding (side face); (f) flexural cracks and debonding (bottom face).

concrete interface could be clearly heard. The number of cracks was also much smaller than that of other failure modes.

Beam WE was strengthened along the full span and not anchored. Its failure began with flexural cracks, followed by CFRP debonding in the middle span and failure by crushing of concrete at the top compression zone (Figure 10(e)). WA failed due to flexural cracks (Figure 10(a)) because the length of CFRP used was shorter than the effective bond length; thus, the CFRP sheet could not effectively improve the beam's load capacity.

4. Conclusions

This study investigated the mechanical properties of aged and seriously damaged RC beams strengthened with different CFRP sheet schemes. The experimental data justify the following conclusions.

Externally bonded CFRP sheet is an effective way to improve the flexural behavior of aged and seriously damaged RC beams. In this study, yield load growth ranges from 5% to 36% after beams were strengthened. The yield load growth is affected by CFRP sheet dimension and position, bond length, and anchoring schemes. The beams strengthened will show higher load capacity, but in the same time will decrease the stiffness of beams.

As strengthening aged and seriously damaged RC beams, CFRP sheets should extend over the shear span to prevent or restrain the formation of shear cracks. Beam side faces also need to be strengthened with the CFRP sheet to prevent propagation of cracks.

The arrangements of external CFRP laminates and end anchorages have a significant effect on beam flexural behavior. In order to have a good strengthening effect, the CFRP bond length must be longer than the effective bond length. Stress concentrates at the CFRP laminate end and decreases rapidly in the midspan direction. U wraps and CF anchors can effectively prevent premature end-peeling of the CFRP sheet, and CF anchors improve beam stiffness more effectively compared with U wraps, especially in the post-yield stage. Beams strengthened with CF anchors can increase ductility in the yielding stage compared to beams with no anchors.

Aged and seriously damaged RC beams strengthened with the CFRP sheet fail in a variety of modes including flexural cracks, shear cracks, shear cracks and debonding, flexural cracks and debonding, and bar bond splitting. The results from this study, and the conclusions they support, can be applied in engineering and construction practice to guide the repair of aged and seriously damaged RC beams.

Notations

α_{yield}: The percent increase of the yield load
f_{str}: The yield load after strengthening (MPa)
f_{bef}: The yield load before strengthening (MPa)
L_e: The effective bond length (m)
E_p: The elastic modulus (MPa) of the bonded plate
t_p: The thickness (mm) of the bonded plate
f'_c: The cylinder compressive strength of the concrete block (MPa)
n: The number of CFRP reinforcements
δ_{ult}: The beam's midspan deflection at ultimate load (mm)
δ_{yield}: The beam's midspan deflection at yield load (mm)
δ_{fail}: The beam's midspan deflection at failure
μ_{ult}: The ductility index at ultimate load and failure
μ_{fail}: The ductility index at failure.

Conflicts of Interest

The authors declare that they have no conflicts of interest.

Acknowledgments

This work was supported by the National Natural Science Foundation of China (51679080 and 51379073), Fundamental Research Funds for the Central Universities (2018B655X14), Postgraduate Research and Practice Innovation Program of Jiangsu Province (KYCX18_0630), and Scientific and Technology Project of Jiangsu Provincial Water Resources Department (2017030) and sponsored by Qing Lan Project.

References

[1] V. Alecci, S. B. Bati, and G. Ranocchiai, "Concrete columns confined with CFRP wraps," *Materials and Structures*, vol. 47, no. 3, pp. 397–410, 2014.

[2] N. Attari, S. Amziane, and M. Chemrouk, "Flexural strengthening of concrete beams using CFRP, GFRP and hybrid FRP sheets," *Construction and Building Materials*, vol. 37, no. 3, pp. 746–757, 2012.

[3] C. Jiang, K. Fan, F. Wu, and D. Chen, "Experimental study on the mechanical properties and microstructure of chopped basalt fibre reinforced concrete," *Materials and Design*, vol. 58, pp. 187–193, 2014.

[4] R. Eid, N. Roy, and P. Paultre, "Normal and high-strength concrete circular elements wrapped with FRP composites," *Journal of Composites for Construction*, vol. 13, no. 2, pp. 113–124, 2009.

[5] H. K. Shehab, A. S. Eisa, and K. A. El-Awady, "Strengthening of cutouts in existing one-way Spanning R. C. Flat slabs using CFRP sheets," *International Journal of Concrete Structures and Materials*, vol. 11, no. 2, pp. 327–341, 2017.

[6] N. Zhuang, Y. Zhou, and H. Sun, "Effects of steel reinforcement corrosion on carbon-fibre-reinforced polymer repaired slabs," *Proceedings of the Institution of Civil Engineers-structures and Buildings*, vol. 169, no. 1, pp. 46–53, 2015.

[7] B. B. Adhikary and H. Mutsuyoshi, "Behavior of concrete beams strengthened in shear with carbon-fiber sheets," *Journal of Composites for Construction*, vol. 8, no. 3, pp. 258–264, 2004.

[8] I. A. Bukhari, R. Vollum, S. Ahmad, and J. Sagaseta, "Shear strengthening of short span reinforced concrete beams with CFRP sheets," *Arabian Journal for Science and Engineering*, vol. 38, no. 3, pp. 523–536, 2013.

[9] T. Norris and H. Saadatmanesh, "Shear and flexural strengthening of R/C beams with carbon fiber sheets," *Journal of Structural Engineering*, vol. 123, no. 7, pp. 903–911, 1997.

[10] M. J. Chajes, T. F. Januszka, D. R. Mertz, T. A. T. Jr, and W. W. F. Jr, "Shear strengthening of reinforced concrete beams using externally applied composite fabrics," *ACI Structural Journal*, vol. 92, no. 3, pp. 295–303, 1995.

[11] P. V. Llauradó, J. Fernández-Gómez, and F. J. G. Ramos, "Influence of geometrical and installation parameters on performance of CFRP anchors," *Composite Structures*, vol. 176, pp. 105–116, 2017.

[12] S. V. Grelle and L. H. Sneed, "Review of anchorage systems for externally bonded FRP laminates," *International Journal of Concrete Structures and Materials*, vol. 7, no. 1, pp. 17–33, 2013.

[13] F. M. Mohee, A. Al-Mayah, and A. Plumtree, "Development of a novel prestressing anchor for CFRP plates: experimental investigations," *Composite Structures*, vol. 176, pp. 20–32, 2017.

[14] J. G. Teng, L. Lam, W. Chan, and J. Wang, "Retrofitting of deficient RC cantilever slabs using GFRP strips," *Journal of Composites for Construction*, vol. 4, no. 2, pp. 75–84, 2000.

[15] C. P. Pantelides and G. Fitzsimmons, "Case study of strategies for seismic rehabilitation of reinforced concrete multicolumn bridge bents," *Journal of Bridge Engineering*, vol. 17, no. 1, pp. 139–150, 2012.

[16] G. Zanardo, H. Hao, Y. Xia, and A. J. Deeks, "Evaluation of the effectiveness of strengthening intervention by CFRP on MRWA bridge no. 3014," *Journal of Composites for Construction*, vol. 11, no. 4, pp. 363–374, 2007.

[17] G. Williams and C. Higgins, "Fatigue of diagonally cracked RC girders repaired with CFRP," *Journal of Bridge Engineering*, vol. 13, no. 1, pp. 24–33, 2008.

[18] I. Ray, G. C. Parish, J. F. Davalos, and A. Chen, "Effect of concrete substrate repair methods for beams aged by accelerated corrosion and strengthened with CFRP," *Journal of Aerospace Engineering*, vol. 24, no. 2, pp. 227–239, 2010.

[19] D. Zhang, S. Shen, Y. Zhao, W. Jin, and T. Ueda, "Cracking behavior of CFRP laminate-strengthened RC beams with premechanical and postmechanical environmental damage," *Journal of composites for construction*, vol. 19, no. 4, article 04014066, 2014.

[20] S. J. E. Dias and J. A. O. Barros, "NSM shear strengthening technique with CFRP laminates applied in high-strength

concrete beams with or without pre-cracking," *Composites Part B: Engineering*, vol. 43, no. 2, pp. 290–301, 2012.
[21] S. B. Zhao, J. F. Li, F. J. Shi, and Z. H. Du, "Shear resistance of pre-cracking reinforced concrete T-beam strengthening with CFRP sheets," *Applied Mechanics and Materials*, vol. 438-439, pp. 472–476, 2013.
[22] ASTM, *G1-90: Standard Profile for Preparing, Cleaning, And Evaluating Corrosion Test Specimens*, ASTM International, West Conshohocken, PA, USA, 1999.
[23] K. Joongkoo, "A study on structural performance evaluation of RC beams strengthened with CFRP plate," *Korean Journal of Construction Engineering and Management*, vol. 5, no. 6, pp. 212–217, 2004.
[24] C. Zeris, J. Anastasakis, and J. Kyriakidis, "Investigation of monotonic and cyclic response of fiber-reinforced polymer strengthened beams," *ACI Structural Journal*, vol. 106, no. 1, pp. 3–13, 2009.
[25] J. F. Chen and J. G. Teng, "Anchorage strength models for FRP and steel plates bonded to concrete," *Journal of Structural Engineering*, vol. 127, no. 7, pp. 784–791, 2001.
[26] M. T. El-Mihilmy and J. W. Tedesco, "Prediction of anchorage failure for reinforced concrete beams strengthened with fiber-reinforced polymer plate," *ACI Structural Journal*, vol. 98, no. 3, pp. 301–314, 2001.
[27] A. U. R. Khan and S. Fareed, "Behaviour of reinforced concrete beams strengthened by CFRP wraps with and without end anchorages," *Procedia Engineering*, vol. 77, pp. 123–130, 2014.
[28] W. Li and C. K. Y. Leung, "Effect of shear span-depth ratio on mechanical performance of RC beams strengthened in shear with U-wrapping FRP strips," *Composite Structures*, vol. 177, no. 1, pp. 141–157, 2017.
[29] M. Kaya and Z. Ç. Kankal, "Effect of anchorage number on behavior of reinforced concrete beams strengthened with glass fiber plates," *International Journal of Concrete Structures and Materials*, vol. 9, no. 4, pp. 415–425, 2015.

6

Biological Resistance and Application Properties of Particleboards Containing Nano-Zinc Oxide

Ladislav Reinprecht, Ján Iždinský, and Zuzana Vidholdová

Faculty of Wood Sciences and Technology, Technical University in Zvolen, T. G. Masaryka 24, 960 53 Zvolen, Slovakia

Correspondence should be addressed to Ladislav Reinprecht; reinprecht@tuzvo.sk

Academic Editor: Mikhael Bechelany

Special particleboards (PBs) proposed for kitchens, bathrooms, hospitals, and some other specific products for interiors should have a sufficient resistance against bacteria, molds, and decaying fungi. This work deals about effects of zinc oxide nanoparticles (nano-ZnO) added into melamine-urea-formaldehyde (MUF) glue in the amounts of 0, 2, 6, 12, or 24% wt. on selected biological, moisture, and strength properties of laboratory-produced one-layer PBs. The nano-ZnO-treated PBs had a higher biological resistance: (1) against the Gram-positive bacterium *Staphylococcus aureus* by up to ca. 70% and the Gram-negative bacterium *Escherichia coli* by up to 50%, since their bacterial activities at using 1.0 McFarland bacterial inoculum decreased from 0.38–0.40 by up to 0.12–0.19 × 10^8 CFU/ml; (2) against the molds *Penicillium brevicompactum* and *Aspergillus niger* by up to ca. 50–63%, since their growth intensities (0–4) on the top surfaces of treated PBs decreased according to a modified EN 15457 from 2.33–2.67 by up to 1.17–1.0; (3) against the brown-rot fungus *Coniophora puteana* by up to 85.7%, since their weight losses reduced according to a modified ENV 12038 from 17.4% by up to 2.5%. The presence of nano-ZnO in PBs uninfluenced their swelling, water absorption, and bending strength; however, it decreased their internal bond strength by up to 38.8%.

1. Introduction

According to Imamura [1] and more other researches (e.g., [2–4]), the biological resistance of particleboards (PBs)—and similar other wood composites, that is, fibreboards, cardboards, oriented strand boards, plywood, laminated veneer lumbers, and glulam—against bacteria, molds, decaying fungi, and insects is in a defined environment affected first of all by (1) the used wood species, (2) the size and shape of wood particles, (3) the type and amount of organic glue or inorganic cement adhesive and used additives including biocides, and (4) the dimension, surface roughness, and other geometric characteristics of the PB and products made from this material. By the abovementioned factors (1) to (4) can also be influenced application properties of PBs, for example, their swelling and strength [4–8].

The common types of PBs produced for a dry inner environment, that is, for the use classes 1 and 2 [9]—it means unlike water-resistant glulam or plywood suitable for the use classes 3 and 4 [9]—must not be resistant to water or to bacterial and fungal attacks. However, the need for a higher biological resistance of PBs in interiors occurs when they have to withstand a shorter-time impact of additional wetting when achieving a moisture content above 20%. So, in the high bacterial, mold, and decay risk environments, for example, if creation of condensate water cannot be excluded, the PBs should have an increased biological durability. It can be achieved (1) by using acceptable-environmentally friendly and technologically suitable and applicable biocides [10–14] and (2) by using particles either from more durable wood species [15] or from cuttings and other residues of thermally and chemically modified woods having a higher resistance to pests as natural woods.

The increase of the biological resistance of PBs against bacteria, molds, and decaying fungi can be realised (1) by their postmanufacture surface treatment (PMST), for example, by lamination with traditional melamine resin-impregnated papers [16], by lamination with impregnated papers containing biocides [17, 18], by painting with acrylic, epoxy, or other coatings unmodified or modified without biocides [19], and by fumigation with creation of stable biocides, for example, with trimethyl borate which is transformed to boric acid [20] or with supercritical carbon dioxide fluid (SCF-CO_2) of organic biocides [21], and (2) by the in-process treatment (IPT) applying biocides directly in the production process, for example, by impregnation of wood particles with thermostable inorganic, heterocyclic, or other organic fungicides [2, 22], by using special bioactive glues, by addition of biocides into glues [23, 24], by addition of biocides when applying a glue, when layering wood particles, or during pressing of particleboards. A more effective biological protection of PBs can usually be achieved in those cases where the biocides are applied by the IPTs. However, when using such direct treatments, some application properties of PBs can more rapidly be worsened.

For interiors, the general recommendation applies that using environmental-harmful and health-harmful substances, including more types of wood preservatives, should be maximally limited. Therefore, in situations, when a treatment of wood or wooden composites with biocides is needed by EN 460 [25], these substances should be applied only on the basis of valid and regularly updated regulations for biocidal products by ECHA [26, 27].

Nano-ZnO can be applied for interiors. Due to a reduced size of particles, it is evidently a more effective biocide compared to classical ZnO powders having larger-sized particles [28–30]. Nano-ZnO is able to interact with bacteria, fungi, insects, and other pests, either with the surfaces of a living cell or after penetration to the living cell with its core [31]. Several authors described a biological activity of ZnO nanoparticles against the Gram-positive and Gram-negative bacteria [32–35], various species of wood-decaying fungi [36, 37], usually more against the white-rot fungi as against the brown-rot ones [38], and molds [39]. For example, in our previous experiments, it was found that zinc oxide nanoparticles, used as 0.1–1% dispersions for treatment of lime and maple woods, suppressed the decay activity of the white-rot fungus *Trametes versicolor* by up to 56.3% and the brown-rot fungus *Coniophora puteana* by up to 39.3% [40], which on the top surfaces of these wood species partly reduced the activity of the molds *Aspergillus niger* and *Penicillium brevicompactum* [41].

The aim of this study was a complete search on selected bactericidal, fungicidal, moisture, and strength properties of nano-ZnO-treated PBs prepared by the IPT method, when zinc oxide nanoparticles were added into melamine-urea-formaldehyde glue.

2. Experimental

2.1. Treatment of MUF Glue with Nano-Zinc Oxide. The basic melamine-urea-formaldehyde (MUF) glue with a molar ratio of formaldehyde and urea of 1 : 1.15 and with 20% wt. addition of melamine was prepared in Kronospan Bučina DDD Co. Ltd., Zvolen, Slovakia. Its hardener, the "catalyst of condensation," was ammonium nitrate (NH_4NO_3), and it was used in the amount of 2.5% wt. (weight of solid hardener per solid weight of catalysed glue).

Table 1: Curing times of MUF glues treated with zinc oxide nanoparticles determined at 100°C.

MUF glue	Addition of nano-ZnO (% wt.)				
	0	2	6	12	24
Curing time (s)	198	261	271	317	384

The basic MUF glue was treated with zinc oxide nanoparticles (nano-ZnO) (Sigma-Aldrich Co. Ltd., Munich, Germany; <100 nm particle size (DLS); <35 nm average particle size (APS); 50% wt. in water) in the amounts of 0, 2, 6, 12, or 24% wt. (weight of solid nano-ZnO per solid weight of catalysed MUF glue). MUF glues treated with nano-ZnO had maximally 1.94 times prolonged curing time (Table 1). However, the viscosity of all modified MUF glues was constant, that is, 62 ± 2 s measured by Ford 4 cup at 20°C.

2.2. Particleboard Preparation. One-layer particleboards (PBs) with the area size of 360 mm × 360 mm and the nominal thickness of 16.0 mm were prepared in laboratories of the Technical University in Zvolen. For their production, these substances were used with recalculation of their parts on the absolute dry mass: (1) 10 parts of wood-chipped particles—9.5 parts of spruce, fir, and pine and 0.5 part of beech and oak—used at 3.5% moisture content; (2) 1 part of MUF glue—0.965 part of the basic MUF glue, 0.025 part of the hardener ammonium nitrate, and 0.010 part of paraffin—used in a form of 67% water solution; and (3) 0 to 0.24 part of nano-ZnO used in a form of 50% water nanosuspension. A uniform homogenization of ZnO nanoparticles in MUF glues was made by continual mixing of modified glues in a laboratory agitator (Diaf, Denmark) until they were applied to the wood particles.

Application of nano-ZnO-modified MUF glues on wood-chipped particles was performed in a laboratory rotary mixing device (VDL-TU Zvolen, Slovakia). Wood particles coated with glues were then loaded into a prepressing form and finally pressed in a laboratory press CBJ 100-11 (TOS Rakovník, Czech Republic). The pressing process was performed according to a three-stage pressing diagram at the maximum temperature of 210°C, maximum specific pressure of 5.33 MPa, and pressing factor of 14 s, which is analogous with preparation of PBs containing polyethylene terephthalate [42]. Totally 15 PBs were prepared, that is, control and four types treated with nano-ZnO (3 pieces from each).

2.3. Biological Resistance of Particleboards. The bacterial, mold, and decay tests of PBs were performed on the samples with a reduced thickness of 7.0 mm—achieved by cutting of PBs to their half of the nominal thickness—with the intention to perform all biological tests in Petri dishes with a diameter of 120 mm and an internal height of 17 mm. Firstly, PB samples were conditioned at a temperature of 20 ± 2°C and a relative humidity of 65 ± 2%, achieving their EMC of 12 ± 1%. Following this, the side and bottom surfaces of PB samples were 2 times coated with epoxy resin CHS-Epoxy 1200 mixed with hardener P11 (Stachema, Mělník, Czech Republic). Finally, all surfaces of PB samples were sterilized for 1 h with a UV light radiator.

A previous leaching of PB samples in distilled water by the Standard EN 84 [43] (or by other leaching methods) was not performed because the laboratory-prepared nano-ZnO-treated PBs contained a partly water unstable MUF glue. It means that these PBs can be recommended only for indoor expositions with an occasional moisture increasing, where the mitigation of nano-ZnO biocidal effect due to leaching processes is not expected.

2.3.1. Bacterial Test. The resistance test of PBs to bacteria was performed with the Gram-positive bacterium *Staphylococcus aureus* ATCC-25923 and the Gram-negative bacterium *Escherichia coli* ATCC-25922.

The top surfaces of the samples with the size of 50 mm × 50 mm × 7 mm were cleaned with alcohol solution (8.8 : 1.2 mixture of ethanol and 2-propanol) and inoculated in Petri dishes with a diameter of 120 mm with 0.1 ml of bacterial suspension. Two types of bacterial suspension were used in the physiological solution, that is, 0.5 and 1.0 of the McFarland scale (1.5×10^8 and 3.0×10^8 CFU/ml). Incubation of bacteria on the top surfaces of the samples was performed at 37°C for 48 hours. Afterwards, bacteria were stripped from the top surfaces of PBs using a sterile swab, then taken up in the liquid culture medium for 48 hours, and finally preinoculated from the liquid medium into the sodium chloride diagnostic soil in Petri dishes.

The activity of bacteria on the top surfaces of PBs was assessed indirectly in these diagnostic soils in CFU/ml.

2.3.2. Mold Test. The mold resistance test of PBs was performed in accordance with a modified Standard EN 15457 [44], that is, using (1) the standard mold *Aspergillus niger* Tiegh. and also the nonstandard mold *Penicillium brevicompactum* Dierckx instead of the usually recommended *P. purpurogenum*, (2) other dimensions of the samples, and (3) other sterilization methods of the samples.

Samples with the size of 40 mm × 20 mm × 7 mm were at sterile conditions placed into Petri dishes with a diameter of 120 mm on plastic mats—two tested samples into one dish filled with a 3-4 mm thick layer of the 4.9% Czapek–Dox agar (HiMedia Ltd., India)—and inoculated with water suspension of mold spores. Incubation of the samples lasted 21 days at a temperature of 24 ± 2°C and a relative humidity of 90–95%.

The growth intensity of molds on the top surfaces of PBs was assessed by the scale from 0 to 4, using the following criteria: 0 = no mold on the surface; 1 = mold up to 10%; 2 = mold up to 30%; 3 = mold up to 50%; and 4 = mold more than 50% on the surface.

2.3.3. Decay Test. The decay resistance test of PBs was carried out according to a modified Standard ENV 12038 [45] with the brown-rot fungus *Coniophora puteana* (Schumacher ex Fries) Karsten, strain BAM Ebw. 15 using (1) other dimensions and sterilization of the samples and (2) a shorter time of their incubation (6 weeks instead of 16 weeks, as the dimension of the tested samples was smaller compared to that of the standardized ones, 50 mm × 50 mm × thickness).

Samples with the size of 40 mm × 40 mm × 7 mm were at sterile conditions placed into Petri dishes with a diameter of 120 mm on plastic mats under which a fungal mycelium has already grown up—one tested sample and one control sample per dish filled with a 3-4 mm thick layer of the 4.5% malt agar (HiMedia Ltd., India). Incubation of the samples lasted 6 weeks at a temperature of 24 ± 2°C and a relative humidity of 65–75%.

The decay attack of PB samples was evaluated from their weight losses (Δm). For these aims, the samples before and after the decay test were air-conditioned at a temperature of 20 ± 2°C and a relative humidity of 65 ± 2%, achieving their constant weight with an accuracy of 0.001 g. The weight losses of PBs were determined from their weights in conditioned states before (m_0) and after the fungal attack ($m_{\text{fungal attack}}$) by

$$\Delta m = \frac{m_0 - m_{\text{fungal attack}}}{m_0} \cdot 100\,(\%). \quad (1)$$

2.4. Application Properties of Particleboards. PB samples, having a nominal thickness of 16.0 mm, were used for testing their density, moisture, and strength properties. The thickness swelling (TS) and water absorption (WA) were determined after 2 and 24 hours by the Standard EN 317 [46]. The internal bond (IB) strength, that is, the tensile strength perpendicular to the plane of PBs, was determined by the Standard EN 319 [47], and the bending strength (BS), that is, the modulus of rupture of PBs, was determined by the Standard EN 310 [48]. For both strength analyses, a universal machine TiraTest 2200 (VEB TIW Rauenstein, Germany) was used.

3. Results and Discussion

3.1. Biological Resistance of PBs. The nano-ZnO-treated PBs had a better biological resistance against bacterial, mold, and decay attacks compared to the control untreated PB (Table 2).

The activity of the Gram-positive bacterium *Staphylococcus aureus* on the top surfaces of the nano-ZnO-treated PBs decreased maximally about 70–74% (Table 2), that is, from 0.19 up to 0.05×10^8 CFU/ml at using a 0.5 McFarland scale of bacterial inoculum (SBI), or from 0.40 up to 0.12×10^8 CFU/ml at using a 1.0 McFarland SBI. The activity of the Gram-negative bacterium *Escherichia coli* decreased as well (Table 2), however milder, maximally about 50% (i.e., from 0.38 up to 0.19×10^8 CFU/ml) at using 1.0

TABLE 2: Biological resistances of the control PB and the nano-ZnO-treated PBs against two bacteria, two molds, and the brown-rot fungus *C. puteana*.

Biocidal properties of PBs	Bacterial activity (10^8 CFU/ml)				Growth intensity of the mold (0–4)		Weight loss (%)
Nano-ZnO	*Staphylococcus aureus*		*Escherichia coli*		*Penicillium*	*Aspergillus*	*Coniophora*
(% wt. in MUF glue)	0.5*	1.0*	0.5*	1.0*	*brevicompactum*	*niger*	*puteana*
0	0.19	0.40	0.15	0.38	2.33 (0.52)	2.67 (0.52)	17.38 (2.96)
2	0.17	0.40	0.13	0.36	1.83 (0.41)^d	2.00 (0.63)^c	11.60 (0.27)^a
6	0.13	0.31	0.15	0.34	1.33 (0.52)^b	2.17 (0.41)^d	6.34 (0.31)^a
12	0.12	0.23	0.14	0.24	1.33 (0.52)^b	1.83 (0.41)^b	4.68 (1.61)^a
24	0.05	0.12	0.12	0.19	1.17 (0.41)^a	1.00 (0.00)^a	2.48 (1.04)^a

Note. Mean values: (1) activity of each bacterium from 3 samples, (2) growth intensity of each mold from 6 samples, and (3) decay attack with *C. puteana* from 6 samples; standard deviations are in parentheses; the Duncan test with levels of significance (^{a} >99.9%, ^{b}>99% to ≤99.9%, ^{c} > 95% to≤ 99%, and ^{d} < 95%) evaluating bioeffects of nano-ZnO on the treated PBs in relation to the control PB; *the values 0.5 and 1.0 represent the density of the bacterial suspension in the McFarland scale.

McFarland SBI, or only about 20% (i.e., from 0.15 up to 0.12×10^8 CFU/ml) at using 0.5 McFarland SBI.

Positive effect of ZnO nanoparticles reducing the growth activity of the molds *Aspergillus niger* and *Penicillium brevicompactum* on the top surfaces of PBs was unambiguous (Table 2). The growth activity of the *P. brevicompactum* significantly decreased from 2.33 up to 1.17, that is, maximally about 49.8%, and the growth activity of the *A. niger* significantly decreased from 2.67 up to 1.0, that is, maximally about 62.6%. These results are consistent with a low mold resistance of commercial untreated PBs [16], as well as with possibilities for improving their insufficient mold resistance by application of suitable fungicides [3, 49].

It is well known that fungicides based on Zn, Cu, and other heavy metals have a better efficiency against white-rot fungi compared to brown-rot fungi [38, 50]. In spite of this fact, zinc oxide nanoparticles used in our experiment apparently improved the resistance of PBs against the brown-rot fungus *Coniophora puteana*, since weight losses of nano-ZnO-treated PBs significantly decreased from 17.38% up to 2.48%, that is, maximally about 85.7% (Table 2; Figures 1 and 2). Weight losses of PBs, caused by the activity of this decaying fungus, reduced exponentially in accordance with a higher content of nano-ZnO in MUF glues: $\Delta m = 3.02 + \exp\ (2.66 - 0.24 \times w_{(ZnO)}/w_{(MUF)})$ (Figure 1). Similarly, Marzbani et al. [24] determined for PBs bonded with nano-ZnO-modified MUF glues an exponential decrease of their weight losses at action of *C. puteana* and also at action of the white-rot fungus *Trametes versicolor*.

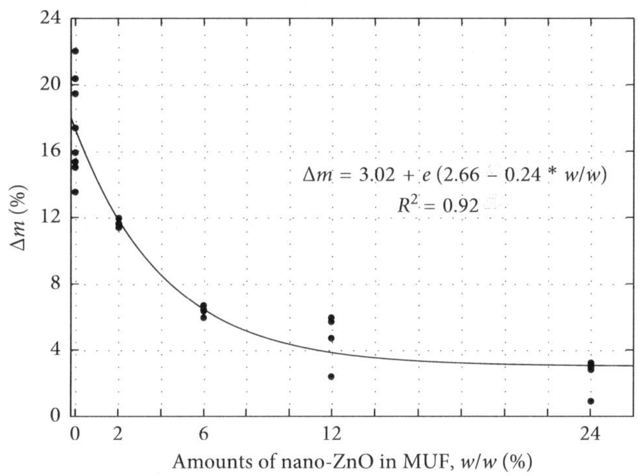

FIGURE 1: Exponential correlation between a higher content of nano-ZnO in MUF glue (w/w, %) used in the production of PBs and a smaller weight loss (Δm, %) of the nano-ZnO-treated PBs exposed to the attack by the brown-rot fungus *Coniophora puteana*.

Biological resistance of chemically preserved PBs depends on more factors, among which the most dominant are the three ones: (1) the kind and amount of biocides; (2) the type and amount of wood particles, glues, waxes, and other additives; and (3) the macro- and microdistribution of biocides, for example, which is dependent on the production technology of PB. However, in practice, environmental conditions and leaching or evaporation of biocides from preserved PBs cannot be neglected [51, 52].

Treatment of wooden composites with biocides was studied by several researchers. For example, Stangierska et al. [3] found that PBs can be completely protected from damage by the brown-rot fungus *Coniophora puteana* with organic fungicides (a mixture of didecyldimethylammonium nitrate and 1,2,4-triazole derivative in amount of 3.0 kg per m^3 of PB in amino glues). Bayatkashkoli et al. [53] showed that the highest biological resistance of PBs against termites was obtained by addition of BFCA (boron-fluorine-chromium-arsenic) or ACQ (alkaline copper quaternary). Sen et al. [54], on testing composites, which have been made from recycled carton, wood veneers, and four glue types, found that the most resistant composite against the fungus *C. puteana* was the one prepared with MUF glue. Evans et al. [55] on visualizing the microdistribution of zinc borate in OSBs with X-ray microcomputed tomography and SEM-EDX pointed out that nonuniform concentration and nonhomogeneous distribution of this biocide may lead to a partial damage of final composite boards by fungi.

Biocidal effect of nano-ZnO on bacteria, molds, and decaying fungi is usually explained by its direct impact on these organisms. He et al. [56] on examining the efficacy of nano-ZnO on the fungus *Botrytis cinerea* with Raman spectroscopy found that this biocide significantly increased the content of nucleic acid and carbohydrate groups, but no significant change occurred in protein and fat groups. Differences in the spectrum were confirmed by principal

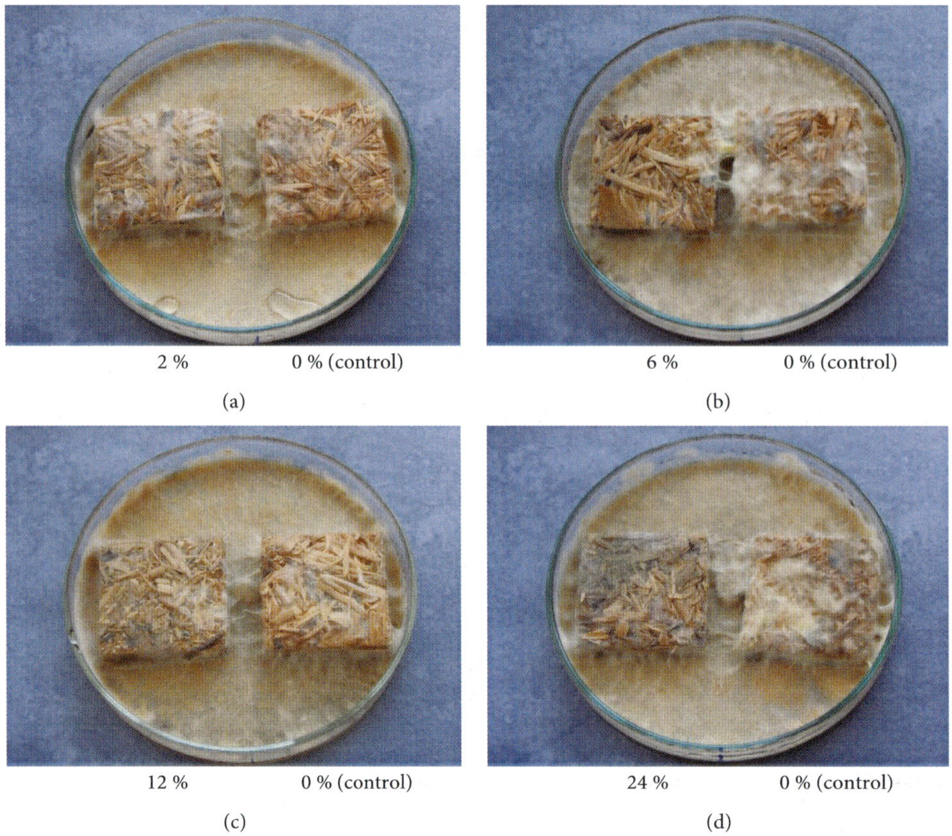

FIGURE 2: Growth of the surface mycelia of the brown-rot fungus *Coniophora puteana* on the top surfaces of the nano-ZnO-treated PBs (prepared by MUF glues modified by addition of 2, 6, 12, or 24% wt. ZnO) and the control PB, after 6 weeks of the decay test.

TABLE 3: Physical and mechanical properties of the control PB and the nano-ZnO-treated PBs.

Physical and mechanical properties of PBs	Nano-ZnO (% wt. in MUF glue)				
	0	2	6	12	24
Density (kg·m^{-3})	644 (19.26)	638 (17.78)	629 (25.66)	653 (15.69)	637 (12.40)
Thickness swelling (TS) after 2 h (%)	5.91 (0.41)	6.88 (0.89)	6.79 (0.36)	6.72 (0.48)	6.61 (1.47)
Thickness swelling (TS) after 24 h (%)	15.51 (1.69)	18.82 (1.33)	19.71 (1.38)	19.33 (2.61)	16.60 (2.29)
Water absorption (WA) after 2 h (%)	15.23 (1.43)	18.25 (2.14)	17.77 (1.60)	17.47 (1.26)	15.73 (2.20)
Water absorption (WA) after 24 h (%)	39.88 (3.88)	46.75 (1.40)	48.83 (2.50)	45.71 (3.35)	39.87 (3.42)
Internal bond (IB) strength (N·mm^{-2})	0.49 (0.05)	0.35 (0.09)	0.30 (0.06)	0.32 (0.05)	0.33 (0.06)
Bending strength (BS) (N·mm^{-2})	9.80 (1.33)	7.80 (0.95)	8.08 (0.47)	9.13 (0.83)	8.71 (0.54)

Note. Mean values: density from 6 samples, TS from 8 samples, WA from 8 samples, IB from 10 samples, and BS from 6 samples. Standard deviations are in the parentheses.

component analysis (PCA). These results indicate that nucleic acids cause a stress response in the fungus mycelia. The growth of carbohydrates in the fungus can be explained by their excessive production in the defense mechanism of the fungus against zinc oxide nanoparticles [57, 58], at which excessive accumulation of nucleic acids and carbohydrates causes deformation of the cell structure in the fungus mycelia.

3.2. Application Properties of PBs. Selected physical and mechanical properties of the nano-ZnO-treated PBs are presented in Tables 3 and 4 and in Figure 3.

A higher proportion of zinc oxide nanoparticles in PBs did not have a more apparent tendentious effect on their density, thickness swelling after 2 or 24 hours, water absorption after 2 or 24 hours, and bending strength. This knowledge was for each of the above PBs property documented by almost the same mean value (Table 3) and also confirmed by minimal value of the coefficient of determination R^2 of linear correlation "Property = $a + b \times (w_{(ZnO)}/w_{(MUF)})$" (Table 4). On the other hand, interesting was an evident decrease of the internal bond at lower proportions of nano-ZnO in PBs (Tables 3 and 4; Figure 3). In summary, nano-ZnO in PBs manifested (1) as a neutral component from the point of view of its effect on their

Table 4: Linear correlation analyses between the proportion of nano-ZnO in MUF glues ($x \rightarrow w_{(ZnO)}/w_{(MUF)}$) and the selected properties of PBs containing nano-ZnO (y).

Properties of PBs	n	R	R^2	T	P	$y = a + b \times x$
Density (kg·m^{-3})	30	−0.03	0.00	−0.16	0.88	$641.3 - 0.06 \times w/w$
Thickness swelling after 2 h (%)	40	0.11	0.01	0.70	0.49	$6.49 + 0.01 \times w/w$
Thickness swelling after 24 h (%)	40	−0.05	0.00	−0.32	0.75	$18.12 - 0.01 \times w/w$
Water absorption after 2 h (%)	40	−0.15	0.02	−0.92	0.36	$17.19 - 0.03 \times w/w$
Water absorption after 24 h (%)	40	−0.27	0.07	−1.70	0.10	$45.47 - 0.14 \times w/w$
Internal bond strength (N·mm^{-2})	50	−0.39	0.16	−3.03	0.00	$0.40 - 0.004 \times w/w$
Bending strength (N·mm^{-2})	30	−0.13	0.02	−0.79	0.43	$9.03 - 0.02 \times w/w$

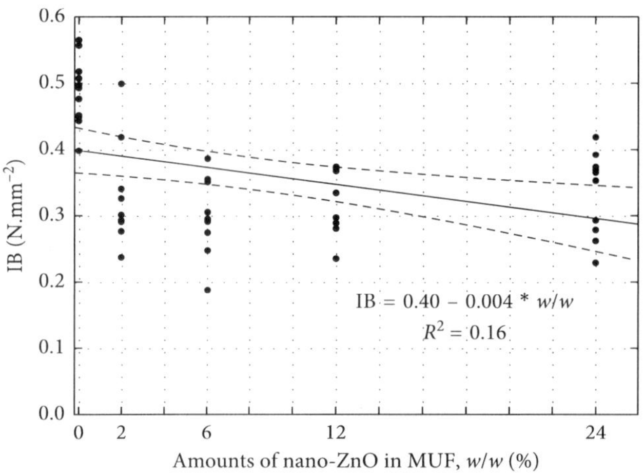

Figure 3: Linear correlation between a higher content of nano-ZnO in MUF glue (w/w, %) used in the production of PBs and a reduced internal bond (IB) strength of the nano-ZnO-treated PBs.

moisture and bending properties and (2) as a negative component for their internal bond strength.

Iždinský and Reinprecht [59] founded that the traditional copper biocide "copper sulphate pentahydrate," applied in the same concentrations as in this work used nano-ZnO, had a negative effect on some application properties of PBs, that is, the thickness swelling increased by about 19.2% after 2 hours and by about 6.7% after 24 hours and the bending strength decreased by about 19.9%. Bayatkashkoli et al. [53] found that the strength properties of PBs decreased or unchanged in the presence of other protective substances: BFCA, ACQ, or CTL (chlorothalonil). According to Guerreiro-Tanomaru et al. [60], nanoparticles of ZnO and ZrO_2 added into Portland cement significantly decreased the compressive strength of silicate-based composite materials.

Rangavar and Hoseiny [61], on the contrary, showed that addition of copper biocides in a new form of nanoparticles into UF glue when manufacturing PBs from palm tree waste improved all their tested physical and mechanical properties. Similarly, according to Yingprasert et al. [62], PBs in the presence of some types of essential oils had better moisture properties (TS and WA); however, their mechanical properties were not affected, except a slight ca. 1.8% reduction of the impact bending strength by cinnamon and clove oils.

In summary, environmental and application properties of wooden materials are more or less influenced by the type of glue, biocide, and other additives [63]. Some of these factors can also affect the adhesion forces on the phase interfaces "wood-glue" and subsequently the moisture and strength properties of wooden composites including PBs.

4. Conclusions

Nano-zinc oxide nanoparticles added into melamine-urea-formaldehyde (MUF) glue in the amounts of 0, 2, 6, 12, or 24% wt. produce the following phenomena:

(1) Increase the bioresistance of one-layer particleboards against

- (a) the Gram-positive bacterium *Staphylococcus aureus* by up to ca. 72% and the Gram-negative bacterium *Escherichia coli* by up to 20–50%;
- (b) the molds Aspergillus *niger* and *Penicillium brevicompactum* by up to ca. 50–63%;
- (c) the brown-rot fungus *Coniophora puteana* by up to 85.7%.

(2) Individual change or unchange the physical and mechanical properties of one-layer particleboards:

- (a) unchange the density, thickness swelling, water absorption, and bending strength;
- (b) decrease the internal bond strength by up to 38.8%.

Generally, special particleboards bonded with MUF glues, and if necessary with addition of nano-ZnO as an ecologically friendly biocide, can be recommended for use in interiors with temporarily increased humidity, for example, for bathroom and kitchen cabinets which are occasionally exposed to bacterial and fungal risks.

Conflicts of Interest

The authors declare that they have no conflicts of interest.

Acknowledgments

This experimental work was supported by the Slovak Research and Development Agency under Contract no. APVV-0200-12.

References

[1] Y. Imamura, "Estimation of the fungal resistance of wood composites for structural use," in *Recent Research on Wood*

and Wood-Based Materials-Current Japanese Materials Research, N. Shiraishi, H. Kajita, and M. Norimoto, Eds., vol. 11, pp. 75–84, Elsevier, London, 2013.

[2] L. Reinprecht and V. Štefka, "Drevotrieskové dosky pojené močovinoformaldehydovým a fenolformaldehydovým lepidlom s prídavkom fungicídov," in *Proceedings of the Pokroky vo výrobe a použití lepidiel v drevopriemysle (9th Symposium)*, pp. 227–238, Chemko Strážske, Czechoslovakia, 1989.

[3] A. Stangierska, J. Zabielska-Matejuk, and I. Frąckowiak, "The assessment of the resistance of particleboard protected with ionic liquids to Basidiomycetes and mould fungi," *Annals of Warsaw University of Life Sciences-SGGW, Forestry and Wood Technology*, vol. 84, pp. 190–193, 2013.

[4] A. N. Papadopoulos, "Mechanical properties and decay resistance of hornbeam cement bonded particleboards," *Research Letters in Materials Science*, vol. 2008, Article ID 379749, 4 pages, 2008.

[5] T. C. Chiang, S. Hamdan, and M. S. Osman, "Urea formaldehyde composites reinforced with sago fibres analysis by FTIR, TGA, and DSC," *Advances in Materials Science and Engineering*, vol. 2016, Article ID 5954636, 10 pages, 2016.

[6] M. J. Saad and I. Kamal, "Mechanical and physical properties of low density kenaf core particleboards bonded with different resins," *Journal of Science and Technology*, vol. 4, no. 1, pp. 17–32, 2012.

[7] H. R. Taghiyari and O. F. Bibalan, "Effects of copper nanoparticles on permeability, physical and mechanical properties of particleboard," *European Journal of Wood and Wood Products*, vol. 71, pp. 69–77, 2013.

[8] C. C. Tay, S. Hamdan, and M. S. B. Osman, "Properties of sago particleboards resinated with UF and PF resin," *Advances in Materials Science and Engineering*, vol. 2016, Article ID 5323890, 12 pages, 2016.

[9] European Committee for Standardization, *EN 335: Durability of Wood and Wood-Based Products–Use Classes: Definitions, Application to Solid Wood and Wood-Based Products*, European Committee for Standardization, Brussels, Belgium, 2013.

[10] T. Godish, *Indoor Environmental Quality*, Lewis Publisher, CRC Press LLC, Boca Raton, Florida, USA, 2001.

[11] J. W. Kirkpatrick and H. M. Barnes, "Biocide treatments for wood composites–a review," IRG/WP 06–40323, in *Proceedings of the Paper prepared for the 37th Annual Meeting*, pp. 1–21, Tromso, Norway, 2006.

[12] H. M. Barnes, "Durable composites: an overview," *Processing of American Wood Protection Association*, vol. 107, pp. 267–279, 2012.

[13] J. Ważny and A. Kundzewicz, "Conditions and possibilities of nanobiocides formulation for wood protection," *Drewno-Wood*, vol. 51, no. 180, pp. 107–115, 2008.

[14] W. R. Smith and Q. Wu, "Durability improvement for structural wood composites through chemical treatment," *Forest Products Journal*, vol. 55, no. 2, pp. 8–17, 2005.

[15] European Committee for Standardization, *EN 350: Durability of Wood and Wood-Based Products-Testing and Classification of the Durability to Biological Agents of Wood and Wood-Based Materials*, European Committee for Standardization, Brussels, Belgium, 2016.

[16] Z. Vidholdová, J. Iždinský, L. Reinprecht, and J. Krokošová, "Activity of bacteria and moulds on surfaces of commercial wooden composites," *Materials Science Forum*, vol. 818, pp. 190–193, 2015.

[17] W. D. Hanraman, L. J. Parlano, and I. W. Ong, "Antimicrobial melamine resin and products made therefrom," U. S. Patent 2006/0068662 A1, 2006.

[18] E. Nosáľ and L. Reinprecht, "Anti-bacterial and anti-mold efficiency of ZnO nanoparticles present in melamine-laminated surfaces of particleboards," *BioResources*, vol. 12, no. 4, pp. 7255–7267, 2017.

[19] J. Iždinský, L. Reinprecht, E. Nosáľ, Z. Vidholdová, and J. Krokošová, "The activity of bacteria on surfaces of wooden composites painted with acrylate coating with addition of silver nanoparticles," in *Proceedings of the Wood Modification Research and Applications, (ModWoodLife) COST FP 1407 Conference*, pp. 130-131, Kuchl, Salzburg, Austria, September 2017.

[20] R. J. Murphy, "Vapour phase treatments for wood products," in *Wood Preservation in 90th and Beyond*, Proceedings no. 7308, pp. 83-88, Forest Product Society, Madison, WI, USA, 1994.

[21] J. J. Morrell, M. N. Acda, and A. R. Zahora, "Performance of oriented strandboard, medium density fiberboard, plywood, and particleboard treated with tebuconazole in supercritical carbon dioxide," IRG/WP 05-30364, pp. 1–11, Bangalore, India, 2005.

[22] L. Reinprecht, Z. Vidholdová, and J. Iždinský, "Particleboards prepared with addition of copper sulphate–part 1: biological resistance," *Acta Facultatis Xylologiae Zvolen*, vol. 59, no. 2, pp. 53–60, 2017.

[23] L. Reinprecht and J. Perlác, "Properties of particleboards protected by TBTN and TCMTB fungicides," in *Proceedings of the Wood modification '95 (10th Symposium)*, pp. 255–264, Poznaň, Poland, 1995.

[24] P. Marzbani, Y. M. Afrouzi, and A. Omidvar, "The effect of nano-zinc oxide on particleboard decay resistance," *Maderas: Ciencia y tecnología*, vol. 17, no. 1, pp. 63–68, 2015.

[25] European Committee for Standardization, *EN 460: Durability of Wood and Wood-Based Products. Natural Durability of Solid Wood. Guide to the Durability Requirements for Wood to be Used in Hazard Classes*, European Committee for Standardization, Brussels, Belgium, 1994.

[26] European Chemicals Agency, *ECHA: Transitional Guidance on the Biocidal Products Regulation-Transitional Guidance on Efficacy Assessment for Product Type 8 Wood Preservatives*, European Chemicals Agency, Helsinki, Finland, 2015.

[27] European Chemicals Agency, *ECHA: Practical Guide on Biocidal Products Regulation*, European Chemicals Agency, Helsinki, Finland, 2016.

[28] L. S. Mende and J. MacManus-Driscoll, "ZnO nanostructures, defects, and devices," *Materials Today*, vol. 10, no. 5, pp. 40–48, 2007.

[29] L. E. Shi, Z. H. Li, W. Zheng, Y. F. Zhao, Y. F. Jin, and Z. X. Tang, "Synthesis, antibacterial activity, antibacterial mechanism and food applications of ZnO nanoparticles: a review," *Food Additives and Contaminants Part A, Chemistry, Analysis, Control, Exposure and Risk Assessment*, vol. 31, no. 2, pp. 173–186, 2014.

[30] A. Sirelkhatim, S. Mahmud, A. Seeni et al., "Review on zinc oxide nanoparticles: antibacterial activity and toxicity mechanism," *Nano-Micro Letters*, vol. 7, no. 3, pp. 219–242, 2015.

[31] J. T. Seil and T. J. Webster, "Antimicrobial applications of nanotechnology: methods and literature," *International Journal of Nanomedicine*, vol. 7, pp. 2767–2781, 2012.

[32] K. R. Raghupathi, R. T. Koodali, and J. T. Manna, "Size-dependent bacterial growth inhibition and mechanism of antibacterial activity of zinc oxide nanoparticles," *Langmuir*, vol. 27, no. 7, pp. 4020–4028, 2011.

[33] C. Wang, L. L. Liu, A. T. Zhang, P. Xie, J. J. Lu, and X. T. Zou, "Antibacterial effects of zinc oxide nanoparticles on

Escherichia coli K88," *African Journal of Biotechnology*, vol. 11, no. 44, pp. 10248–10254, 2012.
[34] S. Ghosh, R. G. Sankar, and V. Vandana, "Curious case of bactericidal action of ZnO," *Journal of Nanoscience*, vol. 2014, Article ID 343467, 8 pages, 2014.
[35] A. Stanković, M. Sezen, M. Milenković, S. Kaišarević, N. Andrić, and M. Stevanović, "PLGA/nano-ZnO composite particles for use in biomedical applications: preparation, characterization, and antimicrobial activity," *Journal of Nanomaterials*, vol. 2016, Article ID 9425289, 10 pages, 2016.
[36] A. Rezazadeh, M. R. M. Farahani, Y. M. Afrouzi, and A. A. D. Khalaji, "Investigation on rot resistance of *Populus deltoides* wood treated with nano-zinc oxide," *World of Science Journal*, vol. 2, no. 2, pp. 19–28, 2014.
[37] M. R. M. Farahani and F. Banikarim, "Effect of nano-zinc oxide on decay resistance of wood-plastic composites," *BioResources*, vol. 8, no. 4, pp. 5715–5720, 2013.
[38] G. Mantanis, E. Terzi, S. N. Kartal, and A. N. Papadopoulos, "Evaluation of mold, decay and termite resistance of pine wood treated with zinc- and copper-based nanocompounds," *International Biodeterioration and Biodegradation*, vol. 90, pp. 140–144, 2014.
[39] N. Bellotti, R. Romagnoli, C. Quintero, C. Domínguez-Wong, F. Ruiz, and C. Deyá, "Nanoparticles as antifungal additives for indoor water borne paints," *Progress in Organic Coatings*, vol. 86, pp. 33–40, 2015.
[40] L. Reinprecht, Z. Vidholdová, and M. Kožienka, "Inhibícia hniloby lipového dreva nanočasticami oxidu zinočnatého v kombinácii s akrylátom," *Acta Facultatis Xylologiae Zvolen*, vol. 57, no. 1, pp. 43–52, 2015.
[41] L. Reinprecht and Z. Vidholdová, "Growth inhibition of moulds on wood surfaces in presence of nano-zinc oxide and its combinations with polyacrylate and essential oils," *Wood Research*, vol. 62, no. 1, pp. 37–44, 2017.
[42] J. Iždinský, V. Tóth, and J. Kúdela, "Polyethylene terephthalate recycling in particle board production," in *Wood the Best Material for Mankind*, pp. 93–97, Arbora, Zvolen, Slovakia, 2013.
[43] European Committee for Standardization, *EN 84: Wood Preservatives. Accelerated Ageing of Treated Wood prior to Biological Testing. Leaching Procedure*, European Committee for Standardization, Brussels, Belgium, 1997.
[44] European Committee for Standardization, *EN 15457: Paints and Varnishes-Laboratory Method for Testing the Efficacy of Film Preservatives in a Coating against Fungi*, European Committee for Standardization, Brussels, Belgium, 2014.
[45] European Committee for Standardization, *ENV 12038: Durability of Wood and Wood-Based Products-Wood-Based Panels-Method of Test for Determining the Resistance against Wood-Destroying Basidiomycetes*, European Committee for Standardization, Brussels, Belgium, 2002.
[46] European Committee for Standardization, *EN 317: Particleboards and Fibreboards: Determination of Swelling in Thickness after Immersion in Water*, European Committee for Standardization, Brussels, Belgium, 1993.
[47] European Committee for Standardization, *EN 319: Particleboards and Fibreboards: Determination of Tensile Strength Perpendicular to the Plane of the Board*, European Committee for Standardization, Brussels, Belgium, 1993.
[48] European Committee for Standardization, *EN 310: Wood-Based Panels: Determination of Modulus of Elasticity in Bending and of Bending Strength*, European Committee for Standardization, Brussels, Belgium, 1993.
[49] W. Y. Chung, S. G. Wi, H. J. Bae, and B. D. Park, "Microscopic observation of wood-based composites exposed to fungal deterioration," *Journal of Wood Science*, vol. 45, pp. 64–68, 1999.
[50] F. Green and C. A. Clausen, "Copper tolerance of brown rot fungi: oxalic acid production in southern pine treated with arsenic-free preservatives," *International Biodeterioration & Biodegradation*, vol. 56, pp. 75–79, 2005.
[51] C. Tascioglu, K. Umemura, T. Yoshimura, and K. Tsunoda, "Biological performance of zinc borate-incorporated particleboard: effects of leaching on efficacy," *Composites: Part B*, vol. 57, pp. 31–34, 2014.
[52] C. Tascioglu, K. Umemura, S. S. Kusuma, and T. Yoshimura, "Potential utilization of sodium fluoride (NaF) as a biocide in particleboard production," *Journal of Wood Science*, vol. 63, no. 6, pp. 652–657, 2017.
[53] A. Bayatkashkoli, B. Kameshki, S. Ravan, and M. Shamsian, "Comparing of performance of treated particleboard with alkaline copper quat, boron-fluorine-chromium-arsenic and chlorotalonil against *Microcerotermes diversus* and *Anacanthotermes vagans* termite," *International Biodeterioration & Biodegradation*, vol. 120, pp. 186–191, 2017.
[54] S. Sen, N. Ayrilmis, and Z. Candan, "Fungicide and insecticide properties of cardboard panels made from used beverage carton with veneer overlay," *African Journal of Agricultural Research*, vol. 5, no. 2, pp. 159–165, 2010.
[55] P. D. Evans, V. Lube, H. Averdunk et al., "Visualizing the microdistribution of zinc borate in oriented strand board using X-ray microcomputed tomography and SEM-EDX," *Journal of Composites*, vol. 2015, Article ID 630905, 9 pages, 2015.
[56] L. He, Y. Liu, A. Mustapha, and M. Lin, "Antifungal activity of zinc oxide nanoparticles against *Botrytis cinerea* and *Penicillium expansum*," *Microbiological Research*, vol. 166, no. 3, pp. 207–215, 2011.
[57] F. J. Alvarez-Peral, O. Zaragoza, Y. Pedreno, and J. C. Argüelles, "Protective role of trehalose during severe oxidative stress caused by hydrogen peroxide and the adaptive oxidative stress response in *Candida albicans*," *Microbiology*, vol. 148, no. 8, pp. 2599–2606, 2002.
[58] K.-J. Kim, W. S. Sung, B. K. Suh et al., "Antifungal activity and mode of action of silver nano-particles on *Candida albicans*," *Biometals*, vol. 22, no. 2, pp. 235–242, 2009.
[59] J. Iždinský and L. Reinprecht, "Particleboards prepared with addition of copper sulphate–part 2: moisture and strength properties," *Acta Facultatis Xylologiae Zvolen*, vol. 59, no. 2, pp. 61–66, 2017.
[60] J. M. Guerreiro-Tanomaru, A. Trindade-Junior, B. C. Costa et al., "Effect of zirconium oxide and zinc oxide nanoparticles on physicochemical properties and antibiofilm activity of a calcium silicate-based material," *Scientific World Journal*, vol. 2014, Article ID 975213, 6 pages, 2014.
[61] H. Rangavar and M. S. F. Hoseiny, "The effect of nanocopper additions in a urea-formaldehyde adhesive on the physical and mechanical properties of particleboard manufactured from date palm waste," *Mechanics of Composite Materials*, vol. 51, pp. 119–126, 2015.
[62] W. Yingprasert, N. Matan, P. Chaowana, and N. Matan, "Fungal resistance and physico-mechanical properties of cinnamon oil- and clove oil-treated rubber-wood particleboards," *Journal of Tropical Forest Science*, vol. 27, no. 1, pp. 69–79, 2015.
[63] A. Pizzi, "Wood products and green chemistry," *Annals of Forest Science*, vol. 73, no. 1, pp. 185–203, 2016.

7

Compression Behavior of Confined Columns with High-Volume Fly Ash Concrete

Sung-Won Yoo,[1] Young Cheol Choi,[2] and Wonchang Choi[2]

[1]*Department of Civil and Environmental Engineering, Gachon University, Seongnam-si, Gyeonggi-do 13120, Republic of Korea*
[2]*Department of Architectural Engineering, Gachon University, Seongnam-si, Gyeonggi-do 13120, Republic of Korea*

Correspondence should be addressed to Wonchang Choi; wonchang.choi@gmail.com

Academic Editor: Prinya Chindaprasirt

The use of fly ash in ordinary concrete provides practical benefits to concrete structures, such as a gain in long-term strength, reduced hydration heat, improved resistance to chloride, and enhanced workability. However, few studies with high-volume fly ash (HVFA) concrete have been conducted that focus on the structural applications such as a column. Thus, there is a need to promote field applications of HVFA concrete as a sustainable construction material. To this end, this study investigated the compressive behavior of reinforced concrete columns that contain HVFA with a 50 percent replacement rate. Six columns were fabricated for this study. The study variables were the HVFA replacement rate, tied steel ratio, and tie steel spacing. The computed ultimate strength by the American Concrete Institute (ACI) code conservatively predicted the measured values, and, thus, the existing equation in the ACI code is feasible for confined RC columns that contain HVFA. In addition, an analysis model was calibrated based on the experimental results and is recommended for predicting the stress-strain relationship of confined reinforced concrete columns that contain HVFA.

1. Introduction

Fly ash is an industrial byproduct of the electrical power industry. Recently, partially recycled fly ash has been used as supplementary cementitious material in the cement industry. Fly ash content that is less than 25 percent of the total cementitious content is now commonly used in concrete mixtures due to its apparent benefits for concrete, such as a gain in long-term strength, reduced hydration heat, improved resistance to chloride, and enhanced workability [1–4]. On a limited basis, an increase up to a 100 percent fly ash replacement rate in concrete mixtures has been attempted [5]. However, this increase in the replacement rate of cement has led to shortcomings, such as difficulties associated with adequate quality control and low early-age strength and inferior material properties [6].

Most of the previous studies on this topic have concentrated on evaluating the material properties of high-volume fly ash (HVFA) concrete and developing mix designs [3, 7]. For example, Huang et al. [8] investigated mix designs that incorporate HVFA and are associated with two types of "loss on ignition" (LOI). Huang et al. [8] confirmed the superior mechanical properties of mixtures with low LOI fly ash and the feasibility of incorporating HVFA in concrete mixtures up to 80 percent. On the other hand, with similar water-to-binder ratio in the mix design of both HVFA concrete mix and the OPC mix, less compressive strength in HVFA concrete mix will be expected [6]. In addition, limited studies have been conducted that target the use of HVFA for structural applications [9–12].

Therefore, this study focused on the compressive behavior of confined reinforced concrete (RC) columns that contain HVFA. Numerous confinement models are available in the literature [13–15] to evaluate the axial capacity and to predict the envelope of the stress-strain relationship for RC columns. However, HVFA in concrete results in inferior structural performance in terms of early-age strength. To develop an accurate stress-strain model, the proper material properties must be considered. This paper presents the results of a seriespagebreak of tests of RC columns that contain HVFA and proposes an analysis model to predict the behavior of RC columns according to HVFA content.

Table 1: Chemical proportion for OPC and FA.

Types	Items								
	Chemical composition [%]							Physical properties	
	SiO_2	Al_2O_3	Fe_2O_3	CaO	MgO	SO_3	Ig. loss	Specific gravity [g/cm^3]	Blaine [cm^2/g]
NPC	21.96	5.27	3.44	63.41	2.13	1.96	0.79	3.16	3,214
FA	55.66	27.76	7.04	2.70	1.14	0.49	4.3	2.19	3,621

Table 2: Mix designs for OPC and HVFAC.

Binder type	W/B (%)	S/a (%)	Unit weight (kg/m^3)					Superplasticizer (% by cement weight)
			W	C	FA	S	G	
OPC	50.0	40.2	175	350	0	710	1,054	0.05
HVFAC	35.0	44.1	120	154	189	821	1,042	1.50

Note. S/a is the sand-to-aggregate ratio; W is water; C is cement; FA is fly ash; S is sand; G is gravel.

Table 3: Testing matrix.

ID	Longitudinal steel ratio (%)	Tied steel ratio (%)	Tied steel spacing (mm)	f_c (MPa)	Fly ash (%)
N-100		0.475	100		
N-150		0.317	150	30.6	0
N-250	2.25	0.190	250		
F-100		0.475	100		
F-150		0.317	150	30	55
F-250		0.190	250		

Six columns were manufactured for this study. The study variables are the replacement rate of the fly ash, the tied steel ratio, and tie steel spacing. In addition, based on the experimental results, an analysis model was developed to predict the compressive behavior of RC columns that contain HVFA. Comparisons of the model results with the test results were used to calibrate the accuracy of the proposed analysis model for HVFA RC columns.

2. Experimental Program

2.1. Test Parameters and Specimen Fabrication. Tests were conducted based on two replacement ratios of fly ash (0% and 50%) as the variables. The design compressive strength of the concrete was 30 MPa. The longitudinal rebar reinforcement ratio was fixed at 2.25 percent with three different spacings (100 mm, 150 mm, and 250 mm) of tie rebar in the column specimens. The rebar sizes of D13 (ϕ = 13 mm) and D10 (ϕ = 10 mm) were used for the longitudinal rebar and tie rebar, respectively. The measured yield strength of D13 rebar is 412 MPa. And the chemical composition of fly ash in the study was presented in Table 1. Extra process was performed to ensure quality control of fly ash and the loss of ignition values was controlled to 5% or less.

Table 2 shows the mix design for ordinary Portland cement (OPC) and high-volume fly ash concrete (HVFAC). Due to the low water-to-binder ratio (W/B), a relatively large amount of superplasticizer was used in the HVFAC mix. The use of fly ash in OPC provides practical benefits to concrete, such as a gain in long-term strength, reduced hydration heat, improved resistance to chloride, and enhanced workability. The mix proportion used in this study has been developed based on the related research results by the authors [16]. To determine material properties of concrete, three cylindrical specimens (100 mm × 200 mm) were made and cured in moisture condition for 28 days with 23°C ± 3.

Table 3 shows the variables for the specimens: tie steel ratio, spacing of the tied steel, and the replacement rate of the fly ash. The specimen identification (ID) indicates the fly ash and tie steel spacing; for example, N-100 means that the specimen is fabricated with OPC with 100 mm of tie steel spacing.

Each specimen cross-section is 150 mm × 150 mm at the midheight of the column and is 1,030 mm in height. The column specimens were cured in moisture for 28 days with 23°C ± 3. Compression tests were performed at the age of 28 days after fabrication. Figure 1 presents the reinforcement details of a specimen, the installed gauge locations, and the test set-up used in this study.

2.2. Test Procedure. A universal testing machine (UTM) with 2,000 kN was used to apply monotonic axial compressive loading with displacement control at a rate of 0.5 mm/min (cross-head speed). To monitor the strain of the concrete

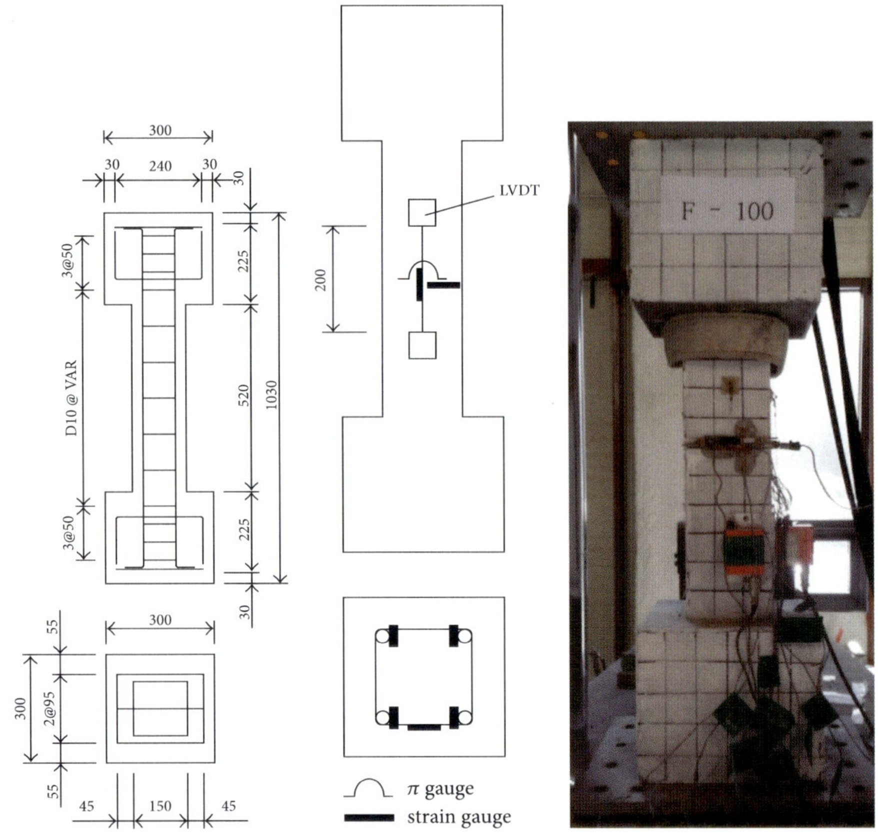

(a) Specimen dimensions (b) Gauge locations (c) Test set-up

FIGURE 1: Specimen design and instrumentation.

TABLE 4: Material properties of concrete mixtures.

ID	f_c (MPa)	ε_c	Unit weight (kg/m^3)	E_c (MPa) Measured	E_c (MPa) ACI 318	Meas./Pred.
OPC	30.6	0.00285	2,304	28,918	26,522	1.09
HVFAC	30.0	0.00287	2,102	21,173	23,058	0.92

surface and the displacement of the column, strain gauges, a pressure indicator (PI) gauge, and a linear variable displacement transducer (LVDT) were installed at the midheight of the column, as shown in Figure 1(b). Strain gauges also were installed on the longitudinal rebar and the tie rebar. Monotonic loading tests were terminated when an abrupt load drop was observed for each specimen.

3. Test Results and Discussion

3.1. Material Properties. Figure 2 presents the stress-strain relationship for the OPC and HVFAC specimens. Table 4 presents a summary of the test results, including the compressive strength, elastic modulus values, and the predicted values in accordance with the current American Concrete Institute (ACI) 318 specifications. The predicted elastic modulus value, $E_c = 0.077(m_c)^{1.5}(f_c)^{0.5}$, is computed based on the unit weight of the concrete mixture (m_c) and the compressive concrete strength (f_c). A similar compressive concrete strength value was obtained for both concrete mixtures. However, a lower elastic modulus value was obtained for the HVFAC mixture than for the OPC mixture. This outcome was due to the low unit weight of the HVFAC mixture that is caused by the low unit weight of fly ash compared to the weight of cement. These results are in good agreement with those found in the literature [1].

3.2. Failure Mode and Load versus Displacement Relationship. Figure 3 presents typical failure modes of the axially loaded column specimens. No significant concrete spalling occurred during the axial load tests. Shear failure in specimens N-250 and F-150 occurred at failure. Similar failure patterns were observed for each specimen regardless of whether HVFA was present.

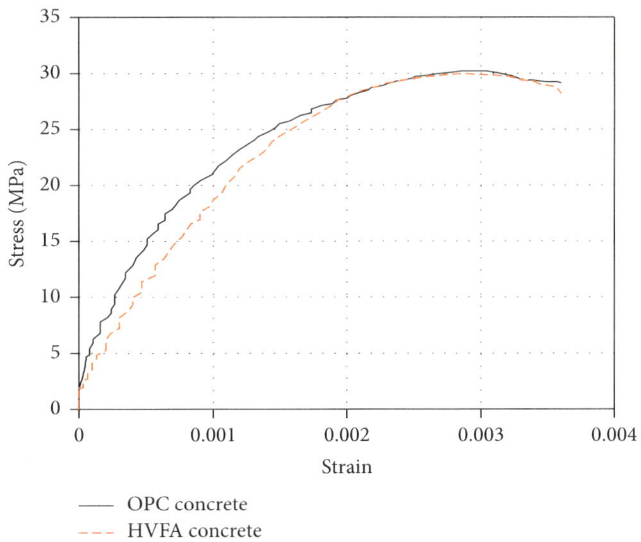

FIGURE 2: Stress-strain relationship in compression.

N-100 N-150 N-250 F-100 F-150 F-250

FIGURE 3: Failure modes of specimens.

Figure 4 shows the load versus displacement relation. The load-carrying capacity of the columns rapidly decreased after reaching the peak load. This behavior is consistent with that observed in similar RC columns with OPC. The ultimate load for the specimens, regardless of whether they contained HVFA, tended to decrease with an increase in the spacing of the transverse steel due to confining pressure.

Table 5 shows the measured axial loads at the yield of the longitudinal reinforcement and at the ultimate points. The ratio of the yield load to the ultimate load for the specimens with OPC is in the range of 0.73 to 0.79, and the ratio of the yield load to the ultimate load for the specimens with HVFA is in the range of 0.75 to 0.84. The ultimate load for both series of tests tended to increase with an increase in the tie reinforcement ratio. The ultimate design axial strength of a column (ϕP_n) was computed using the equation found in ACI 318-16: $P_n = 0.8 * (0.85f'_c(A_g - A_s) + f_y \cdot A_s)$. The measured longitudinal reinforcement strain levels reached the yield strain at the ultimate point, so the measured yield strength of the longitudinal reinforcement was used for the calculation. The predicted ultimate design strength of the RC columns with HVFA led to failure at loads ranging from 54 percent to 57 percent of the measured peak strength. The computed values conservatively predicted the measured values, and, thus, the existing equation in the ACI code is feasible for confined RC columns that contain HVFA.

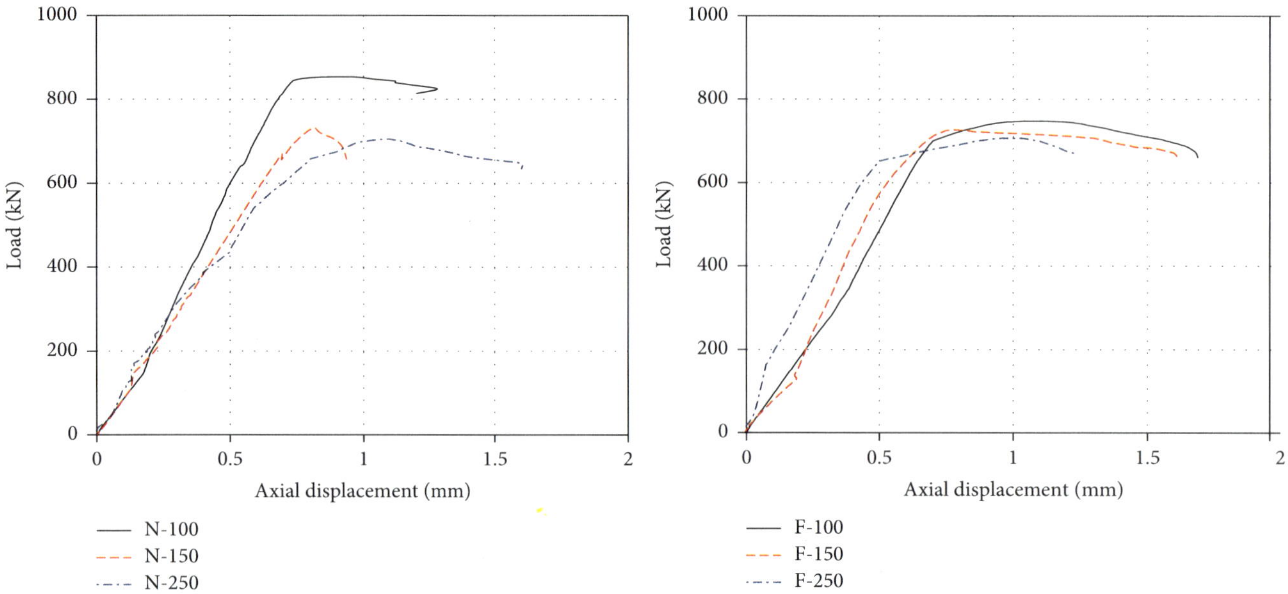

FIGURE 4: Load versus displacement relationship.

TABLE 5: Measured axial strength values at yield and ultimate points.

ID	P_y (kN)	$P_{\max}$ (kN)	$P_y/P_{\max}$	ϕP_n (kN)	$\phi P_n/P_{\max}$
N-100	626.8	854.0	0.73	406	0.48
N-150	548.8	732.8	0.75	406	0.55
N-250	559.0	705.8	0.79	406	0.58
F-100	625.5	747.2	0.84	400	0.54
F-150	673.5	727.1	0.93	400	0.55
F-250	531.6	707.7	0.75	400	0.57

3.3. Concrete Strain of Specimens. Figure 5 shows the measured concrete strain values in the axial and lateral directions. In the early stages, the response was a relatively linear relationship between the axial strain and the lateral strain of the concrete; subsequently, nonlinear responses in the lateral strain beyond $\varepsilon_l = 0.0005$ were experienced. The axial strain at peak stress was assumed to be axial strain of 0.002 [17]; however, this lateral strain value varied depending on the tie reinforcement ratio.

3.4. Effects of Tie Reinforcement Ratio. Figure 6 shows the relation between the load and strain for the longitudinal steel, including the load when the longitudinal steel yielded. For all the tested columns, the longitudinal steel yielded. However, the strain of the transverse rebar for all the specimens did not reach yield strain, so a confinement effect for the columns hardly was observed in the study. Similar patterns were observed irrespective of the presence of HVFA.

4. Analytical Stress-Strain Models for Confined RC Columns

The modified Kent and Park model [18], which was developed based on the second-order parabola model, allows simplicity and accuracy and thus is widely adopted in the literature. In the modified Kent and Park model, the monotonic concrete stress-strain relationship in compression is described as presented in

$$f_c = Kf_{cc}\left[2\left(\frac{\varepsilon_c}{\varepsilon_{cc}}\right) - \left(\frac{\varepsilon_c}{\varepsilon_{cc}}\right)^2\right] \quad \varepsilon_c \le \varepsilon_{cc}$$
$$f_c = Kf_{cc}\left[1 - Z\left(\varepsilon_c - \varepsilon_{cc}\right)\right] \ge 0.2Kf'_c \quad \varepsilon_c \ge \varepsilon_{cc}, \qquad (1)$$

where the concrete strain at the peak compressive stress is $\varepsilon_{cc} = 0.002K$; $K = 1 + \rho_s f_{yh}/f_{cc}$ is a factor that accounts for the increase in strength that is due to confinement; the strain softening slope is $Z = 0.5/(((3 + 0.29f_{cc})/(145f_{cc} - 1000)) + 0.75\rho_s\sqrt{h'/S_h} - 0.002K)$; f_{cc} is the ultimate concrete compressive strength (MPa); f_{yh} is the yield strength of the transverse steel (MPa); ρ_s is the ratio of the volume of the transverse steel to the volume of the concrete core measured to the outside of the stirrups; h' is the width of the concrete core measured to the outside of the stirrups; and s_h is the center-to-center spacing of the stirrups.

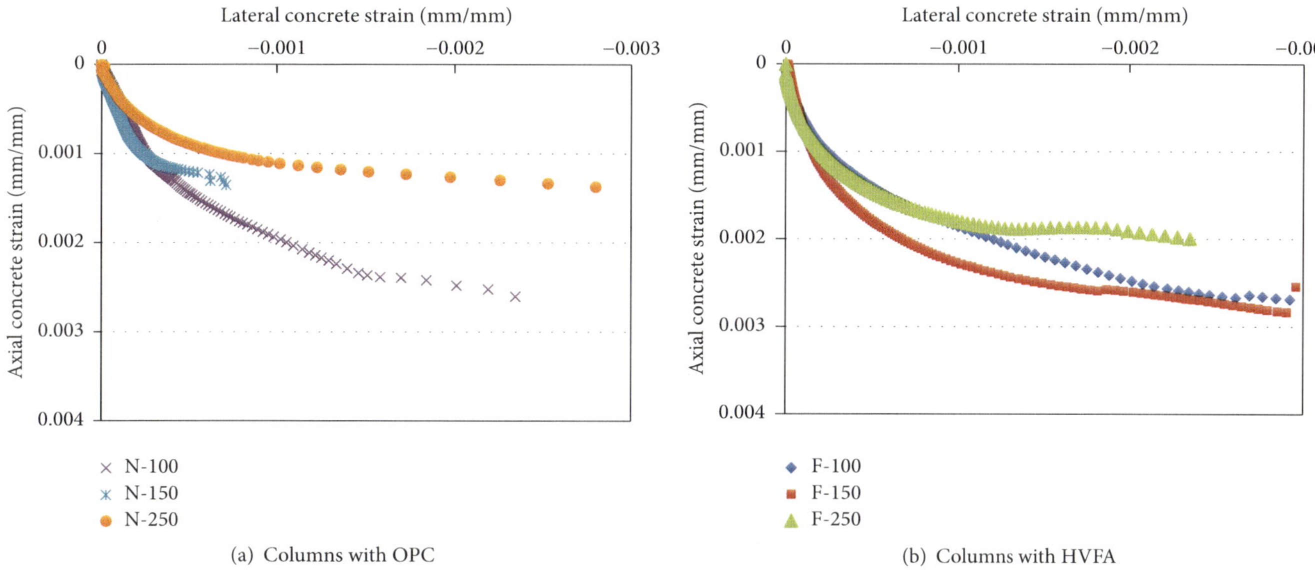

FIGURE 5: Axial versus lateral concrete strain of specimens.

Hoshikuma et al. [19] adopted an exponential function to simplify the expression for the ascending part, shown here as

$$f_c = E_c \varepsilon_c \left\{ 1 - \frac{1}{n} \left(\frac{\varepsilon_c}{\varepsilon_{cc}} \right)^{n-1} \right\}, \tag{2}$$

where ε_{cc} is the strain at peak stress, E_c is the initial stiffness (MPa), and n is a coefficient.

Also, the descending part of the stress curve can be modeled from the test results by a straight line, as expressed in

$$f_c = f_{cc} - E_{\text{des}} \left(\varepsilon_c - \varepsilon_{cc} \right), \tag{3}$$

where E_{des} is the gradient at the descending part (MPa) ($E_{\text{des}} = (11.2 \cdot f_{co}{}^2)/(\rho_s \cdot f_{yh})$).

Figure 7 presents comparisons of the experimental results and the analytical model results. The experimental results for the specimens with OPC are consistent with the results obtained from the analytical modified model proposed by Kent and Park [18] and the Hoshikuma et al. model [19] irrespective of the confinement spacing. On the other hand, the experimental results of the specimens with HVFAC were slightly overestimated by the analytical modified model of Kent and Park but matched relatively well with the proposed equation by Hoshikuma. Therefore, the Hoshikuma equation was employed in this study to develop a modified analytical model for confined RC columns with HVFAC.

The adopted Hoshikuma equation was calibrated by adjusting the initial stiffness value (E_c) of the HVFAC and the peak concrete strain that is due to the low unit weight of HVFA. Equations (4) and (5) are proposed for the initial stiffness value (E_c) of RC columns with HVFA and the strain at peak stress (ε_{cc}), respectively.

$$E_c = 0.077 m_c^{1.5} \cdot \sqrt[3]{f_c} \tag{4}$$

$$\varepsilon_{cc} = 0.002 \left(\frac{2{,}300}{m_c} \right)^{1.5} + 0.033 \beta \frac{\rho_{sh} f_{yh}}{f_c}, \tag{5}$$

where β is a modification factor that depends on the section shape; for a square section, 0.4 is used by Hoshikuma et al. [19].

Figure 8 presents comparisons between the measured stress-strain values of the specimens using the proposed equations with modification of the elastic modulus values for the HVFAC mixture. The ascending part and descending part of the proposed stress-strain curves show good agreement with the measured envelopes.

Figure 9 shows that available experimental results in the literature [20] were compared with the proposed equation and confirmed that the stiffness of the specimen containing fly is less than that of the specimen without fly ash. The proposed equation properly predicts the stress-strain envelope for the column specimens containing fly ash.

The proposed equation was driven by limited experimental tests, and the results of the previous studies on HVFA concrete were insufficient for verification. Further study on the compressive behavior of HVFA concrete is needed. If a lot of experimental data are accumulated, it will be necessary to do further studies on the basis of the formula proposed.

5. Conclusions

Six specimens were tested in this study to develop an analytical stress-strain model for confined RC columns that contain HVFA. Based on the limited experimental results,

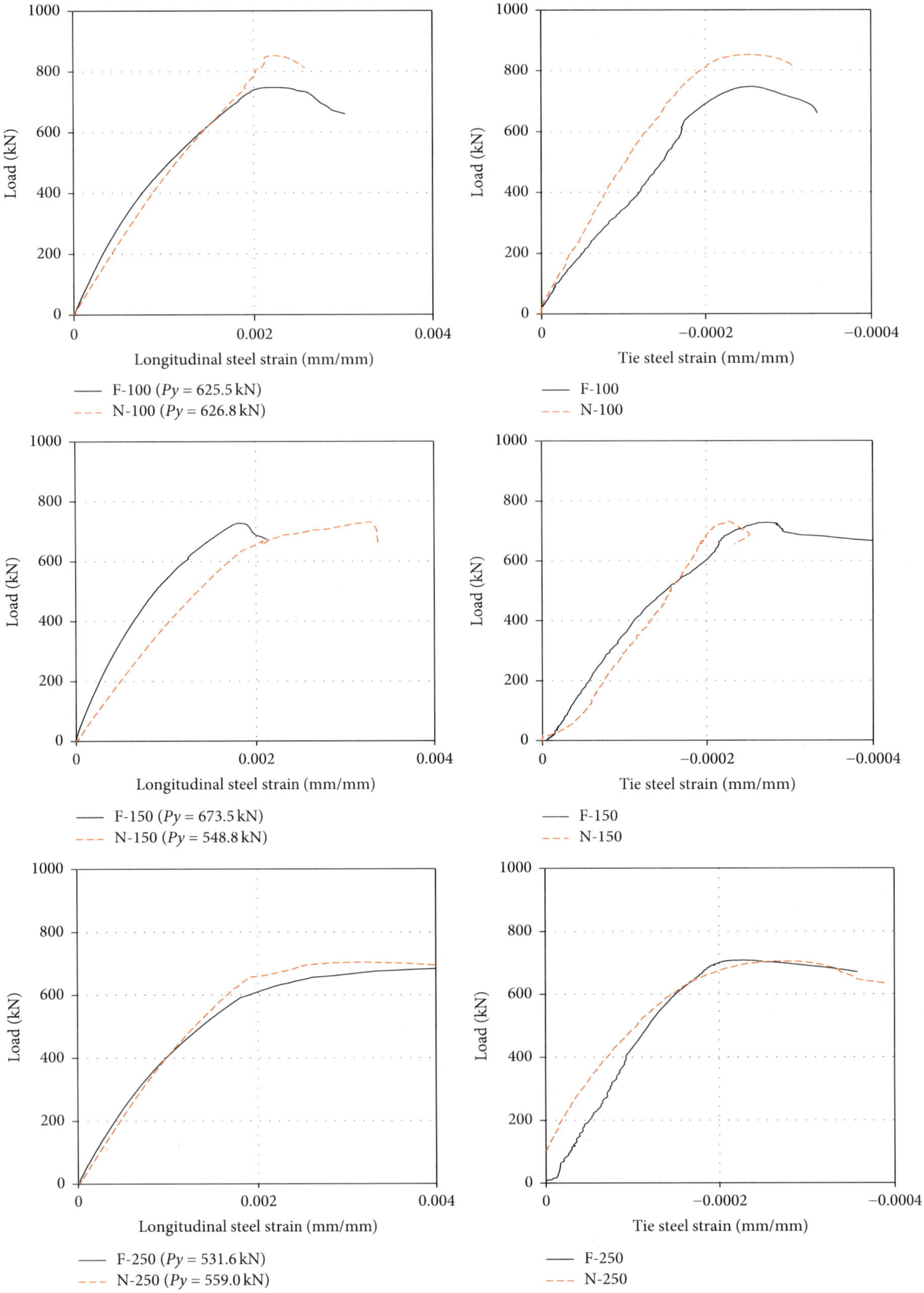

FIGURE 6: Load versus strain relationship of longitudinal steel for the specimens.

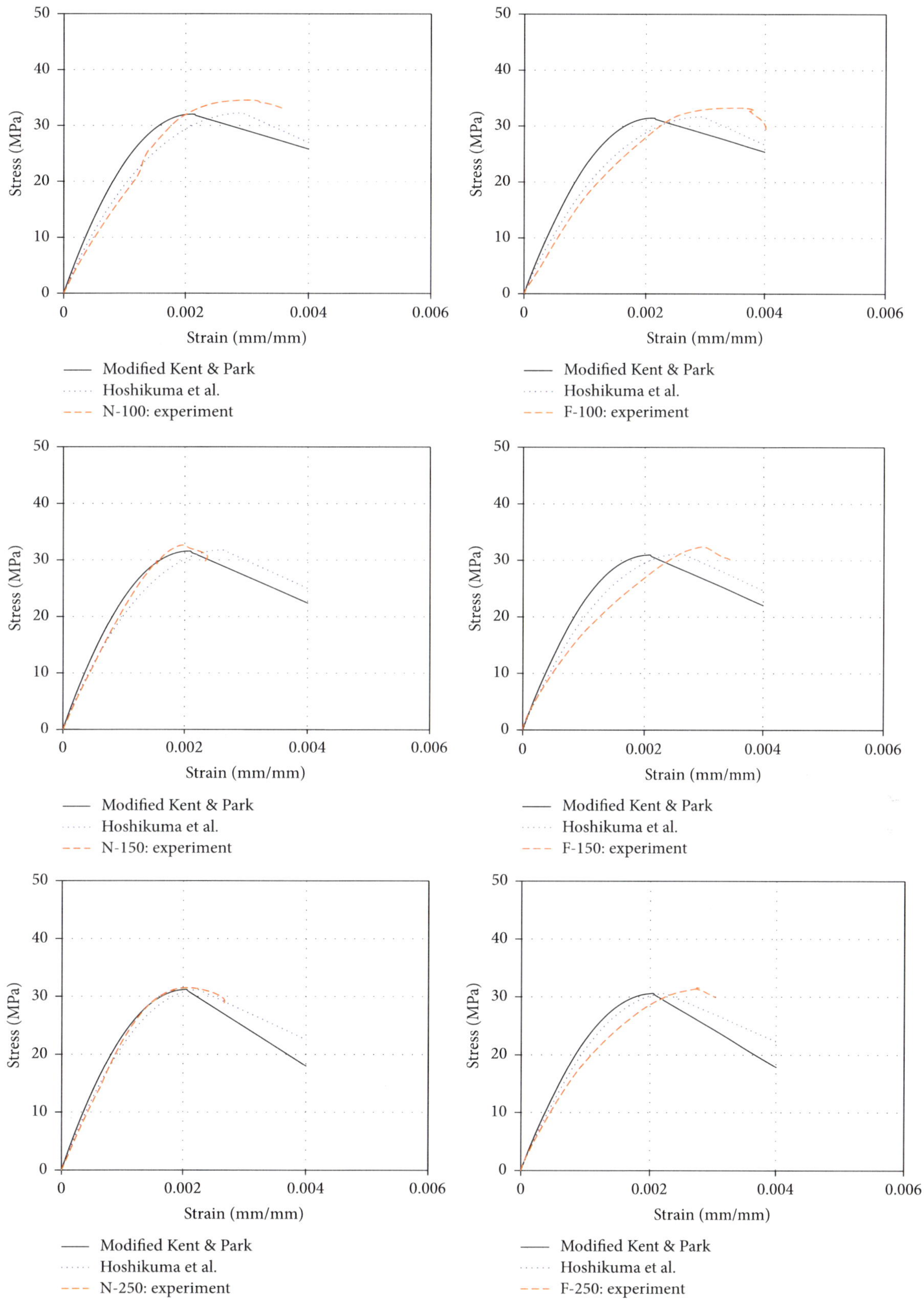

FIGURE 7: Comparisons of the experimental results and analytical model results.

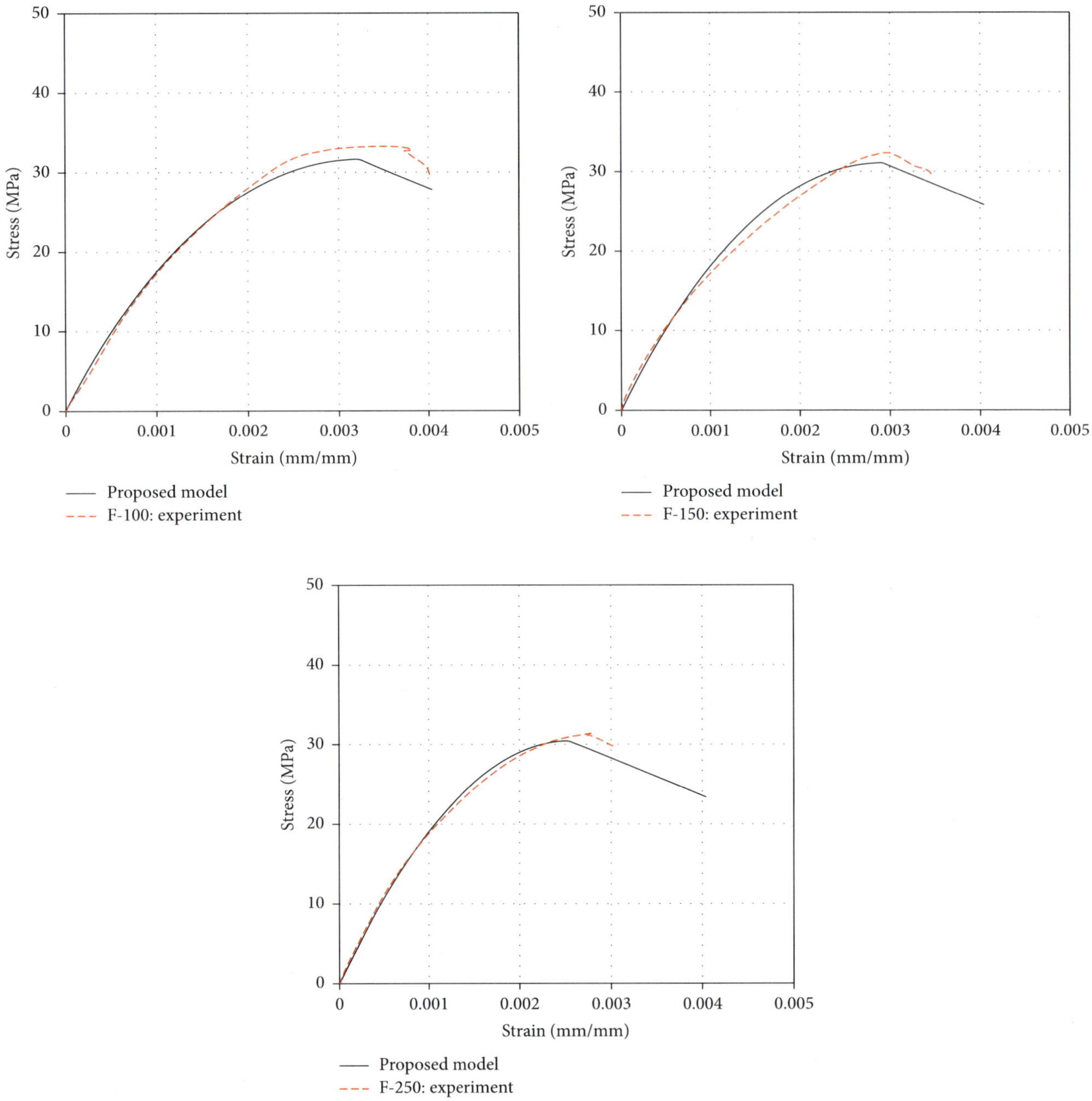

FIGURE 8: Comparisons between the experimental results and the proposed equation.

a stress-strain envelope for RC columns with HVFA was derived analytically and then calibrated using the experimental results. The following conclusions can be drawn with regard to the compressive behavior of the OPC and HVFAC mixes.

The elastic modulus values of the HVFAC mix were lower than those of the OPC mix. This outcome was due to the different water-to-binder ratio and low unit weight of the HVFAC mixture that is caused by the low specific gravity of its inherent fly ash compared to the weight of cement. Meanwhile, the compressive behavior of the HVFAC and OPC columns was similar with regard to crack morphology and crack progression. The maximum load-carrying capacity of the columns with HVFA tended to decrease up to 14 percent compared to the columns with OPC. The RC column elements made with HVFA behaved similarly, as expected, based on the equation used to predict OPC column behavior. Existing design standards will conservatively predict the ultimate capacity of HVFAC columns.

The initial stiffness of a column affects the elastic modulus of HVFAC; therefore, the proposed modified elastic modulus equation allows for adjustment of the initial stiffness value of concrete columns that contain HVFA. The modified

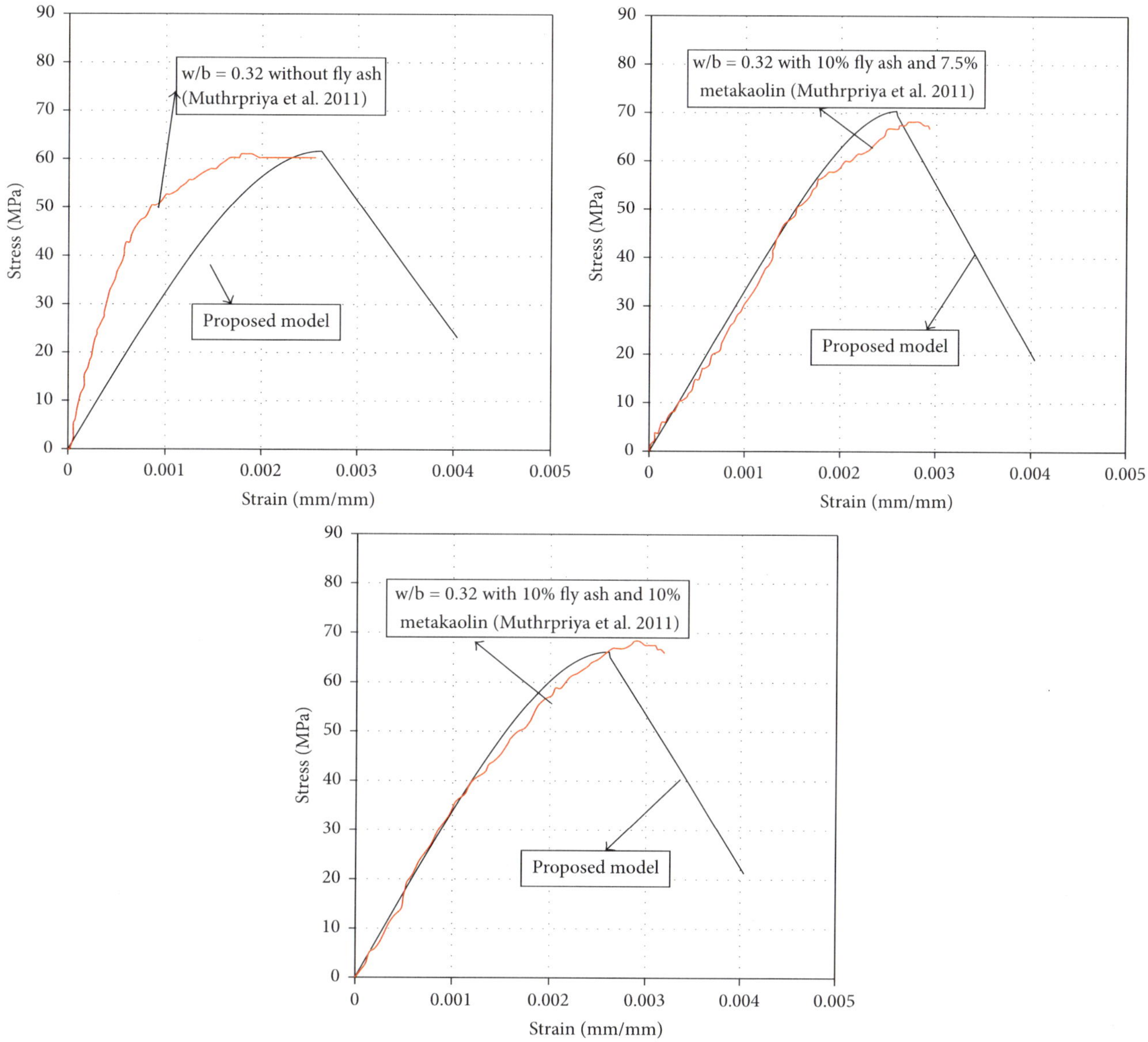

FIGURE 9: Comparisons between the available experimental result in the literature and the proposed equation.

stress-strain model can predict both the ascending and descending branches of the curve for confined RC columns with HVFA.

Conflicts of Interest

The authors declare that there are no conflicts of interest regarding the publication of this paper.

Acknowledgments

This research was financially supported by the Korean Ministry of Environment as "Public Technology Program Based on Environmental Policy" (no. 2016000700003).

References

[1] R. Siddique, "Performance characteristics of high-volume Class F fly ash concrete," *Cement and Concrete Research*, vol. 34, no. 3, pp. 487–493, 2004.

[2] K. Mala, A. K. Mullick, K. K. Jain, and P. K. Singh, "Effect of relative levels of mineral admixtures on strength of concrete with ternary cement blend," *International Journal of Concrete Structures and Materials*, vol. 7, no. 3, pp. 239–249, 2013.

[3] V. M. Malhotra, "Durability of concrete incorporating high-volume of low-calcium (ASTM Class F) fly ash," *Cement and Concrete Composites*, vol. 12, no. 4, pp. 271–277, 1990.

[4] T. R. Naik, B. W. Ramme, and J. H. Tews, "Use of high volumes of Class C and Class F fly ash in concrete," *Cement, Concrete and Aggregates*, vol. 16, no. 1, pp. 12–20, 1994.

[5] D. Cross, J. Stephens, and J. Vollmer, "Structural applications of 100 percent fly ash concrete," in *Proceedings of the World of Coal Ash (WOCA)*, pp. 1–9, Springer, Lexington, Kentucky, USA, 2005.

[6] J.-E. Kim, W.-S. Park, Y.-I. Jang et al., "Mechanical Properties of Energy Efficient Concretes Made with Binary, Ternary, and Quaternary Cementitious Blends of Fly Ash, Blast Furnace Slag, and Silica Fume," *International Journal of Concrete Structures and Materials*, vol. 10, no. 3, pp. 97–108, 2016.

[7] T. R. Naik and S. S. Singh, "Influence of fly ash on setting and hardening characteristics of concrete systems," *ACI Materials Journal*, vol. 94, no. 5, pp. 355–360, 1997.

[8] C. H. Huang, S. K. Lin, C. S. Chang, and H. J. Chen, "Mix proportions and mechanical properties of concrete containing very high-volume of Class F fly ash," *Construction and Building Materials*, vol. 46, pp. 71–78, 2013.

[9] G. Venkatesan, S. Reghu Raman, and M. Chandra Sekaran, "Flexural behavior of reinforced concrete beams using high volume fly ash concrete confinement in compression zone," *Journal of Civil Engineering (IEB)*, vol. 41, no. 2, pp. 87–97, 2013.

[10] M. Arezoumandi, C. A. Ortega, and J. S. Volz, "Flexural behavior of high-volume fly ash concrete beams-experimental study," *Transportation Research Record*, vol. 2508, pp. 22–30, 2015.

[11] M. Arezoumandi, J. S. Volz, and J. J. Myers, "Shear behavior of high-volume fly ash concrete versus conventional concrete," *Journal of Materials in Civil Engineering*, vol. 25, no. 10, pp. 1506–1513, 2013.

[12] R. Thangaraj and R. Thenmozhi, "Experimental study on rc beams using high volume fly ash," *IRA-International Journal of Technology & Engineering*, vol. 3, no. 2, pp. 71–85, 2016.

[13] V. V. Cao and H. R. Ronagh, "A model for damage analysis of concrete," *Advances in Concrete Construction*, vol. 1, no. 2, pp. 187–200, 2013.

[14] M. K. M. Reddar, *Stress-strain model of Unconfined and Confined Concrete and Stress-Block Parameters, Master Thesis [Master, thesis]*, University of Texas A & M, College Station, Tex, USA, 2009.

[15] T. Tavio and A. Tata, "Predicting Nonlinear behavior ad stress-strain relationship of rectangular confined reinforced concrete columns with ANSYS," *Civil Engineering Dimension*, vol. 11, no. 1, pp. 23–31, 2009.

[16] S. W. Yoo, S. H. Jung, and S.-J. Kwon, "Evaluation of compressive and shear strength characteristics in high volume fly ash concrete," *Applied Mechanics and Materials*, vol. 584-586, pp. 1282–1288, 2014.

[17] M. H. Harajli, E. Hantouche, and K. Soudki, "Stress-strain model for fiber-reinforced polymer jacketed concrete columns," *ACI Structural Journal*, vol. 103, no. 5, pp. 672–682, 2006.

[18] D. C. Kent and R. Park, "Flexural members with confined concrete," *Journal of the Structural Division*, vol. 97, no. 7, pp. 1969–1990, 1971.

[19] J. Hoshikuma, K. Kawashima, K. Nagaya, and A. W. Taylor, "Stress-strain model for confined reinforced concrete in bridge piers," *Journal of Structural Engineering*, vol. 123, no. 5, pp. 624–633, 1997.

[20] P. Muthupriya, K. Subramanian, and B. G. V. Vishnuram, "Investigation of Behaviour of high performance reinforced concrete columns with metakaolin and fly ash as admixture," *International Journal of Advanced Engineering Technology*, vol. 2, no. 1, pp. 190–202.

8

Improvement of Mechanical Properties in Polypropylene- and Glass-Fibre-Reinforced Peach Shell Lightweight Concrete

Fan Wu,[1] Changwu Liu,[1,2] Zhaofeng Diao,[2] Bo Feng,[2] Wei Sun,[2] Xiaolong Li,[2] and Shuang Zhao[2,3]

[1]*Institute of Disaster Management and Reconstruction, Sichuan University and The Hong Kong Polytechnic University, No. 1 Huanghe Road, Chengdu 610065, China*

[2]*College of Water Resource and Hydropower, Sichuan University, No. 24 South Section 1, Yihuan Road, Chengdu 610065, China*

[3]*China Yangtze Power Co., Ltd., China Three Gorges Corporation, No. 1 Construction Road, Yichang, China*

Correspondence should be addressed to Changwu Liu; liuchangwu@scu.edu.cn

Academic Editor: Nadezda Stevulova

The use of the polypropylene fibre and glass fibre with different volume fractions to improve the mechanical properties of peach shell lightweight concrete was investigated in this study. The volume fractions of 0.25%, 0.50%, and 0.75% were used for each fibre. The results showed that, as the polypropylene fibre and glass fibre were added into peach shell concrete, the density was reduced by up to 6.1% and the compressive strength, splitting tensile strength, and flexural strength were increased by 19.1%, 54.3%, and 38.6%, respectively. The highest compressive strength, splitting tensile strength, and flexural strength of 29.3 MPa, 2.87 MPa, and 3.09 MPa, respectively, were produced by peach shell concrete with 0.75% glass fibre. The results indicated that the incorporation of fibres significantly enhanced the postfailure toughness of peach shell concrete. It was found that the glass fibre was more effective than the polypropylene fibre in improving the mechanical properties of peach shell concrete. Although the incorporation of fibres slightly increased the water absorption and porosity, the type and content of fibres had no significant effect on water absorption and porosity. Therefore, the mechanical properties of peach shell lightweight concrete can be improved by adding polypropylene fibres and glass fibres.

1. Introduction

The use of agricultural or industrial wastes in concrete to replace the conventional raw material has achieved environmentally friendly and sustainable development by reducing the production costs of the raw material and contributing to waste disposal. The expanding extraction of conventional aggregates such as granite aggregate and sand in concrete production has caused serious environmental pollution problems [1]. The use of lightweight aggregates such as expanded clay, foamed slag, and natural pumice instead of conventional coarse aggregates can produce lightweight aggregate concrete (LWAC) [2]. LWAC has a lot of advantages including better fire resistance, heat insulation, sound absorption, and frost resistance [3]. Recently, the utilization of agricultural wastes such as coconut shell, oil palm shell, and apricot shell as a substitute for normal weight aggregates (NWAs) is gradually rising. On the other hand, the additional advantages of using agricultural wastes instead of normal aggregates to produce LWAC are to reduce environmental pollution and to deal with wastes [4]. Agricultural waste is rapidly storing in developing countries, and peach shell (PS) is the seed of peach fruits, which causes environmental problem around their cultivated areas. More than thousand tons of PS wastes need to be discharged each year in China. The lightweight and regenerative characteristic of PS compared to the NWAs can make PS as one of the potential lightweight aggregates and sustainable building materials in the production of LWAC. This could reduce the environmental pollution caused by PS and contribute to the reutilization of PS.

The improvement in the mechanical properties of LWAC depends on the oven-dried density, water-to-cement ratio, addition of other binder materials, aggregate content, and particle size [5, 6]. Previous studies showed that LWAC with higher compressive strength can be produced, but this concrete was prone to brittle and poor tensile strength due to the fact that the lightweight aggregate is usually weaker under tension [7, 8]. The tensile strength of LWAC is about one-tenth of its compressive strength [9]. The use of PS instead of NWA has a positive impact on density, and a reduction in density up to 30% can be achieved in the peach shell concrete (PSC). However, the splitting tensile and flexural strengths and modulus of elasticity (MOE) of PSC are lower than those of LWAC made of other lightweight aggregates such as pumice and expanded clay. The low tensile strength tends to cause significant tensile cracking to occur under lower tensile loads [10]. Hence, the improvement of the mechanical properties of PSC requires more attention.

Incorporation of fibres is an effective way to improve the mechanical properties of concrete, such as splitting tensile and flexural strengths and other related characteristics [11]. In fibre-reinforced concrete, the role of fibres is to improve the performance of concrete through the crack bridge effect of fibres and the fibre-cement-aggregate interfacial bond [5]. Previous research showed that fibre-reinforced concrete with two types of different fibres had better ductility, crack growth resistance, and impact resistance. In the composite fibre material system, the harder fibres enhance the first crack stress and ultimate strength, and the softer fibres control the crack propagation and increase the ductility [12]. Yap et al. [13] reported that oil palm shell lightweight concrete with 0.9% steel and 0.1% polypropylene (PP) fibres had better improvement in flexural toughness characteristics.

Although a lot of study works are ongoing on the addition of fibres in LWAC, there is only little literature on the incorporation of fibres in PSC. In order to make PSC like other LWAC applied to practical building structure elements such as sidewalks, parking lots, roadblocks, and partition walls, the mechanical properties and durability of PSC need to be further improved. The addition of fibres into PSC can make it as a ductile and sustainable building material. Commonly used fibres are the steel fibre, PP fibre, and nylon fibre, and there are few literatures available on the use of glass (G) fibres in LWAC. The main disadvantages of adding steel fibres to LWAC are a significant increase in density and reduced workability. The PP fibre is mainly used in concrete to enhance the ductility, toughness, and impact resistance, but do not expect them to improve the strength [12]. Mastali et al. [14] reported that adding G fibres led to a significant improvement in compressive strength, tensile strength, and impact resistance of G fibre-reinforced concrete. Compared to steel fibres, the advantage of adding PP fibres and G fibres to PSC is that the density is less than 2000 kg/m^3 and lower than the upper limit of the density set for LWAC. However, there are few literatures on the comparison of PP fibre- and G fibre-reinforced peach shell lightweight concrete.

The purpose of the present study is to investigate the influence of two types of fibres (PP fibre and G fibre) and each fibre at a volume fraction of 0.25%, 0.50%, and 0.75% on the mechanical properties of peach shell lightweight concrete. A comparison of the performance of PP fibre- and G fibre-reinforced peach shell lightweight concrete is investigated and reported, including workability, density, compressive strength, splitting tensile strength, flexural strength, modulus of elasticity, residual compressive strength, water absorption, and porosity. This work helps to expand the knowledge on the effect of PP fibres and G fibres with different contents on the mechanical properties of peach shell fibre-reinforced concrete (PSFRC).

2. Materials and Methods

2.1. Materials

2.1.1. Cement and Silica Fume. Type I 42.5 grade ordinary Portland cement was used in this study and obtained from a local cement company. The Blaine specific surface area and specific gravity are 3532 cm^2/g and 3.14 g/cm^3, respectively. In addition, silica fume (SF) was added in concrete as an additional mineral binder to improve the performance of concrete. SF content at 10% of the cement weight was used in all mixes.

2.1.2. Coarse and Fine Aggregates. Peach shell (PS) was used as an alternative coarse aggregate, and crushed pebble was used as a normal weight coarse aggregate. The density of PS is 63% lower than that of normal weight aggregates (NWAs), which is significant for reducing dead loads of building structures. River sand was used as a fine aggregate and obtained from a local sand company. The physical properties of the aggregates are shown in Table 1. PS was collected from a local peach plantation. Before using, they were washed to remove the residual dried peach pulp and the dust from the PS surface. Then, the dry PS was crushed by a crushing machine in the laboratory. Crushed PS was sieved with 4.75 mm and 9.5 mm sieves, respectively. Particles of size 4.75–9.5 mm were considered as coarse aggregates. Uncrushed and crushed PS and scanning electron microscopic (SEM) image at 5000 magnification times of crushed PS are shown in Figure 1. As can be seen from Figure 1, crushed PS has an irregular and rough surface, and the PS aggregate has a lot of circular porosity on the surface, and the diameter of the micropore is about 10–20 μm. Although a large number of porosity on the surface of PS reduces the density, it leads to the fact that PS has a higher water absorption compared to NWA. Therefore, PS was submerged in water for 24 h and kept in an internal saturated and surface dry condition before mixing.

2.1.3. Water and Superplasticizer. The water used in this study was normal tap water. A high-range naphthalene sulphonated superplasticizer (SP) was used in the study to increase the workability of fresh concrete. SP content at 1% of the cement weight was used in all mixes.

2.1.4. Fibres. Two different fibres used in this study are the (i) polypropylene (PP) fibre and (ii) glass (G) fibre. The properties of fibres are shown in Table 2.

Table 1: Physical properties of the aggregates.

Physical property	Coarse aggregate		Fine aggregate
	PS	NWA	River sand
Specific gravity (g/cm^3)	1.26	2.66	2.65
Fineness modulus	5	4.9	2.7
Bulk density (kg/m^3)	536	1449	1538
Water absorption (24 h) (%)	26.84	0.5	1.2
Aggregate impact value (%)	1.95	16.72	—
LA abrasion value (%)	6	25	—
Flakiness index (%)	48	24	—
Elongation index (%)	61	34	—
Grading			
Sieve size (mm)	Cumulative % by weight passing		
16	100		100
9.5	98.2		100
4.75	24.8		99
2.36	0		86
1.18	0		74.7
0.6	0		48.3
0.3	0		16.3
0.15	0		0

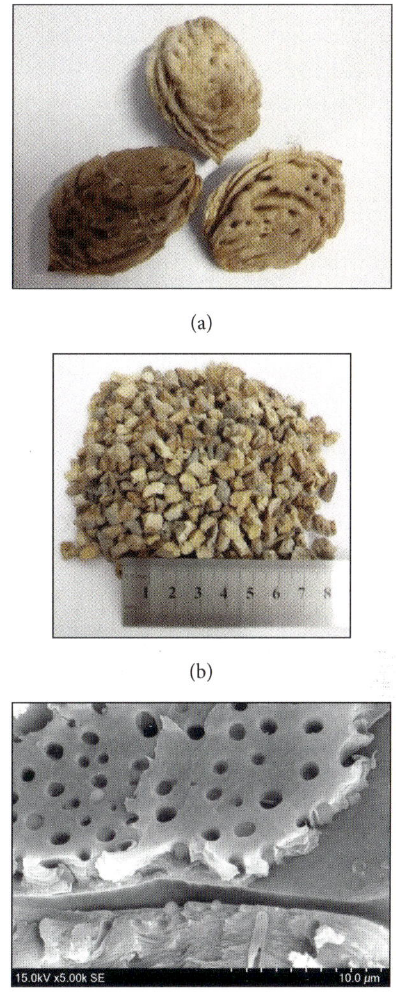

(a)

(b)

(c)

Figure 1: (a) Uncrushed PS, (b) crushed PS, and (c) SEM image of the crushed PS particle at 5000 magnification times.

2.2. Mix Proportioning. The mix proportion of concretes is presented in Table 3. A total of six concrete mixes were prepared with different fibre contents, and a control mix without the addition of fibres was also produced for comparison purpose. The total amount of materials for all mixes was the same except in the case of two different fibres and three fibre volumes. The volume fractions of fibres were 0.25%, 0.50%, and 0.75%, respectively.

2.3. Specimen Preparation and Test Methods. The procedure of specimen preparation of peach shell fibre-reinforced concrete (PSFRC) was detailed as follows: firstly, PS and river sand were poured into a blender and dry mixed for 2 min. Then, fibres and cement and SF were added to the mixture and dry mixed for 1 min. Before the fibres were mixed, fibres were manually distributed. Following this, 60% water mixed with SP was added to the mixture and mixed for 3 min. Finally, 40% water was added to the mixture and mixed for 5 min. After mixing was complete, the slump test was carried out immediately, and then, the specimens were cast in oiled plastic moulds. All specimens were compacted by the vibrating table. Immediately after compaction, all specimens along with the moulds were covered by a plastic sheathing to prevent moisture evaporation. Specimens were removed from the moulds after 24 h. Finally, specimens were stored in a controlled room with a relative humidity of 95% ± 5% and a temperature of 20°C ± 2°C until the test age.

The workability of all mixes was measured by the slump test according to ASTM C143/C143M-12 (Figure 2). The density of all specimens was measured in accordance with ASTM C138. 100 mm cube specimens were used for the compressive strength test (BS EN12390-2:200 9 and BS EN12390-3:200 9) at the age of 3, 7, 28, and 56 days (Figure 2).

The splitting tensile and flexural strengths and modulus of elasticity of all mixes were measured at the age of 28 days according to ASTM C496/C496M-11, ASTM C78-10, and ASTM C469-10 (Figure 2), respectively. The average value of at least three specimens is taken for each test result.

The water absorption test was carried out according to the procedures prescribed in ASTM C642. All specimens were oven-dried at 110 ± 5°C for not less than 24 h to remove moisture and achieve constant before testing. The specimens were then submerged in water for 24 h. The water absorption values of the specimen were calculated using the following formula:

$$\text{Water absorption (\%)} = \frac{M_2 - M_1}{M_1} \times 100, \tag{1}$$

where M_1 is the mass of the oven-dried specimen in air (g) and M_2 is the mass of the surface-dried specimen in air after immersion (g).

The porosity of all specimens was determined in accordance with ASTM C642. Absolute density is necessary to determine the total porosity by the method prescribed in ASTM C642; hence, the pycnometry method was applied for

Table 2: Properties of the PP fibre and G fibre.

Fibre		Fibre type	Length (mm)	Diameter (μm)	Specific gravity (g/cm^3)	Tensile strength (MPa)
PP	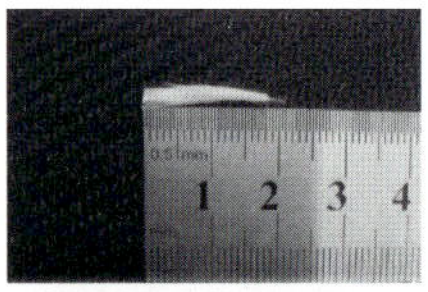	Monofilament	19	40	0.91	500
G	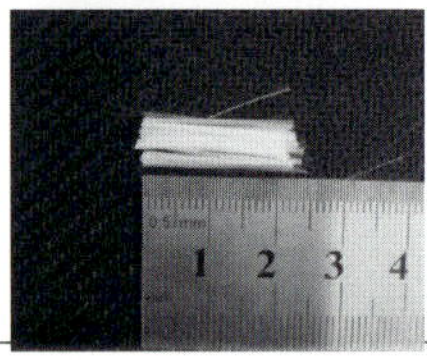	Monofilament	22	10	2.65	1200

Table 3: Mix proportion of concretes.

Mix code	Cement (kg/m^3)	SF (kg/m^3)	Water (kg/m^3)	Sand (kg/m^3)	SP (kg/m^3)	PS (kg/m^3)	PP fibre (%)	G fibre (%)
Control	550	55	180	780	5.5	360	0	0
PP25	550	55	180	780	5.5	360	0.25	0
PP50	550	55	180	780	5.5	360	0.50	0
PP75	550	55	180	780	5.5	360	0.75	0
G25	550	55	180	780	5.5	360	0	0.25
G50	550	55	180	780	5.5	360	0	0.50
G75	550	55	180	780	5.5	360	0	0.75

(a) (b) (c) (d)

Figure 2: (a) Slump test, (b) compressive test, (c) splitting tensile test, and (d) flexural strength test.

the determination of absolute density. The total porosity and open porosity values of the specimen can be calculated by (2) and (3), respectively. The closed porosity volume can be obtained by subtracting the open porosity volume from the total porosity volume.

$$\text{Total porosity volume}\,(\%) = \frac{\rho_a - \rho_b}{\rho_a} \times 100, \qquad (2)$$

where ρ_a is the absolute density of the specimen (g/cm^3) and ρ_b is the bulk density of the specimen (g/cm^3).

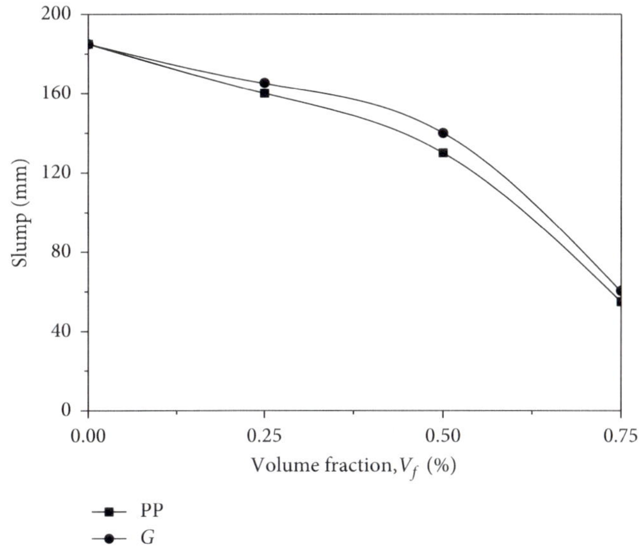

FIGURE 3: Relationship between the fibre content and slump value.

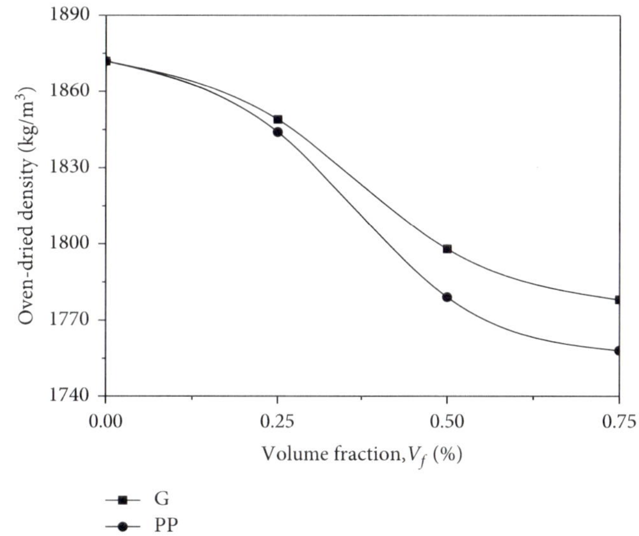

FIGURE 4: Relationship between the fibre content and oven-dried density.

$$\text{Open porosity volume}\,(\%) = \frac{M_3 - M_1}{M_3 - M_4} \times 100, \qquad (3)$$

where M_1 is the mass of the oven-dried specimen in air (g); M_2 is the mass of the surface-dried specimen in air after immersion (g); M_3 is the mass of the surface-dried specimen in air after immersion and boiling (g); and M_4 is the apparent mass of the specimen in water after immersion and boiling (g).

3. Results and Discussion

3.1. Workability (Slump). Relationship between the fibre content and slump value is shown in Figure 3. Yew et al. [9] reported that the incorporation of different types of PP fibres in oil palm shell concrete caused a considerable decrease in the slump value. As expected, the results showed that the slump values of the fresh PSC with PP fibres and G fibres were reduced to 55 mm and 60 mm, respectively. The reason for this phenomenon can be explained by the fact that the fibre-cement paste interfacial bond in concrete restricts the dispersion and flow of the cement paste and improves the viscosity of the mixtures. With the increase in fibre volume fraction, the capability of the fibre-cement paste interfacial bond in concrete becomes stronger as more fibres absorb the cement paste to wrap around it. Therefore, the workability of concrete decreased as the fibre volume (V_f) increased from 0% to 0.75%. Mehta and Monteiro [15] reported that the LWAC with a slump value of 50–75 mm is similar to normal weight concrete with a 100–125 mm slump value. In the case of PSC, the slump value of PP fibre- and G fibre-reinforced PSC between 55 mm and 60 mm can easily be compacted.

Figure 3 also shows that the two types of fibres have an impact on the workability of concrete. For all V_f, the G fibre produced slightly higher slump values than the PP fibre. The increase in slump values at a range of 5–10 mm and the biggest difference between the G fibre and PP fibre were observed at $V_f = 0.50\%$, and the slump value of G50 was 7.8%

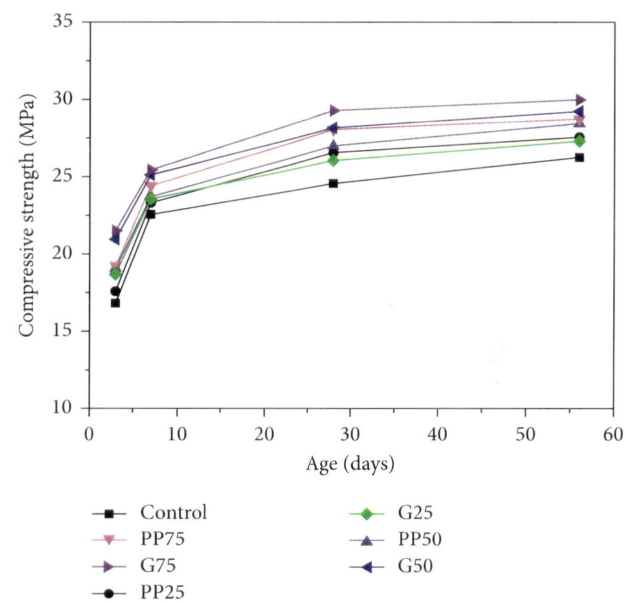

FIGURE 5: The development of the compressive strength of concretes.

higher than that of PP50. Song et al. [16] reported that the workability of fibres with a shorter length was lower than that of longer fibres. The shorter fibres have more surface area for the reinforcing fibre-cement paste bond [9]. Since the length of the G fibre is longer than that of the PP fibre and the surface of the G fibre is smoother, G fibres have lower effective surface area to develop a fibre-cement paste bond compared to PP fibres. Hence, the slump value of PSC incorporated with G fibres was slightly higher than that of PP fibres.

3.2. Density. Lightweight aggregate concrete (LWAC) is defined as concrete with a compressive strength of more than 15 MPa and an oven-dried density (ODD) in the range

TABLE 4: Compressive strength of concretes.

Mix	Compressive strength (MPa)				% increase in compressive strength (28 days)
	3 days	7 days	28 days	56 days	
Control	16.8 (68.3%)	22.6 (91.9%)	24.6	26.3 (106.9%)	—
PP25	17.6 (66.2%)	23.3 (87.6%)	26.6	27.6 (103.8%)	8.1
PP50	19.1 (70.7%)	23.7 (87.8%)	27.0	28.5 (105.6%)	9.8
PP75	19.2 (68.3%)	24.4 (86.8%)	28.1	28.8 (102.5%)	14.2
G25	18.7 (71.6%)	23.6 (90.4%)	26.1	27.3 (104.6%)	6.1
G50	21.0 (74.5%)	25.1 (89.0%)	28.2	29.3 (103.9%)	14.6
G75	21.5 (73.4%)	25.5 (87.0%)	29.3	30.2 (103.1%)	19.1

Note. Values in brackets denote the percentages of compressive strength to the corresponding 28-day compressive strength.

of 1600–2000 kg/m^3 [17]. Relationship between the fibre content and ODD is shown in Figure 4. The results showed that the ODD of all PSC ranged from 1758 kg/m^3 to 1872 kg/m^3 and fulfilled the requirement of LWAC because PS is about 63% lighter than NWA. The addition of PP fibres and G fibres in concrete reduced the ODD of PSC because fibres have a low specific gravity. When the volume fraction of PP fibres is added at percentages of 0.25%, 0.50%, and 0.75%, the ODD of these concretes decreased by 1.5%, 5.0%, and 6.1%, respectively, compared to the control mix.

Figure 4 also showed that the different types and volume fractions of fibres had an influence on the ODD of PSC. It was found that concrete with G fibres produced a slightly higher ODD than PP fibres, and the difference in the ODD of PSC is more significant as fibre volume fraction ranges from 0.25% to 0.75%. At $V_f = 0.75\%$, G75 had the lowest ODD of 1778 kg/m^3, and it was only 1.1% higher than PP75. It should be noted that although the specific gravity of PP fibres and G fibres is significantly different, its impact on the ODD of PSC is not obvious. However, compared to the density of approximately 2400 kg/m^3 for ordinary weight concrete, the ODD of PSC is reduced by about 26%, which is a significant reduction in the weight of concrete.

3.3. Compressive Strength. The development of the compressive strength of concretes up to 56 days is shown in Figure 5. The test results revealed that the compressive strength of all concretes increased as the curing age increased. As can be seen from Figure 5, there is a significant difference in the compressive strength between PSC containing different types and contents of fibres and control mix without any fibres. The enhancement of the ultimate compressive strength may be attributed to the fact that fibres arrest the growth of cracks due to the fibre-cement paste interfacial bond and the crack bridging effect of fibres [18].

The addition of PP fibres and G fibres had significantly increased the compressive strength of PSC at all ages, and the compressive strength increased as an increase in fibre content. The 28-day compressive strength of all mixes varied between 24.6 MPa and 29.3 MPa, which fulfilled the requirements for the strength and density of LWAC. It can be observed that compared to the control mix, the 28-day compressive strength of PP25, PP50, and PP75 increased by 8.1%, 9.8%, and 14.2%, respectively, and the 28-day compressive strength of G25, G50, and G75 increased by 6.1%, 14.6%, and 19.1%, respectively. Except for $V_f = 0.25\%$, the 28-day compressive strength of G fibre-reinforced PSC was slightly higher than that of PP fibre. This phenomenon indicated that the use of G fibre-reinforced PSC is a good choice.

Table 4 shows the compressive strength of concretes. All concretes attained approximately 66%–75% of compressive strength at 3 days and 87%–92% of compressive strength at 7 days. This indicated that all PSC developed high early compressive strength because the incorporation of SF increased the cohesiveness of cement paste and reduced the advance of microcracks, and eventually, the early compressive strength of PSC is increased. The addition of PP fibres and G fibres in concrete increased the 28-day compressive strength by about 6%–19%. At $V_f = 0.75\%$, G75 produced the highest 28-day compressive strength of 29.3 MPa, which is 19.1% higher than the control mix of 24.6 MPa. Although not every kind of lightweight aggregate is suitable for the production of LWAC, the results based on this study indicate that it is feasible to produce LWAC using PS as a lightweight aggregate.

Generally, the initiation and advance of cracks in concrete are due to the continuous increase in compressive loading. Since the tensile stress generated by the fibre is perpendicular to the crack propagation path, the debonding begins at the interface between the fibre and cement paste. Once the advancing cracks reach the fibre-cement paste interface, the stress concentration at the crack tip is released, and as a result, the development of the advancing crack is blunted and blocked. The fibre is like a bridge in this process that arrests the advancing cracks and consequently improves the strength of concretes [19, 20]. Therefore, the addition of PP fibres and G fibres enhanced the bonding of the fibre-cement paste interface and improved the compressive strength [16]. The compressive strength of G fibre-reinforced PSC is higher than that of PP fibre-reinforced PSC. This reason can be explained by the following: firstly, the tensile strength of G fibres in this study is 2.4 times that of PP fibres, and thus, compared to the PP fibre, the G fibre is easier to transfer higher tensile stress from a cracked fibre-cement interface to fibre. Secondly, under the same fibre volume fraction, the density of G fibre-reinforced PSC in this study is higher than that of PP fibre-reinforced PSC. Although the addition of fibres in this study reduced the density of PSC,

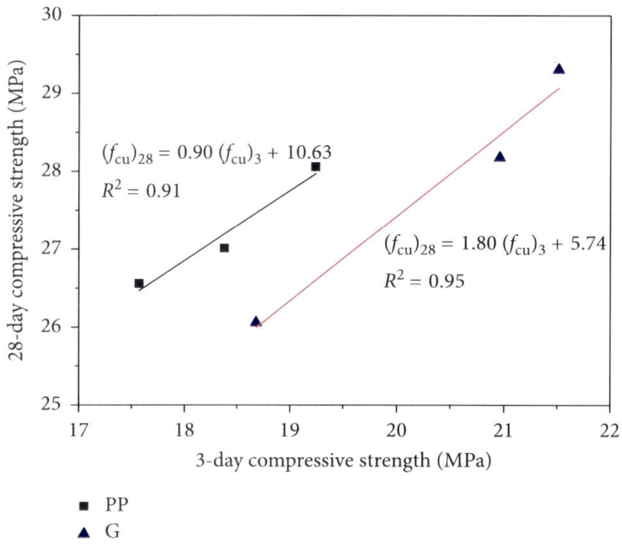

FIGURE 6: Relationship between 3-day and 28-day compressive strengths.

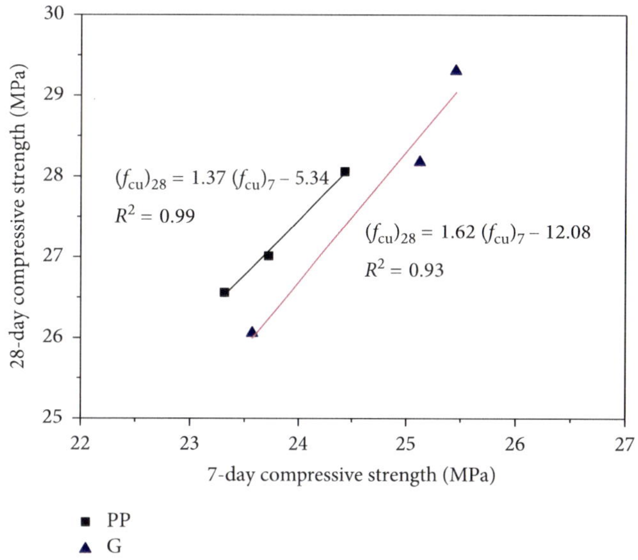

FIGURE 7: Relationship between 7-day and 28-day compressive strengths.

for normal weight concrete, the higher the density the higher the compressive strength. Thus, the compressive strength of higher density of G fibre-reinforced PSC is higher than that of PP fibre-reinforced PSC.

Due to the importance of knowing the 28-day compressive strength from the early age compressive strength, Figures 6 and 7 show the relationship between the 28-day compressive strength and compressive strength at early ages of 3 days and 7 days, respectively. As can be seen from Figures 6 and 7, there is a linear relationship with the high correlation coefficient between the 28-day compressive strength and the early compressive strength, and the 28-day compressive strength of PSC can be predicted by the early compressive strength.

TABLE 5: Splitting tensile and flexural strengths and modulus of elasticity of concretes (28 days).

Mix	Splitting tensile strength (MPa)	Flexural strength (MPa)	Modulus of elasticity (GPa)
Control	1.86	2.23	8.71
PP25	2.05	2.38	9.63
PP50	2.15	2.50	10.15
PP75	2.54	2.63	10.85
G25	2.00	2.35	9.79
G50	2.52	2.69	10.75
G75	2.87	3.09	11.32

3.4. Splitting Tensile and Flexural Strengths. Splitting tensile strength and flexural strength of all mixes are shown in Table 5. Comiloli et al. [21] reported that the incorporation of PP fibres in concrete can slightly enhance the flexural toughness of concrete. The fibre enhances the strength of concretes by the crack bridging effect and carrying part of load [12]. The results showed that the incorporation of PP fibres and G fibres in concrete increased both splitting tensile strength and flexural strength of PSC, and the more the fibre content, the higher the splitting tensile strength and flexural strength. The addition of PP fibres from 0.25% to 0.75% and glass fibres from 0.25% to 0.75% increased the splitting tensile strength up to 10.2%–36.6% and 7.5%–54.3%, respectively. Furthermore, the flexural strength of PP fibre- and G fibre-reinforced PSC was also improved. The flexural strength of PSC increased as an increase in fibre volume fraction. When the fibre volume fraction varies from 0.25% to 0.75%, the flexural strength for PP fibre and G fibre reinforced PSC increased by 6.7%–17.9% and 5.4%–38.6%, respectively. The advantage of PSC could be further explained based on Figure 8. From Figure 8, it was evident that the control concrete without any fibres failed once the ultimate splitting tensile strength was reached due to low ductility. Furthermore, the fibre-reinforced PSC showed excellent crack arresting performance as the specimen was not broken into two pieces even after the failure load. This phenomenon attributed to the crack bridging effect of fibres, and fibres in concrete provide additional strength to the fibre-cement interfacial bond which carries a partial load.

Although the ODDs of PP75 and G75 were lower in this study, the splitting tensile strength and flexural strength of PP75 and G75 were higher than those of the control mix. Since the tensile stress of G fibres is greater than that of PP fibres, glass-fibre reinforced PSC produced a higher splitting tensile strength and flexural strength. At $V_f = 0.50\%$ and 0.75%, splitting tensile strength and flexural strength of PSC containing G fibres were found to be about 12.5%–13.0% and 7.6%–17.5%, respectively, and higher than those of PP fibres. Splitting tensile strength and flexural strength of G75 were 2.87 MPa and 3.09 MPa, respectively, which increased by 54.3% and 38.6%, respectively, compared to the control mix. Compared to the PP fibre, the higher tensile stress of the G fibre is more effective in transferring the tensile stress. At the

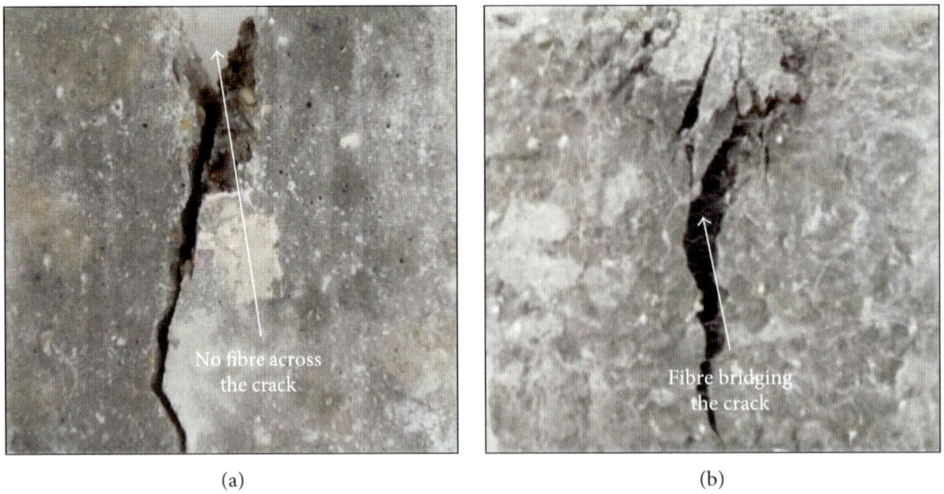

FIGURE 8: Specimen after the splitting tensile test: (a) control concrete without fibres and (b) concrete containing fibres.

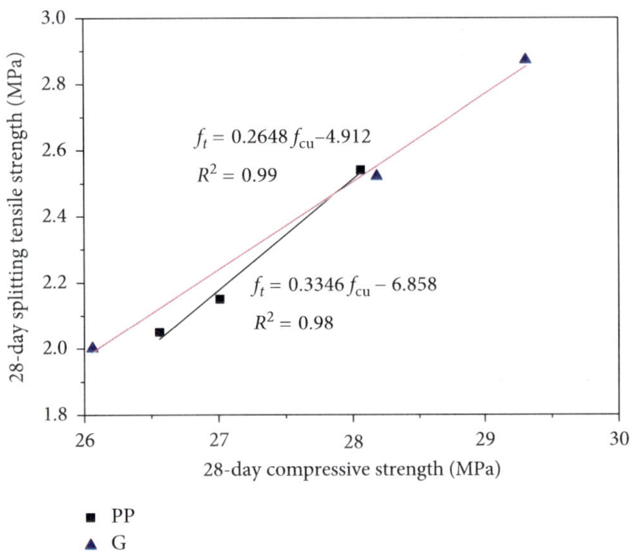

FIGURE 9: Relationship between the compressive strength and splitting tensile strength.

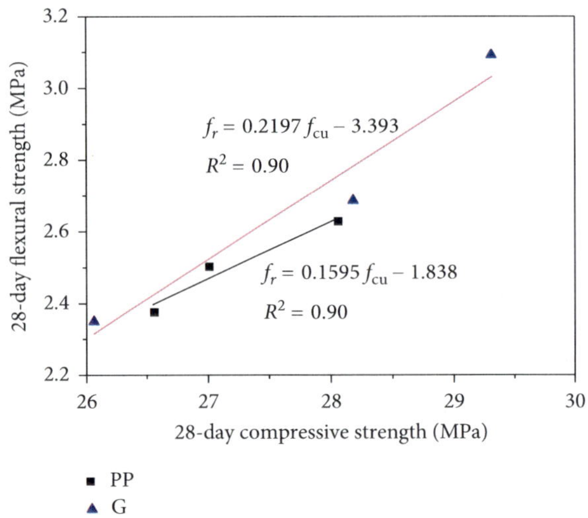

FIGURE 10: Relationship between the compressive strength and flexural strength.

same fibre volume fraction, the G fibre has a smaller diameter, leading to more fibres to join the cracks [22]. Due to the bridging effect of fibres, the stress concentration around cracks is reduced and the development of cracks is hindered. Eventually, the tensile strength of PSC is improved.

The relationship between the compressive strength and splitting tensile strength and flexural strength for PSC at 28 days is shown in Figures 9 and 10, respectively. As can be seen from Figures 9 and 10, the splitting tensile strength and flexural strength of PSFRC increase as an increase in compressive strength.

3.5. Modulus of Elasticity. Table 5 shows the modulus of elasticity (MOE) of all concretes, which ranges between 8.71 and 11.32 GPa. The control concrete in this study produced a minimum MOE of 8.71 GPa. The results showed that the types of G fibres and PP fibres had a significant effect on MOE. At $V_f = 0.25\%$, 0.50%, and 0.75%, the MOE of PP fibre- and G fibre-reinforced PSC increased by 10.6% and 12.4%, 16.5% and 23.4%, and 24.6% and 30.0%, respectively. The incorporation of fibres enhances the MOE of PSC because fibres arrest the initial cracks caused by shrinkage, and the crack bridging effect reduces the strain caused by the compression loading and consequently improves the MOE.

The MOE of PSFRC was found to depend on the fibre volume fraction, while the type of fibres had an insignificant influence on it. The volume fraction of G fibres increased from 0.25% to 0.75%, the MOE of G fibre-reinforced PSC increased by 15.6% from 9.79 GPa to 11.32 GPa, and the MOE of G fibre-reinforced PSC was 1.7%, 5.9%, and 4.3%, respectively, slightly higher than that of PP fibre.

Figure 11 shows the relationship between the compressive strength and MOE. As can be seen from Figure 11, there is a good relationship between the MOE and 28-day

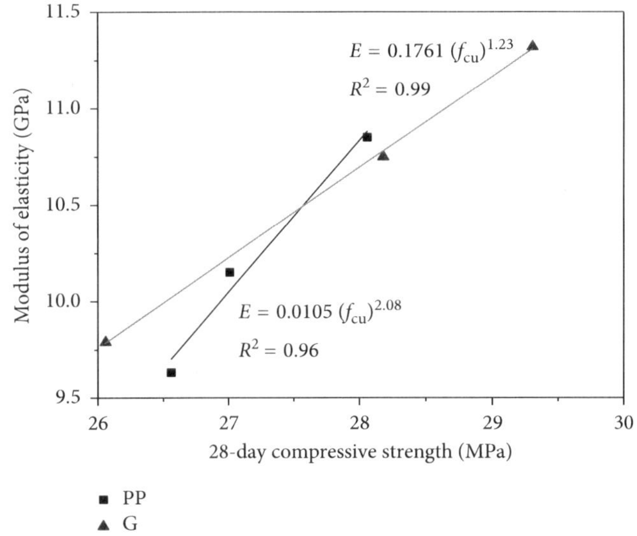

FIGURE 11: Relationship between the compressive strength and MOE.

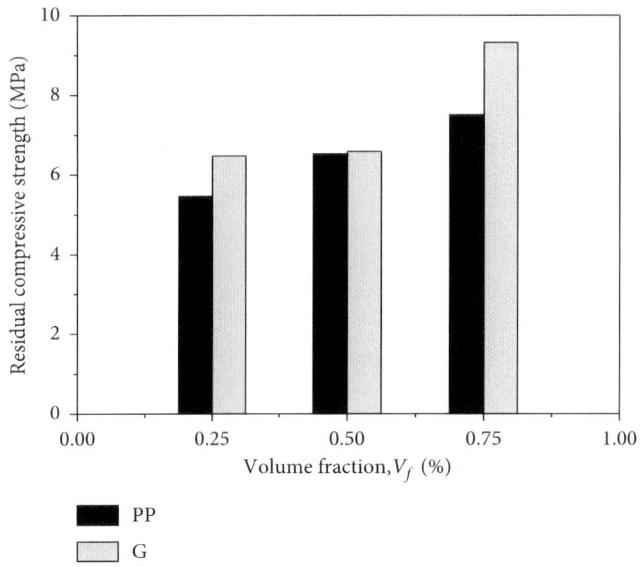

FIGURE 12: Residual compressive strength of concretes.

compressive strength of PSFRC, and the MOE of PP fibre-reinforced PSC and glass fibre-reinforced PSC can be predicted by the 28-day compressive strength, respectively.

3.6. Residual Compressive Strength. Residual compressive strength (RCS) can be used to evaluate the residual strength toughness of concretes. Figure 12 shows residual compressive strength of concretes. Because the crack bridging effect of fibres existed at two crack surfaces, in the fibre-reinforced concrete, additional loading forces are required for further growth of the crack. As expected, the incorporation of PP fibres and G fibres enhanced the residual strength toughness of PSC. The control concrete without any fibres had no RCS as the control concrete was immediately failed when the ultimate loading strength is reached. In

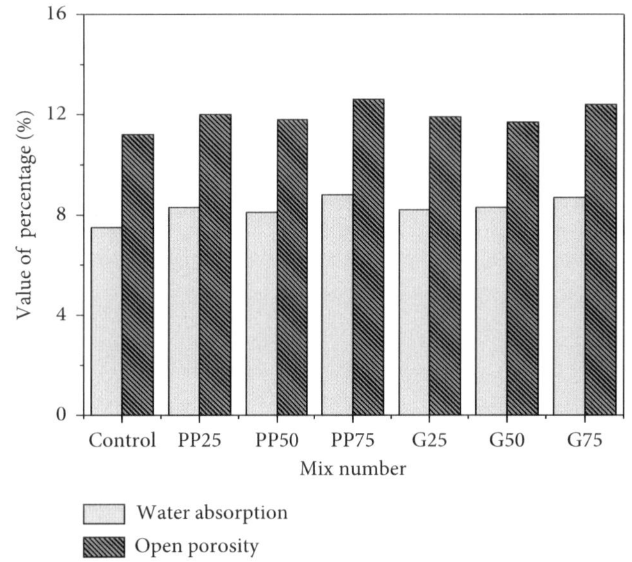

FIGURE 13: Percentage value of water absorption and open porosity of concretes.

addition, the more the fibre content, the higher the RCS value. At $V_f = 0.75\%$, PP fibre- and G fibre-reinforced PSC had the highest RCS value. It might be attributed to the crack bridging effect of fibres that existed at two crack surfaces, which can hinder further propagation of cracks, even after microcrack cracking. The results indicated that the influence of adding fibres on the improvement of postfailure toughness of PSC was very significant.

3.7. Water Absorption and Porosity. Twenty-four-hour water absorption and open porosity for all mixes are shown in Figure 13. Lo et al. [23] studied that the water absorption of the lightweight aggregate affected the internal microstructure of the hardened mortar and the interfacial zone of concrete, and an increase in the water absorption of the lightweight aggregate resulted in an increase in the number of pores in the interfacial zone of concrete. Figure 13 clearly showed that the incorporation of fibres slightly increased the water absorption and open porosity of PSC, and the water absorption values of all concretes varied from 7.5% to 8.8%. However, the type and content of fibres had an insignificant effect on water absorption and open porosity, and the more the number of open pores, the higher the water absorption for all concretes. The addition of fibres might affect the number and area of micropores in PSC, which resulted in a small increase in water absorption and open porosity. Neville and Brooks [24] reported that although the water absorption cannot be used to determine the quality of concrete, the water absorption of most of the good-quality concretes usually is much lower than 10% by mass. It can be seen from Figure 13 that all concretes have less than 10% water absorption and meet the requirements of good concrete.

Total porosity and open porosity of concretes are shown in Figure 14. The result showed that the total porosity of PSC was between 15.1% and 16.5%, and the open porosity

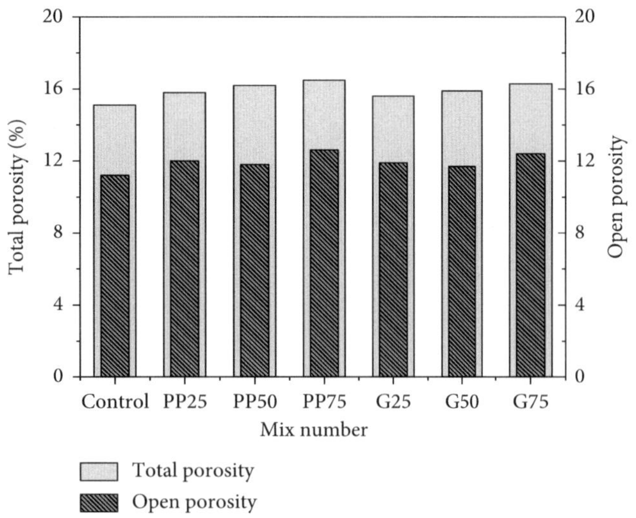

FIGURE 14: Total and open porosity of concretes.

varied from 11.2% to 12.6%. Furthermore, most of the total porosity in PSC was open porosity, and closed porosity accounted for a small portion of the total porosity because the surface of PS aggregates contains a lot of microscopic and connective pore structures. Results also showed that the type and content of fibres had an insignificant influence on the porosity of PSC, and the addition of fibres slightly increased the porosity of PSC. In addition, a slight increase in the total porosity was another reason that adding fibres reduced the density of PSC.

4. Conclusions

The influences of incorporating two types of fibres (polypropylene fibre and glass fibre) and different fibre volume fractions (0.25%, 0.5%, and 0.75%) on the mechanical properties of peach shell lightweight concrete have been investigated in this study. The following conclusions can be drawn based on the present study:

(1) The polypropylene fibre and glass fibre decrease the slump value of peach shell concrete. The slump values of the fresh peach shell concrete with polypropylene fibres and glass fibres are reduced to 55 mm and 60 mm, respectively. However, the concrete containing glass fibres produced slightly higher slump values than polypropylene fibres.

(2) The addition of polypropylene fibres and glass fibres reduces the oven-dried density of peach shell concrete. The oven-dried density of all concretes ranges from 1758 kg/m^3 to 1872 kg/m^3 and fulfills the requirement of lightweight aggregate concrete. In addition, the oven-dried density of peach shell concrete reduces with an increase in fibre content.

(3) All concretes have high early compressive strength. Polypropylene fibres and glass fibres significantly increase the compressive strength of peach shell concrete, and the compressive strength of peach shell concrete increases as an increase in fibre content. In addition, an increase in the 28-day compressive strength of peach shell concrete is about 6%–19%.

(4) The addition of polypropylene fibres and glass fibres in concrete increases the splitting tensile strength of peach shell concrete. Furthermore, the more the fibre content, the higher the splitting tensile strength. The addition of polypropylene fibres and glass fibres from 0.25% to 0.75% increases the splitting tensile strength up to 10.2%–36.6% and 7.5%–54.3%, respectively.

(5) The flexural strength of peach shell concrete increases as an increase in fibre volume fraction. The flexural strength for polypropylene fibres and glass fibres from 0.25% to 0.75% reinforced the peach shell concrete increase by 6.7%–17.9% and 5.4%–38.6%, respectively.

(6) The glass fibre-reinforced peach shell concrete produces a higher splitting tensile strength and flexural strength. At $V_f = 0.50\%$ and 0.75%, the splitting tensile strength and flexural strength of peach shell concrete containing glass fibres are found to be about 12.5%–13.0% and 7.6%–17.5%, respectively, and higher than those of polypropylene fibres.

(7) The effect of incorporating polypropylene fibres and glass fibres enhances the modulus of elasticity of peach shell concrete. Modulus of elasticity of all concretes ranges between 8.71 and 11.32 GPa. At $V_f = 0.25\%$, 0.50%, and 0.75%, the modulus of elasticity of polypropylene fibre- and glass fibre-reinforced peach shell concrete increases by 10.6% and 12.4%, 16.5% and 23.4%, and 24.6% and 30.0%, respectively.

(8) The control peach shell concrete produced no residual compressive strength. However, the addition of polypropylene fibres and glass fibres enhances the residual strength toughness of peach shell concrete. In addition, the higher the fibre volume content, the higher the residual compressive strength. At $V_f = 0.75\%$, the polypropylene fibre- and glass fibre-reinforced peach shell concrete have the highest residual compressive strength values and increase by 46% and 82%, respectively.

(9) The incorporation of fibres slightly increases the water absorption and porosity of peach shell concrete. However, the type and content of fibres have an insignificant effect on water absorption and porosity.

Conflicts of Interest

The authors declare that they have no conflicts of interest.

Acknowledgments

This work was financially supported by the Institute of Disaster Management and Reconstruction, Sichuan University-The Hong Kong Polytechnic University and College of Water Resource and Hydropower, Sichuan University. The authors would also like to thank Dr. Ai Ting for her assistance in conducting some of the tests in this paper.

References

[1] A. Short and W. Kinniburgh, *Lightweight Concrete*, Applied Science Publishers, London, UK, 3rd edition, 1978.
[2] U. J. Alengaram, B. A. A. Muhit, and M. Z. Jumaat, "Utilization of oil palm kernel shell as lightweight aggregate in concrete-a review," *Construction and Building Materials*, vol. 38, pp. 161–172, 2013.
[3] P. Shafigh, M. Z. Jumaat, and H. Mahmud, "Mix design and mechanical properties of oil palm shell lightweight aggregate concrete: a review," *International Journal of Physical Sciences*, vol. 5, no. 14, pp. 2127–2134, 2010.
[4] E. Aguirre-Maldonado and F. Hernández-Olivares, "Study of fine mortar powder from different waste sources for recycled concrete production," in *Sustainable Development and Renovation in Architecture, Urbanism and Engineering*, pp. 253–262, Springer, Berlin, Germany, 2017.
[5] S. P. Yap, U. J. Alengaram, and M. Z. Jumaat, "Enhancement of mechanical properties in polypropylene- and nylon-fibre reinforced oil palm shell concrete," *Materials and Design*, vol. 49, pp. 1034–1041, 2013.
[6] M. K. Yew, H. Mahmud, B. C. Ang, and M. C. Yew, "Effects of heat treatment on oil palm shell coarse aggregates for high strength lightweight concrete," *Materials and Design*, vol. 54, pp. 702–707, 2014.
[7] U. J. Alengaram, M. Z. Jumaat, H. Mahmud, and S. M. Shirazi, "Effect of aggregate size and proportion on strength properties of palm kernel shell concrete," *International Journal of Physical Sciences*, vol. 5, no. 12, pp. 1848–1856, 2010.
[8] A. R. Khaloo and M. Sharifian, "Experimental investigation of low to high-strength steel fibre reinforced lightweight concrete under pure torsion," *Asian Journal of Civil Engineering*, vol. 6, no. 6, pp. 533–547, 2005.
[9] M. K. Yew, H. B. Mahmud, B. C. Ang, and M. C. Yew, "Influence of different types of polypropylene fibre on the mechanical properties of high-strength oil palm shell lightweight concrete," *Construction and Building Materials*, vol. 90, pp. 36–43, 2015.
[10] S. P. Yap, U. J. Alengaram, M. Z. Jumaat, and K. R. Khaw, "Torsional behaviour of steel fibre-reinforced oil palm shell concrete beams," *Materials and Design*, vol. 87, pp. 854–862, 2015.
[11] S. Hamoush, T. Abu-Lebdeh, and T. Cummins, "Deflection behaviour of concrete beams reinforced with PVA microfibres," *Construction and Building Materials*, vol. 24, no. 11, pp. 2285–2293, 2010.
[12] C. X. Qian and P. Stroeven, "Development of hybrid polypropylene-steel fibre reinforced concrete," *Cement and Concrete Research*, vol. 30, no. 1, pp. 63–69, 2000.
[13] S. P. Yap, C. H. Bu, U. J. Alengaram, K. H. Mo, and M. Z. Jumaat, "Flexural toughness characteristics of steel-polypropylene hybrid fibre-reinforced oil shell concrete," *Materials and Design*, vol. 57, pp. 625–659, 2014.
[14] M. Mastali, A. Dalvand, and A. R. Sattarifard, "The impact resistance and mechanical properties of reinforced self-compacting concrete with recycled glass fibre reinforced polymers," *Journal of Cleaner Production*, vol. 124, pp. 312–324, 2016.
[15] P. K. Mehta and P. J. M. Monteiro, *Concrete: Microstructure, Properties and Materials*, McGraw-Hill, New York, NY, USA, 3rd edition, 2006.
[16] P. S. Song, S. Hwang, and B. C. Sheu, "Strength properties of nylon- and polypropylene-fibre-reinforced concretes," *Cement and Concrete Research*, vol. 35, pp. 1546–1550, 2005.
[17] J. Newman and P. Owens, *Properties of Lightweight Concrete. Advanced Concrete Technology Set*, pp. 3–29, Butterworth-Heinemann, Oxford, UK, 2003.
[18] J. M. Gao, W. Sun, and K. Morino, "Mechanical properties of steel fibre-reinforced high-strength lightweight concrete," *Cement and Concrete Composites*, vol. 19, pp. 307–313, 1997.
[19] S. Kakooei, H. M. Akil, M. Jamshidi, and J. Rouhi, "The effects of polypropylene fibres on the properties of reinforced concrete structures," *Construction and Building Materials*, vol. 27, pp. 73–77, 2012.
[20] C. G. Han, Y. S. Hwang, S. H. Yang, and N. Gowripalan, "Performance of spalling resistance of high performance concrete with polypropylene fibre contents and lateral confinement," *Cement and Concrete Research*, vol. 35, no. 9, pp. 1747–1753, 2005.
[21] L. Comiloli, C. Failla, and G. A. Plizzari, "Steel and synthetic fibres for enhancing concrete toughness and shrinkage behaviour," in *Proceedings of International Conference Sustainable Construction Materials and Technologies*, pp. 231–240, Coventry, UK, 2007.
[22] N. Banthia and R. Gupta, "Influence of polypropylene fibre geometry on plastic shrinkage cracking in concrete," *Cement and Concrete Research*, vol. 36, no. 1, pp. 1263–1267, 2006.
[23] T. Y. Lo, H. Z. Cui, W. C. Tang, and W. M. Leung, "The effect of aggregate absorption on pore area at interfacial zone of lightweight concrete," *Construction and Building Materials*, vol. 22, no. 4, pp. 623–628, 2008.
[24] A. M. Neville and J. J. Brooks, *Concrete Technology*, Pearson Education Asia Pvt. Ltd., Malaysia, 2008.

Study on Mechanics-Based Cracking Resistance of Semiflexible Pavement Materials

Duanyi Wang, Xiayi Liang, Danning Li, Hehao Liang, and Huayang Yu

School of Civil Engineering and Transportation, South China University of Technology, Guangzhou 510641, China

Correspondence should be addressed to Huayang Yu; huayangyu@scut.edu.cn

Guest Editor: Munir Nazzal

Semiflexible mixture is a composite paving material combining the advantages of both asphalt and cement concrete materials. It consists of matrix asphalt skeleton and cement mortar. Due to the different volume characters between asphalt structure and cement mortar, stress concentration always happens in this semiflexible mixture, leading to internal cracking. The objective of this study is to alleviate the internal cracking concern of the semiflexible mixture by adjusting the material components. To this end, optimal material design and numerical simulation have been conducted. Matrix asphalt structures with four different air voids were incorporated with different dosages of cement mortar. The contraction strain and expansion strain of cement mortar as well as the indirect tensile strength of matrix asphalt structure were measured. The results were input into ABAQUS for numerical simulation. The results indicated that (1) the internal stress in this semiflexible mixture is mainly determined by the contraction of cement mortar, rather than expansion; (2) larger air void of matrix asphalt structure and less volumetric variation of cement mortar reduce the internal stress; (3) once the air void of matrix asphalt structure is decided, both maximum contraction and expansion deformations of cement mortar should meet specific requirement to ensure less internal cracking. This is a practical-ready paper that provides reference for the anticracking design of semiflexible pavement.

1. Introduction

Semiflexible pavement material is a composite pavement material, which consists of porous asphalt structure and cement mortar. The main advantages of semiflexible pavement include high carrying capacity, high wearing resistance, high shear resistance, high water stability, and driving comfort. Semiflexible pavement material has very bright application prospects and is suitable for intersection turner, gas station, parking, bus station, airport runways, and other sections [1–6]. However, the insufficient volume stability resulted from the defects in the material composition and structure of semiflexible pavement leads to frequent cracking. The cracks occurring in semiflexible pavement materials can be divided into three types based on their causes: fatigue cracking under the action of traffic load, temperature contraction cracking and dry contraction cracking caused by temperature and humidity variation, and reflection cracks of semirigid base cracking [7]. Fatigue crack and reflection crack are mainly caused by traffic load. In addition, temperature contraction cracking and dry contraction crack result from the different volume stability of asphalt mixture matrix and cement mortar. The volume of cement mortar can be greatly changed when the water reaction period or ambient temperature varies greatly. Consequently, the volume deformation of the asphalt mixture matrix and cement mortar cannot be coordinated. This incoordination leads to the stress concentration in semiflexible pavement, which makes the pavement more prone to crack. Thus, internal stress control of semiflexible pavement is particularly critical for crack resistance.

The effects of asphalt, aggregate, and cement mortar on the volumetric stability and cracking resistance of semiflexible pavement were studied by Huang [7], who researched and developed an easy-to-pour small contraction deformation polymer cement mortar. Wang and Wu [8] evaluated the cracking resistance of semiflexible pavement materials based on the fracture toughness using the

J-integral theory and three-point bending semicircular beam test. Huang et al. [9] studied the cracking resistance of semiflexible pavement materials based on cement-emulsified asphalt mortar. At present, studies of the anticracking of semiflexible pavement have been mainly focused on controlling the volumetric stability of the cement mortar and the overall cracking resistance of semiflexible pavement materials. However, few researchers have mechanically studied the internal cracking resistance and anticrack design of semiflexible pavement. Therefore, to improve the cracking resistance of semiflexible pavement from the material composition, the control of the internal stress of the cement mortar and matrix asphalt mixture in semiflexible pavement is particularly critical. In other words, the prerequisite for the anticracking design of semiflexible pavement materials is that the expansion stress and contraction stress of the cement mortar must be lower than the ultimate tensile strength of the asphalt mixture matrix.

In this paper, the expansion strain and contraction strain of cement mortar and the matrix were determined firstly. The ultimate tensile strength of the asphalt mixture matrix was measured. Then digital images of semiflexible pavement cross section were collected using a CCD digital camera. The locations of the cement mortar, asphalt, aggregate, and voids were obtained using AutoCAD software. The finite element analysis method was employed for the mechanical analysis of the digital image results. Taking the internal stress peak as the control index, this paper put forward the standard of strain control of cement mortar, which can effectively control the cracking of semiflexible pavement materials.

2. Materials and Methods

2.1. Materials. The performance indicators of materials used to prepare the cement mortar and the matrix of semiflexible pavement are shown in Tables 1–4.

TABLE 1: Gradation of standard sand.

Sieve size (mm)	1.18	0.6	0.3	0.15	0.075	Content of clay (%)
Passing rate (%)	100	98.5	92.7	51.5	2.2	0.5

TABLE 2: Technical indexes of mineral filler.

Testing indexes	Expressway or first grade highway requirement	Other grade highway requirement	Value
Apparent relative density (g/cm^3)	≥2.50	≥2.45	2.761
Water content (%)	≤1	≤1	0.27
Hydrophilic coefficient (%)	<1	<1	0.33
Plasticity index	≤4	≤4	2.4

TABLE 3: Technical indexes of asphalt.

Testing indexes		Technical requirement	Value
Penetration 25°C, 100 g, 5 s, 0.1 mm		60–80	68
Softening point TR&B (°C) min		≥46	49.5
Ductility 5°C, 5 cm/min, cm		>15	65
Ductility 15°C, 5 cm/min, cm		≥100	>100
Solubility (%) min		≥99.5	99.6
Flash point (°C) min		≥260	285
Wax content (%)		≤2.2	1.8
Dynamic viscosity 60°C		≥180	215
Rotate the film to heat the test residue (163°C, 85 min)	Quality change	±0.8	−0.02
	Penetration ratio: 25°C	≥61	76.5
	Ductility: 10°C, cm/min	≥6	25.2

TABLE 4: Technical indexes of aggregate.

Testing indexes	Value	Technical requirement
Crushing value (%)	11	≤26
Los Angeles wear value (%)	12	≤28
Stone polishing value (BPN)	55	≥42
Coarse aggregate ruggedness (%)	4	≤12
Coarse aggregate needle sheet content (%)	4	≤15
Sand equivalent (%)	53	≤65
Methylene blue value (g/kg)	1.1	≤2.0

2.2. The Expansion Strain and Contraction Strain of Cement Mortar. The expansion strain and contraction strain of cement mortar are the main reasons for the internal stress in semiflexible pavement. In order to explore the anticracking performance of semiflexible pavement, this paper measured the maximum expansion strain and contraction strain of cement mortar.

Referring to "Explain for General Code of Asphalt Pavement Engineering" [10], the cement mortar fluidity, compressive strength, flexural strength, and amount of contraction were used as the evaluation index. Based on the orthogonal test method, taking water-cement ratio and dosage of mineral filler and sand as the influencing factors, the indoor relevant experimental study would be processed to measure the flow performance, bending resistance, and volume stability of cement mortar. Through the range analysis, the best mix ratio of cement mortar can be determined, which is shown in Table 5.

TABLE 5: Optimal mix ratio of cement mortar.

Cement mortar indicators	Water-cement ratio	Dosage of mineral filler (%)	Dosage of sand (%)
Value	0.61–0.63	10	20

2.2.1. The Expansion Strain of Cement Mortar. It has been shown that the thermal expansion coefficient of cement mortar is time-dependent, whose largest value is 48 $\mu\varepsilon$/°C during the initial setting [11]. After the initial setting, the coefficient of thermal expansion decreases and reaches the lowest value of approximately 11 $\mu\varepsilon$/°C during the final setting. As time goes by, the coefficient of thermal expansion will increase slowly to the second largest value of 13.6 $\mu\varepsilon$/°C and will then gradually stabilize at 13.1 $\mu\varepsilon$/°C. Therefore, the thermal expansion deformation becomes the largest when the cement mortar undergoes a hydration reaction. According to the actual project time, semiflexible pavement is generally constructed during the summer. Using the most unfavorable climatic conditions as a reference, the maximum surface temperature difference in Guangdong Province of China is about 40°C in summer, and the maximum expansion strain of cement mortar may be 1.92×10^{-3}.

2.2.2. The Contraction Strain of Cement Mortar. Contraction deformation of cement mortar can be divided into dry contraction, temperature contraction, plastic contraction, self-contraction (hardening contraction), and carbonation contraction. Among them, dry contraction is not only the most common deformation type of cement mortar but also the main reason for the cracking of materials [12]. In this paper, the maximum contraction strain of cement mortar is determined by a contraction test. According to the optimal mix ratio of cement mortar described in Table 5, a triple-test-molded dry contraction specimen with a standard size of 25 mm × 25 mm × 250 mm was created. After the specimen was formed, it was maintained for 24 hours in a curing room at a temperature of 20 ± 3°C and a relative humidity of 98%. Then, the specimens were removed and their initial lengths were measured. The specimens were moved into a box with constant temperature and humidity (20 ± 1°C and relative humidity of 50 ± 3%). The contraction deformations of the cement mortar were measured at 1, 3, 7, 14, 28, 35, 60, 90, 120, 150, and 180 days. The average contraction strains were calculated, and the results of which are shown in Figure 1.

As shown in Figure 1, the average contraction strain of cement mortar increases with time and finally tends to be stable. After 180 days of hardening, the maximum contraction strain is 1.33×10^{-3}.

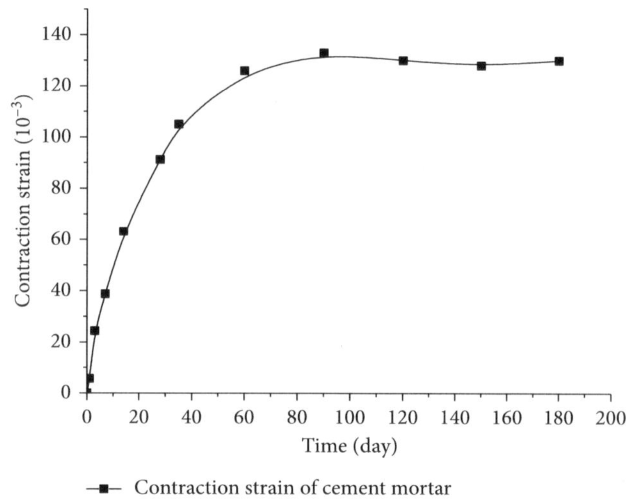

FIGURE 1: Contraction curve of cement mortar.

TABLE 6: Measured coefficients of thermal contraction and expansion at different temperature ranges.

Temperature range (°C)	The coefficients of thermal contraction and expansion ($\mu\varepsilon$/°C)	
	The coefficient of thermal contraction	The coefficient of thermal expansion
−20 to −5	3.3	31.2
−5 to 10	19.2	28.1
10–25	29.4	23.8
25–40	31.1	25.6
40–55	25.7	26.3

Based on the research results [13], gradation of aggregate and air void of the sample do not affect the coefficients of thermal contraction and expansion significantly. Therefore, the maximum contraction strain and expansion strain of the asphalt matrix are referenced to 1.24×10^{-3} and 1.25×10^{-3} in this paper.

2.3. The Expansion Strain and Contraction Strain of the Matrix. The expansion strain and contraction strain of the asphalt matrix can go a long way to influence the internal stresses of the semiflexible pavement and therefore must be accounted for in the study. According to prior research [13], the coefficients of thermal contraction and expansion of AC are nonlinear with temperature, as shown in Table 6.

According to data of Table 6, the largest coefficients of thermal contraction and expansion of AC were determined to be 31.1 and 31.2 $\mu\varepsilon$/°C, respectively. Using the most unfavorable climatic conditions as a reference, the maximum surface temperature difference in Guangdong Province of China is about 40°C in summer, and the maximum contraction strain and expansion strain of AC may be 1.24×10^{-3} and 1.25×10^{-3}.

2.4. Indirect Tensile Test of the Matrix. This study adopted the multistage aggregate skeleton space-filling method [14, 15] to design the gradation of the matrix under the target air void (15%, 20%, 25%, and 30%) and combined the drainage test and the Cantabro test to determine the optimum asphalt content. The grading design results are shown in Table 7.

The ultimate tensile strength of the matrix is directly related to the cracking resistance of the semiflexible pavement materials. In this study, the MTS testing machine was used for the indirect tensile strength test to measure the indirect tensile strength of the matrix specimens with different air voids. Then, the splitting strength of the matrix would be calculated. This study used the indirect tensile strength of the matrix as the ultimate tensile strength. The ultimate tensile strength of the matrix is shown in Table 8.

Table 7: The grading design results of the asphalt mixture matrix with different target air voids.

Sieve size (mm)		Percent passing										Asphalt content (%)
		19	16	13.2	9.5	2.36	1.18	0.6	0.3	0.15	0.075	
Air voids (%)	15	100	81.9	60.1	29.2	23.6	14.5	13.1	8.6	5.1	3.4	4.6
	20	100	78.7	57.3	25.4	19.4	13.6	12.5	8.4	4.8	3.2	4.2
	25	100	74.1	54.0	17.2	15.1	13.7	11.4	8.2	4.5	3.0	3.8
	30	100	71.4	51.4	15.2	13.7	12.4	10.6	7.6	4.2	2.7	3.4

Table 8: Ultimate tensile strength of the matrix.

The air void of the matrix (%)	Ultimate tensile strength of the matrix (MPa)
15	0.36
20	0.34
25	0.31
30	0.28

2.5. *Semiflexible Pavement Molding.* After the specimens were molded, they would be placed statically for 24 hours. The specimens were fixed to the vibration platform, and the cement mortar was poured into the matrix mixture until the surface voids were filled under multiple vibrations. The specimens were then stored in a curing box at constant temperature and humidity (temperature 20°C ± 3°C and humidity >90%) for 7 days. To identify the grouting depth of cement mortar, a small amount of Sudan dye was added to color the cement mortar.

The voids of the matrix can be divided into closed voids and connected voids. It is difficult to pour cement mortar into the closed voids. In order to ensure the accuracy, the amount of cement mortar grouting would be measured. As shown in the formula (1), the actual air void would be calculated to select the specimens whose actual air void is close to target air void. The density of cement mortar is 1.87 g/cm^3. The test results are shown in Table 9.

$$V_{\text{actual air void}} = \frac{m}{\rho \times v}, \tag{1}$$

where $V_{\text{actual air void}}$ is the actual air void of the matrix, m is the amount of cement mortar grouting quantity, ρ is the density of cement mortar, $\rho = 1.87\,\text{g/cm}^3$, and v is the volume of the specimen.

As shown in Table 9, by comparing the actual air void and target air void, the specimens 15-2, 15-3, 15-4, 20-2, 20-3, 20-6, 25-3, 25-4, 25-5, 30-2, 30-5, and 30-6 were selected for the follow-up study.

2.6. *Finite Element Simulation.* In this study, the semiflexible pavement structure was used as the simulation prototype, and the internal stress of the semiflexible pavement was analyzed by using the two-dimensional plane strain finite element method. The peak stress and stress level of the matrix were calculated considering the expansion and contraction deformations of the cement mortar. Asphalt mixture is a typical viscoelastic material, but in order to simplify the calculation, the linear elastic model is used to characterize the constitutive model of the asphalt mixture matrix and cement mortar in the semiflexible pavement structure. The basic assumptions are as follows:

Table 9: The quantity of grout to the asphalt mixture matrix.

Target air void (%)	Specimen number	Cement mortar grouting volume (g)	Actual air void (%)
15	15-1	124.16	13.2
	15-2	133.57	14.2
	15-3	134.51	14.3
	15-4	132.63	14.1
	15-5	126.04	13.4
	15-6	121.34	12.9
20	20-1	171.19	18.2
	20-2	181.54	19.3
	20-3	178.72	19.0
	20-4	167.43	17.8
	20-5	163.67	17.4
	20-6	177.78	18.9
25	25-1	217.28	23.1
	25-2	210.70	22.4
	25-3	225.75	24.0
	25-4	224.81	23.9
	25-5	227.63	24.2
	25-6	216.34	23.0
30	30-1	269.01	28.6
	30-2	273.72	29.1
	30-3	249.26	26.5
	30-4	259.61	27.6
	30-5	272.78	29.0
	30-6	272.78	29.0

(1) The semiflexible pavement structure consists of the asphalt mixture matrix and cement mortar. The asphalt mixture matrix is a homogeneous and isotropic material.

(2) As cement mortar and matrix asphalt mixture both produce contraction deformation and expansion deformation, in order to facilitate the finite element calculation, the volume of the matrix asphalt mixture is locked in the finite element model, and the volumetric deformation of cement mortar is defined as the relative volumetric deformation of the matrix asphalt mixture and cement mortar.

(3) The influence of the road surface weight is not considered.

(4) The adhesion between matrix and cement mortar is good, without any separation phenomenon.

To obtain a semiflexible pavement cracking resistance model, this study cut semiflexible pavement specimens with different air voids and then used a CCD digital camera to collect the screenshots of the specimens, as shown in Figure 2.

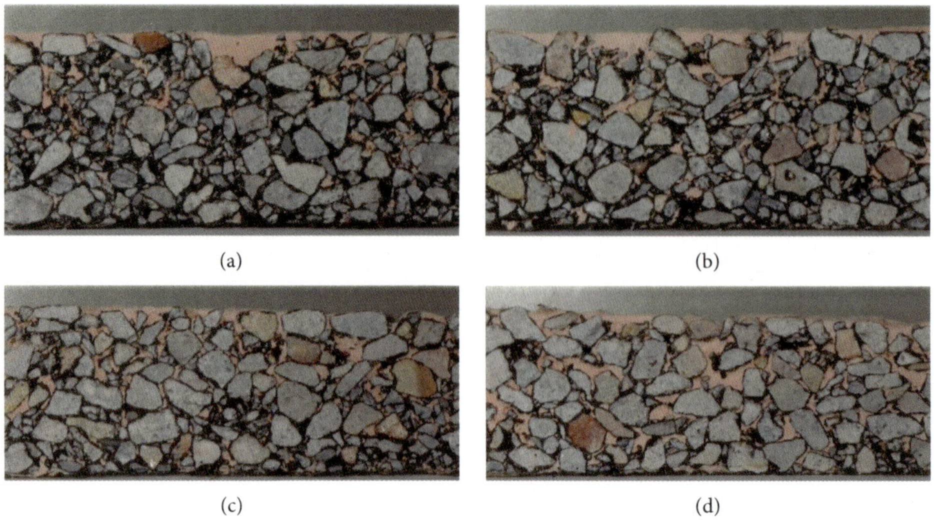

FIGURE 2: Cross-sectional images of semiflexible pavement. (a) 15% air void of the matrix; (b) 20% air void of the matrix; (c) 25% air void of the matrix; (d) 30% air void of the matrix.

The images of the semiflexible pavement were imported into AutoCAD software, and sectional information view was extracted to indicate the distributions of cement mortar, matrix, and air, as shown in Figure 3.

Taking the CAD images of the semiflexible pavement as the calculation model, the CAD images were imported into the finite element software. The material parameters were input, and the contact type was defined. For the boundary conditions, the upper boundary was bounded, the lower boundary was fixed (U1 = UR1 = UR2 = 0), the left and right boundaries were horizontally constrained (U1 = UR1 = UR2 = 0), the meshing partition was controlled using a triangular element (Tri), and the model had about 300,000 microcells. The meshing partition results are shown in Figure 4.

In Figure 4, the green part represents the aggregate, the white part indicates the cement mortar, and the red part is the asphalt. By changing the expansion and contraction of cement mortar, the stress distributions of aggregate and cement mortar were calculated. Due to the large size of the model, the computation takes a great deal of time. Therefore, five cross sections of each semiflexible pavement material with different air voids would be chosen to calculate the average value.

According to prior research [16–19], the material parameters of the semiflexible pavement were set up using the finite element analysis, as shown in Table 10.

Based on Sections 2.2 and 2.3, the contraction strain and expansion strain range of cement mortar is from -1.33×10^{-3} to 1.92×10^{-3}, and the contraction strain and expansion strain range of the matrix is from -1.24×10^{-3} to 1.25×10^{-3}. Therefore, the relative deformation range of the internal structure of the matrix and cement mortar is -2.58×10^{-3} to 3.16×10^{-3}. In this paper, the volume of the matrix was locked, and the volume deformation rate range of cement mortar during the finite element simulation was set from -4×10^{-3} to 4×10^{-3}, with an analysis calculation every 0.5×10^{-3}. The peak stress of the matrix was calculated under the conditions of contraction and expansion of the cement mortar.

Taking a 20% air void of the matrix as an example, the finite element analysis charts of the semiflexible pavement are as shown in Figure 5.

3. Results and Discussion

3.1. Stress Peak Analysis. The finite element simulation results show that the internal stress of the matrix mixture is related to the contraction and expansion deformations of cement mortar, as indicated in Figure 6.

Figure 6 shows the following:

(1) Contraction and expansion deformations of the cement mortar both produce tensile stress in the matrix. In addition, the stress level of the contraction deformation is significantly higher than that of the expansion deformation. Taking a matrix mixture of 15% air void as an example, when the cement mortar shrinks by 0.2%, the internal stress of the matrix asphalt mixture is 0.46 MPa, but when the cement mortar expands by 0.2%, the internal stress of the matrix asphalt mixture is only 0.16 MPa; that is, the contraction stress is 2.8 times greater than the expansion stress. Meanwhile, the internal stress of the matrix mixture under the contraction of cement mortar grows obviously faster than that under the expansion deformation. It can be seen that the contraction deformation of cement mortar is the dominant factor that affects the stress growth of semiflexible pavement materials. Therefore, if the stress peak of semiflexible pavement is taken as an anticracking indicator, the cement mortar contraction stress should be controlled.

(2) When the cement mortar expands, the internal stress of the matrix mixture has a good linear relationship with the expansion deformation. When the contraction deformation of the cement mortar is greater than 0.2%, the internal stress and contraction

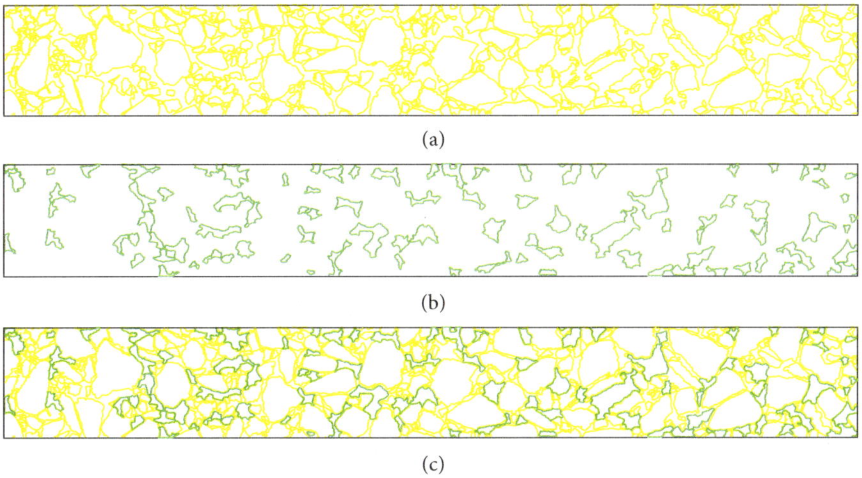

FIGURE 3: Sectional information view of the semiflexible pavement (15% air void). (a) Aggregate distribution; (b) cement mortar distribution; (c) distribution of materials.

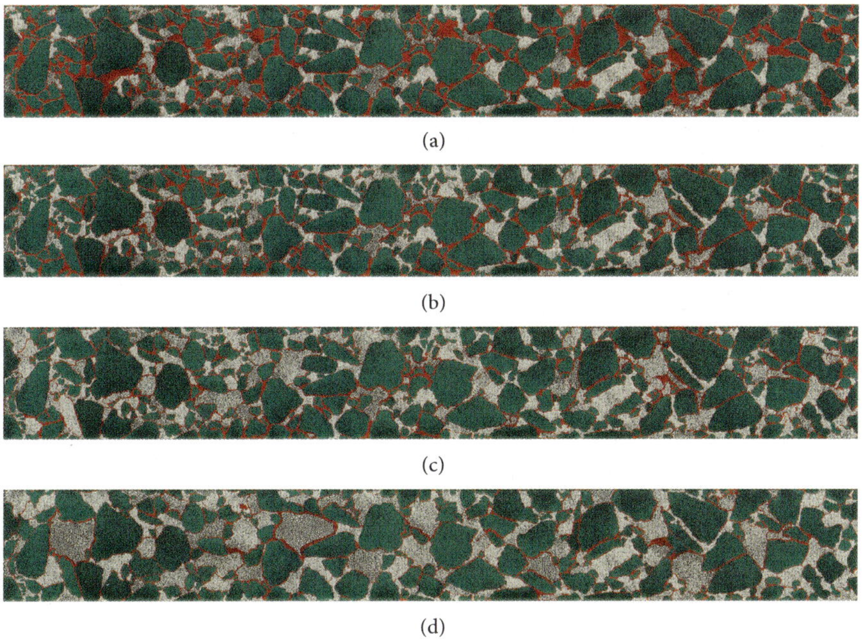

FIGURE 4: Finite element model of the semiflexible pavement. (a) 15% air void of the matrix; (b) 20% air void of the matrix; (c) 25% air void of the matrix; (d) 30% air void of the matrix.

TABLE 10: Material parameters of the semiflexible pavement.

Material	Poisson's ratio	Elastic modulus (MPa)
Asphalt (25°C)	0.35	1250
Aggregate (granite)	0.28	120000
Cement mortar	0.30	15000

deformation of the matrix mixture show a linear growth relationship. When the contraction deformation is less than 0.2%, the internal stress of the matrix mixture is a nonlinear growth model, and the internal stress growth rate decreases with the increase of the deformation amount. It is shown that the matrix mixture is more sensitive to the change in stress when the contraction of the cement mortar is small.

When the cement mortar shrinks and expands, the relationship between air void of the matrix and stress level is as shown in Figure 7.

It can be seen from Figure 7 that the internal stress of the matrix increases with the increase of the contraction of cement mortar. And the air void of the matrix affects the growth rate of stress. Under the same deformation conditions, the smaller the air void of the matrix mixture is, the greater the stress and stress growth rate is. For example, the stress growth rate is about 1.7 MPa/% for 15% air void of the

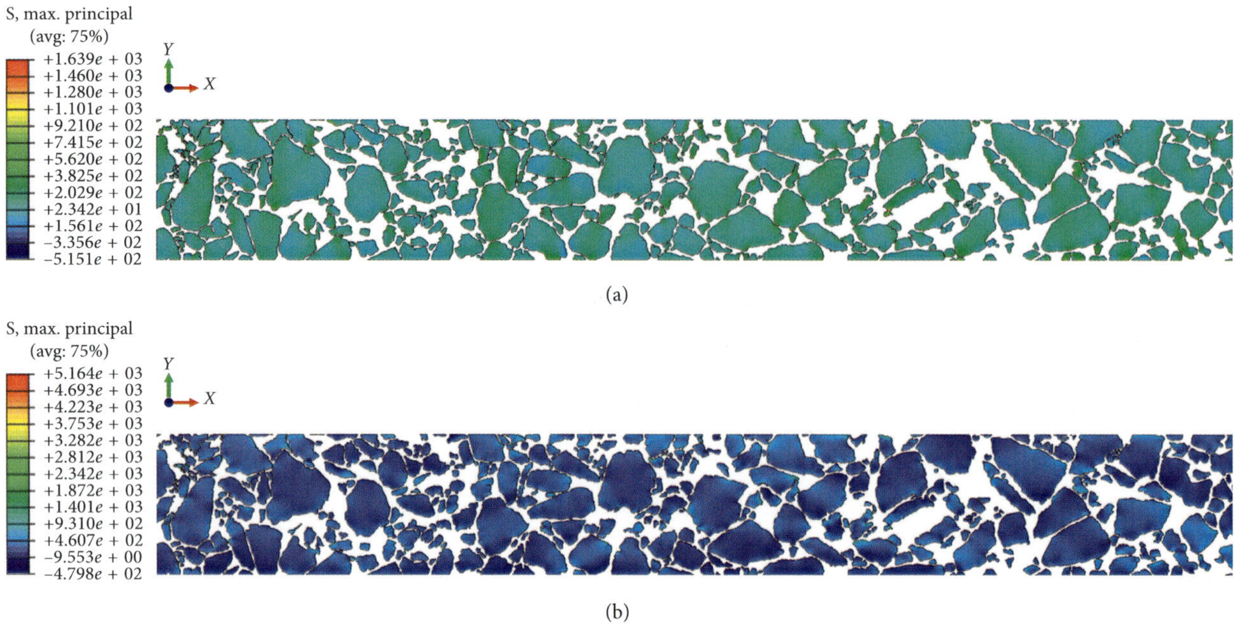

FIGURE 5: Matrix mixture stress cloud. (a) 20% air void, 0.2% expansion; (b) 20% air void, 0.2% contraction.

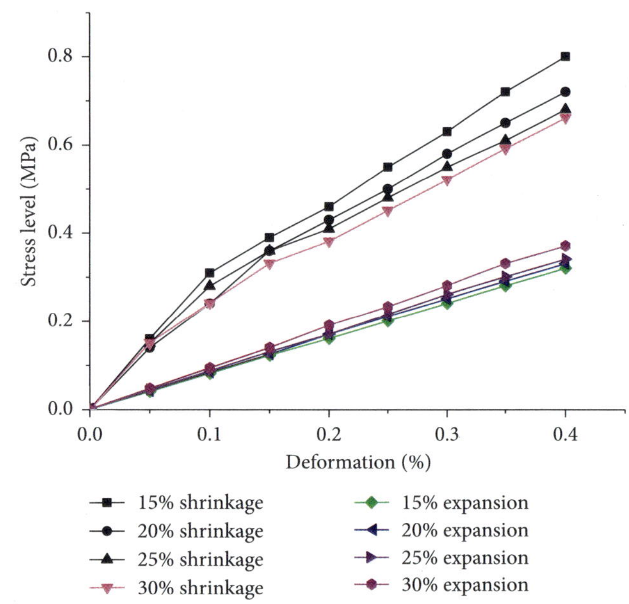

FIGURE 6: The relationship between internal stress of the matrix and deformation of cement mortar.

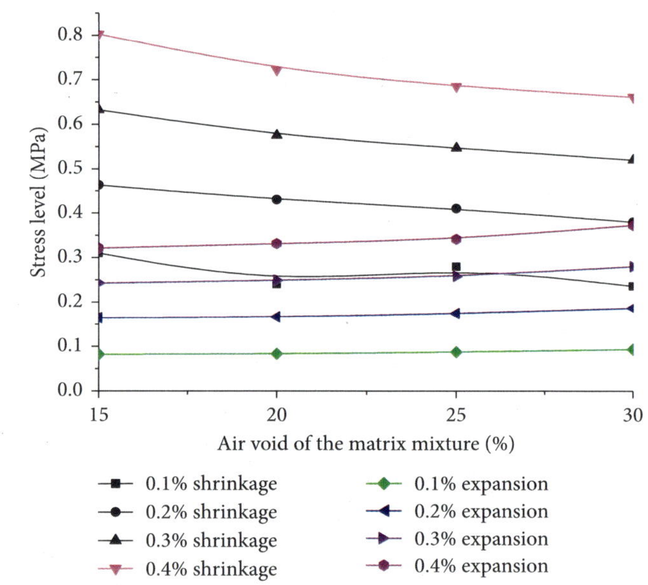

FIGURE 7: Relationship between air void and stress of the asphalt mixture matrix.

asphalt mixture matrix, whereas the stress growth rate is 1.3 MPa/% for 30% air void of the asphalt mixture matrix. Thus, the lower the air void of the matrix mixture is, the more sensitive the stress generated by the contraction of the cement mortar becomes. It can be seen that an appropriate increase in the air void of the matrix, to a certain extent, can reduce the cracking caused by the contraction of cement mortar.

In contrast with the contraction of cement mortar, the larger the air void of the matrix mixture is, the higher the internal stress level under an expansion deformation becomes. In addition, the internal stress produced by expansion deformation is much lower than that by contraction deformation. When the air void of the matrix is between 15% and 25%, the internal stress under the same deformation is basically the same. When the air void of the matrix asphalt mixture is more than 25%, the internal stress of the matrix mixture slowly increases. In general, an increase in the air void of the matrix has little effect on the internal stress growth caused by expansion deformation.

According to the analysis above, larger air void of matrix asphalt structure and less volumetric variation of cement mortar reduce the internal stress. In the actual

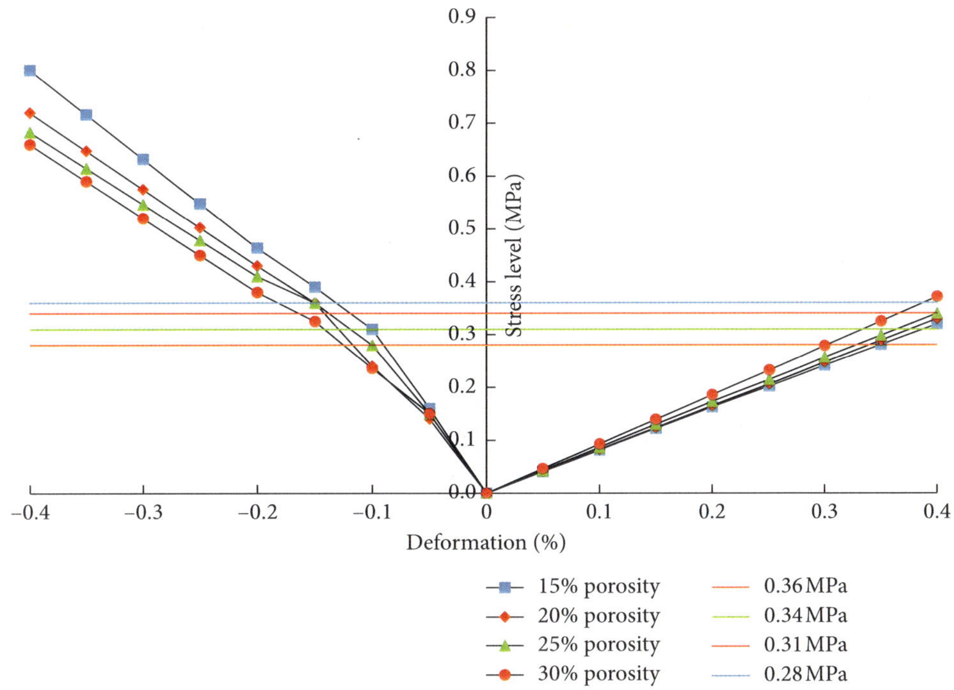

FIGURE 8: Relationship between stress and deformation of the matrix mixture.

project, after measuring the ultimate tensile strength of matrix asphalt structure with target air void, the cement mortar with suitable volume stability can be selected by the peak stress.

3.2. Strain Control Analysis of Cement Mortar. As a prerequisite for the anticracking design of semiflexible pavement materials, the contraction stress and expansion stress of cement mortar should be less than the ultimate tensile strength of the matrix asphalt mixture. Therefore, a threshold value of stress can be set to control the peak stress of cement mortar. As described in Table 7, the threshold value of the ultimate deformation of the cement mortar can be obtained by controlling the ultimate tensile strength of the matrix mixture. When the deformation is negative, it indicates a contraction deformation of the cement mortar, whereas a positive value indicates an expansion deformation.

The relationship between the stress curve caused by the contraction and expansion deformations of cement mortar under different air voids and the corresponding ultimate tensile strength is shown in Figure 8. The results calculated by the interpolation method show that when the air void of the matrix mixture is 15%, the maximum contraction of cement mortar should be less than 0.13%, and the maximum expansion should be less than 0.4%. When the air void of the matrix is 20%, the maximum contraction of cement mortar should be less than 0.14%, and the maximum expansion should be less than 0.4%. When the air void is 25%, the maximum contraction of cement mortar should be less than 0.12%, and the maximum expansion should be less than 0.38%. When the air void is 30%, the maximum contraction should be less than 0.12%, and the maximum expansion should be less than 0.31%.

Therefore, for the matrix and cement mortar used in this paper, the maximum allowable deformation values of cement mortar are summarized in Table 11.

According to the above analysis, this study proposes an anticracking design method of semiflexible pavement materials. Firstly, the ultimate tensile strength of matrix asphalt structure with target air void is measured, and the results are input into a mechanical model for numerical simulation to determine the maximum allowable strain of cement mortar. To avoid the cracking of semiflexible pavement, cement mortar whose maximum contraction strain and expansion strain are both below the maximum allowable strain is selected to alleviate the internal cracking concern of the semiflexible mixture.

4. Conclusions

This study proposed an anticracking design method of semiflexible pavement materials. The ultimate tensile strength of matrix asphalt structure with target air void was measured, and the results were input into a mechanical model for numerical simulation to determine the maximum allowable strain of cement mortar. Based on the laboratory test results and FEM simulation, the following findings were obtained:

(1) Both contraction and expansion deformations of cement mortar produce tensile stress in the asphalt mixture matrix. The stress of cement mortar from contraction deformation is significantly higher than that from expansion deformation. The contraction deformation of cement mortar is the dominant factor

causing internal stress in semiflexible pavement. In addition, the matrix mixture is more sensitive to the changes in stress when the less contraction deformation of the cement mortar occurs.

Table 11: The maximum allowable deformation values of cement mortar.

The air void of the matrix (%)	Maximum allowable contraction (%)	Maximum allowable expansion (%)
15	0.13	0.40
20	0.14	0.40
25	0.12	0.38
30	0.12	0.31

(2) The expansion deformation of cement mortar shows a good linear relationship with the internal stress of the matrix, and the tensile stress produced by the expansion of cement mortar is much lower than that caused by contraction. In addition, the increase in air void of the matrix has little effect on the internal stress growth caused by expansion deformation.

(3) Larger air void of matrix asphalt structure and less volumetric variation of cement mortar reduce the internal stress. For real projects, the ultimate tensile strength of the matrix with target air void can be measured and the peak stress can be used as the control index to select cement mortar with suitable volume stability.

This paper mainly puts forward the research method of semiflexible pavement internal crack resistance. As the deformation of asphalt matrix and cement mortar changes dynamically with temperature and time, and the variation trend is different, the relative deformation of cement mortar and asphalt matrix considered in this paper is not very accurate. In future research, the expansion strain and contraction strain-time curve of cement mortar and asphalt matrix will be further studied, and the expansion and contraction deformations of cement mortar and asphalt matrix will be simulated by the finite element method, and the stress change of semiflexible pavement internal structure will be accurately calculated.

Conflicts of Interest

The authors declare that there are no conflicts of interest regarding the publication of this paper.

Acknowledgments

The authors sincerely acknowledge the funding support from the National Natural Science Foundation of China (NSFC 51808228) and the Fundamental Research Funds for the Central Universities (2018MS92).

References

[1] C. Pratelli, G. Betti, T. Giuffrè, and A. Marradi, "Preliminary in-situ evaluation of an innovative, semi-flexible pavement wearing course mixture using fast falling weight deflectometer," *Materials*, vol. 11, no. 4, p. 611, 2018.

[2] G. Berardi and P. Di Mascio, "Le pavimentazioni aeroportuali in conglomerato bituminoso open grade intasato di Malta cementizia-open grade asphalt concrete filled with cement mortar for airport pavements," *Rassegna del Bitume*, vol. 61, pp. 29–36, 2009.

[3] S. E. Zoorob, K. E. Hassan, and A. Setyawan, "Cold mix, cold laid semi-flexible Grouted Macadams mix design and properties," in *Proceedings of the Performance of Bituminous and Hydraulic Materials in Pavements*, CRC Press, Nottingham, UK, April 2002.

[4] K. E. Hassan, A. Setyawan, and S. E. Zoorob, "Effect of cementitious grouts on the properties of semi-flexible bituminous pavements," in *Proceedings of the Performance of Bituminous and Hydraulic Materials in Pavements*, CRC Press, Nottingham, UK, April 2002.

[5] D. V. Ramsamooj, J. Ramadan, and G. S. Lin, "Model prediction of rutting in asphalt concrete," *Journal of Transportation Engineering*, vol. 124, no. 5, pp. 448–456, 1998.

[6] I. J. Huddleston, H. Zhou, and R. G. Hicks, "Evaluation of open-graded asphalt concrete mixtures used in Oregon," Transportation Research Record 1427TRB, National Research Council, Washington, DC, USA, 1993.

[7] C. Huang, "*Study on volume stability and crack resistance of semi-flexible pavement*," M.S. thesis, Wuhan University of Technology, Wuhan, China, 2010.

[8] W. Wang and K. Wu, "Study on crack resistance of semi-flexible pavement based on J-integral," *Highway*, vol. 61, no. 11, pp. 63–66, 2016.

[9] C. Huang, J. P. Liu, J. X. Hong, and Z. F. Liu, "Improvement on the crack resistance property of semi-flexible pavement by cement-emulsified asphalt mortar," *Key Engineering Materials*, vol. 509, pp. 26–32, 2012.

[10] Japan Association of Roads, *Explain for General Code of Asphalt Pavement Engineering*, Maruzen Co., Ltd., Tokyo, Japan, 1992.

[11] D. Shen, J. Shen, J. Huang et al., "Study on thermal expansion coefficient of cement - based materials in early age and hardening stage," *Journal of Hydraulic Engineering*, vol. 43, no. 1, pp. 153–160, 2012.

[12] S. Li, "*Cement mortar contraction research*," M.S. thesis, Nanjing University of Technology, Nanjing, China, 2004.

[13] M. R. Islam and R. A. Tarefder, "Coefficients of thermal contraction and expansion of asphalt concrete in the laboratory," *Journal of Materials in Civil Engineering*, vol. 27, no. 11, article 04015020, 2015.

[14] G. Lees, "The rational design of aggregate gradings for dense asphaltic compositions," in *Proceedings of Asphalt Paving Technologies*, pp. 60–90, Kansas City, MO, USA, 1970.

[15] X. Cai, D. Wang, K. Wu, and G. Yang, "Design of anti-rutting asphalt mixture by simplified classification method," *Journal of Highway and Transportation Research and Development*, vol. 33, no. 3, pp. 13–17, 2016.

[16] A. Epps, T. Ahmed, D. C. Little et al., "Performance assessment with the MMLS3 at WesTrack," *Asphalt Paving Technology: Association of Asphalt Paving Technologists Proceedings of the Technical Sessions*, vol. 70, pp. 509–542, 2002.

[17] R. B. Mallick and J. A. Bergendahl, *An Evaluation of Rutting and Shoving Potential of, and Chloride Diffusion through Hot Mix Asphalt (HMA) with Rosphalt*, Civil and Environmental Engineering Department, Worcester Polytechnic Institute, Worcester, MA, USA, 2004.

[18] L. Shao, S. Chi, and Y. Jia, "Mesoscopic simulation of large triaxial test of rockfill material," *Rock and Soil Mechanics*, vol. 30, no. S1, pp. 239–243, 2009.

[19] X. Shao, M. Shao, Y. Bi et al., "Asphalt mixture Poisson's ratio test method," *Journal of Tongji University (Natural Science Edition)*, vol. 34, no. 11, pp. 1470–1474, 2006.

10

Adhesive Through-Reinforcement Improves the Fracture Toughness of a Laminated Birch Wood Composite

Wenchang He[1] and Philip D. Evans[1,2]

[1] *Centre for Advanced Wood Processing, University of British Columbia, 2424-2900 Main Mall, Vancouver, British Columbia, Canada V6T 1Z4*

[2] *Department of Applied Mathematics, Research School of Physics and Engineering, The Australian National University, Canberra, ACT 0200, Australia*

Correspondence should be addressed to Wenchang He; hewenvictory@gmail.com

Academic Editor: Peter Chang

In this paper we test the hypothesis that adhesive through-reinforcement in combination with glass-fibre reinforcement of adhesive bond lines will significantly improve the fracture toughness of a laminated birch wood composite. We test this hypothesis using a model composite consisting of perforated veneer that allowed a polyurethane adhesive to penetrate and reinforce veneers within the composite. Model composite specimens were tested for mode I fracture properties, and scanning electron microscopy was used to examine the microstructure of fracture surfaces. Our results clearly show that through-reinforcement, and also reinforcing adhesive bond lines with glass-fibre, significantly improved fracture toughness of the birch wood composite. Our results also indicate that improvements in fracture toughness depended on the level of reinforcement. Improvements in fracture toughness were related to the ability of the reinforcement to arrest crack development during fracture testing and the fibre bridging effect of glass-fibre in adhesive bond lines. We conclude that through-reinforcement is an effective way of improving the fracture toughness of laminated wood composites, but further research is needed to develop practical ways of creating such reinforcement in composites that more closely resemble commercial products.

1. Introduction

Fracture toughness is an indicator of the stress required to propagate a preexisting flaw and is a critical property for materials such as laminated composites used in demanding structural applications [1]. Delamination of laminated composites reduces their stiffness and strength and can lead to loss of structural integrity [2]. Hence, there is significant interest in improving the fracture toughness of laminated composites, particularly those used for aircraft construction [1, 3]. Modern high performance aircraft uses a range of components made from laminated carbon-fibre/epoxy composites [4], but in less demanding aerospace applications the laminated wood composite plywood is still used (Euro Plywood 2016) [5]. Plywood consists of thin layers of wood veneer that are glued together, with adjacent layers having their wood grain rotated up to 90 degrees to one another [6]. The glue lines of plywood and also other laminated wood composites such as laminated veneer lumber (LVL) and glue-laminated lumber (glulam) often contain flaws and they can fail by delamination [7]. Hence, there has also been interest in improving their fracture toughness [8].

The susceptibility of laminated composites to delamination depends on a variety of intrinsic and extrinsic factors, but the lack of reinforcement of adhesive bond lines is a fundamental reason why cracks propagate between laminae [4]. Therefore, one obvious route to increasing the fracture toughness of laminated composites is to provide such reinforcement. For example, numerous studies have shown that the susceptibility of laminated composites to delamination can be reduced by reinforcing adhesive bond lines with fibres that act as bridges to inhibit crack opening [9]. Wegst et al. [10] reviewed this approach, and others used to improve fracture toughness of materials. Another approach

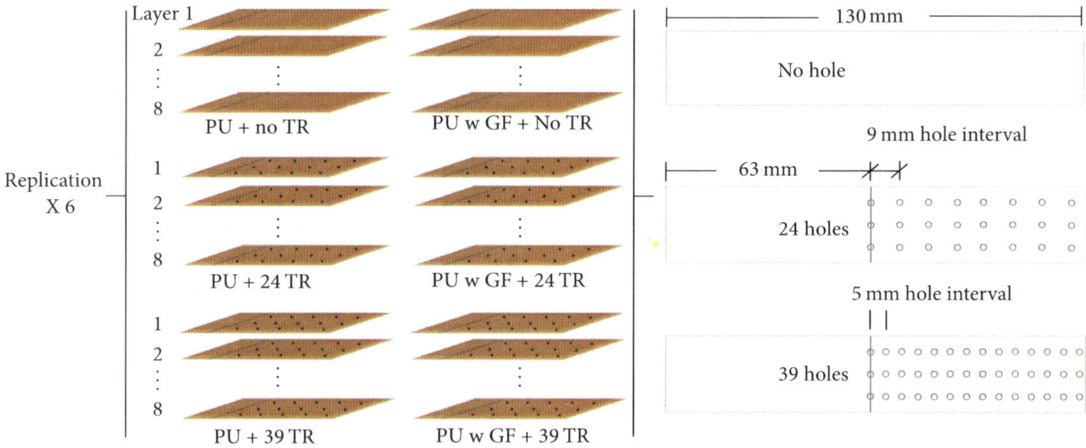

Figure 1: Preparation of laminated birch wood composite specimens with two levels of through-reinforcement (TR) of laminae and polyurethane (PU) adhesive bond reinforcement with glass-fibre (GF).

to improving the fracture toughness of laminates, which we explore here, is to alter the geometry of the adhesive network to provide through-reinforcement across multiple laminae. Such an approach is related although not identical to those used to provide through-thickness reinforcement of advanced composites, for example, 3D-weaving, stitching, braiding, embroidery, tufting, and z-anchoring [3].

In this paper we test adhesive through-reinforcement across multiple laminae in combination with glass-fibre reinforcement of adhesive bond lines as a means of improving the fracture toughness of a model laminated birch wood composite. Our results demonstrate that the fracture toughness of the wood composite can be significantly improved as a result of the introduction of adhesive through-reinforcement. Further significant increases in fracture toughness occurred when glass-fibre was added to the adhesive.

2. Materials and Methods

2.1. Preparation of a Laminated Birch Wood Composite with Through-Reinforcement. Our main experiment examined the effect of adhesive through-reinforcement (two levels) and the effect of glass-fibre reinforcement of adhesive bond lines on different measures of fracture toughness of a model laminated birch wood composite (Figure 1). Two different types of samples were prepared: One contained a higher level of reinforcement than the other (39 Z-direction reinforcements versus 24 reinforcements). There was only one level of glass-fibre reinforcement of adhesive bond lines (5.5% w/w). Birch was chosen as the test substrate because composites made from this wood species were widely used for the manufacture of aircraft during WWII, and there is ongoing interest in using birch plywood for the construction of unmanned aerial vehicles (drones) [11, 12]. Eight birch wood veneer sheets were purchased from a retailer (ENE Wood Products Surrey, British Columbia, Canada) and stored in a conditioned room (20 ± 1°C and 65 ± 5% relative humidity) for 1 month to ensure they reached equilibrium moisture content and stable dimensions before the experiment.

Six different types of specimen were made, and there were six replications of each specimen type (Figure 1). Manufacture of specimens was as follows. Each specimen contained 8 veneer layers, individually grain matched and selected from 8 veneer sheets. Strips measuring 130 mm × 25 mm were cut from veneer sheets using a paper guillotine (Boston™ 2658). These dimensions were chosen because they are the same as those recommended by ASTM standard D5528-13 [13]. Veneer strips used for specimens with adhesive through-reinforcement were perforated with a high-speed dental drill (W&H® Trend WD-56) and a 1 mm diameter drill bit (Dentsply® TN burr) operating at 6000 rpm and 100 g/cm of torque. The drill produced accurately sized, smooth, cylindrical 1 mm diameter holes in each veneer strip. The 8 veneer strips used to make each laminated composite were placed in a customized mould to align them vertically, and all 8 strips were perforated together in the mould. Veneer for specimens with higher levels of adhesive through-reinforcement contained 39 holes (3 rows × 13 columns [density of through-reinforcement = 2.33/cm^2]) with a spacing of 5 mm (Figure 1). Veneer for specimens with lower levels of adhesive through-reinforcement contained 24 holes (3 rows × 8 columns [density of through-reinforcement = 1.43/cm^2]) with a spacing of 9 mm (Figure 1). Control specimens were not perforated.

A one-component polyurethane adhesive (Gorilla Glue Co., USA) was used to bond veneer strips. The glue was either used in an unmodified form or modified by the addition of 5.5% w/w milled (60 mesh) glass-fibre (Fiber-Tek®, Burnaby, BC, Canada). Adhesive (0.187 g ± 0.0005 g) was applied to each veneer strip using a syringe (BD® 1 mL syringe) and spread evenly across each veneer using a glass coverslip (Matsunami®). The *same amount* of adhesive was applied to perforated and unperforated veneer strips. Veneers were then laid up with their grain direction parallel to each other (like the commercial composite, laminated veneer lumber) to make specimens consisting of either perforated or unperforated veneer strips. A strip of aluminum foil measuring 63 mm × 25 mm was inserted into the end of

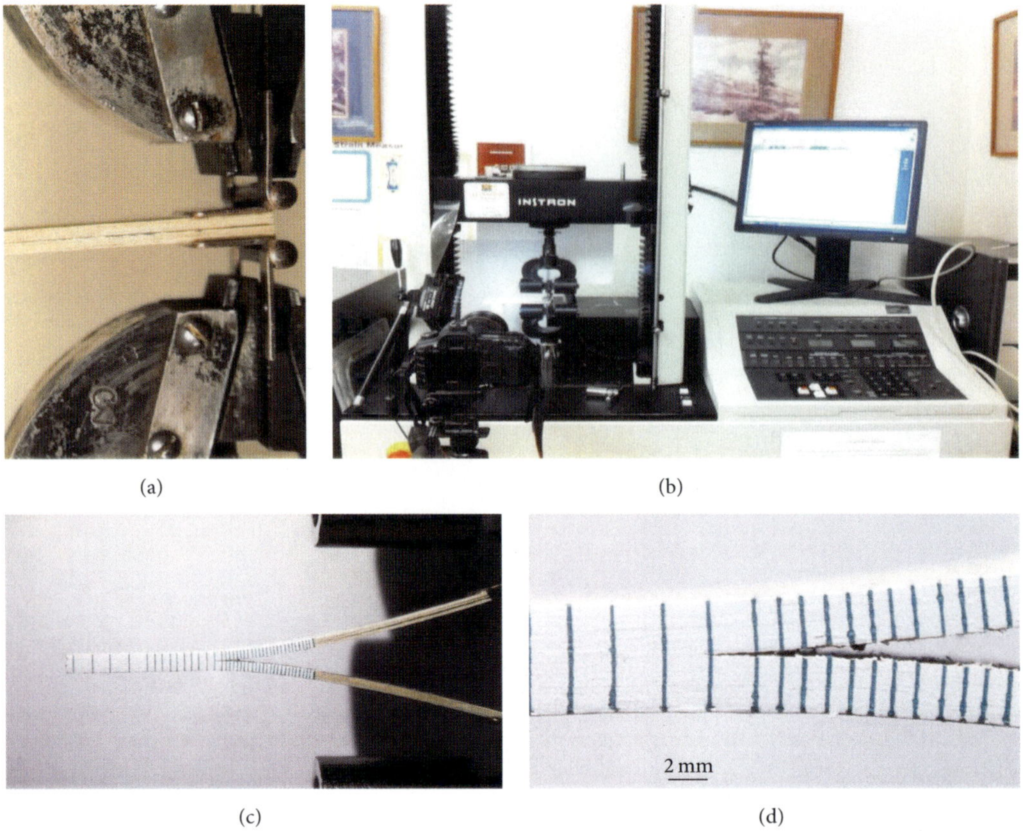

Figure 2: Mode I (crack opening mode) fracture toughness testing of a laminated birch wood composite: (a) Hinges used to attach one end of a specimen to the two cross-heads; (b) universal tensile testing machine with attached computer, and also showing the position of the DSLR camera for measurement of crack propagation and crack length; (c) white coated edge of one side of a specimen with scale markers; (d) higher magnification image of (c) showing the position of a crack tip front.

the middle two veneer strips in each specimen to cover an area of 63 mm × 25 mm. This foil insert served as a crack initiator during fracture testing as recommended by ASTM D5528-13 [13]. Specimens were placed in a small laboratory press (Carver® hydraulic press 3912) and pressed at room temperature at 3 MPa for 8 minutes. The adhesive penetrated through the Z-direction holes in each veneer, creating adhesive through-reinforcement across multiple laminae [14]. No such reinforcement occurred in the unperforated veneers [14]. The resulting laminates were conditioned at $20 \pm 1^{\circ}$C and $65 \pm 5\%$ r.h. for 7 days.

A second experiment examined the effects of different levels of adhesive through-reinforcement (0, 6, 12, 18, 51, and 66 holes) on the fracture toughness of a model laminated birch wood composite bonded with unmodified PU adhesive. This experiment sought to find the lower and upper limits for the ability of adhesive through-reinforcement to improve the fracture toughness of the birch wood composite. The manufacture and testing of duplicate specimens for each level of reinforcement in this second experiment was exactly the same as those used in the first experiment.

2.2. Fracture Toughness Testing of Laminated Wood Composites. The geometry of the specimens described above conforms to the requirement of the standard ASTM D5528-13 test for crack opening mode (mode I) interlaminar fracture toughness testing of unidirectional polymer composites [13]. The edges of specimens were covered with white correcting tape and marked with length scales (1 mm marks for the first 25 mm followed by 2 mm marks for a further 20 mm, and then 5 mm marks for an additional 20 mm, Figure 2). Opening forces were applied to specimens via hinges attached to the cross-head of a universal tensile testing machine (Instron® M3502) (Figures 2(a) and 2(b)). A cross-head speed of 1 mm/min was used, and crack length propagation was measured using a digital SLR camera (Canon® EOS 5D Mark II) with 24–105 mm zoom lens as shown in Figure 2(b). The camera captured an image as soon as the specimen opened. Subsequent images were captured every 20 seconds (until failure) using the tethering photo-capture software, Breeze® DSLR Remote Pro, on a PC. Each time-lapse image was saved in RAW format so the crack front position could be easily seen at high magnification within the software Adobe Photoshop® Raw processer (Figures 2(c) and 2(d)).

2.3. Fracture Toughness Calculation and Analysis of Data. Modified beam theory was used to calculate fracture toughness according to ASTM D5528-13 [13]. It should be noted that test results are primary "specimen properties" rather than "material properties" because calculated fracture toughness

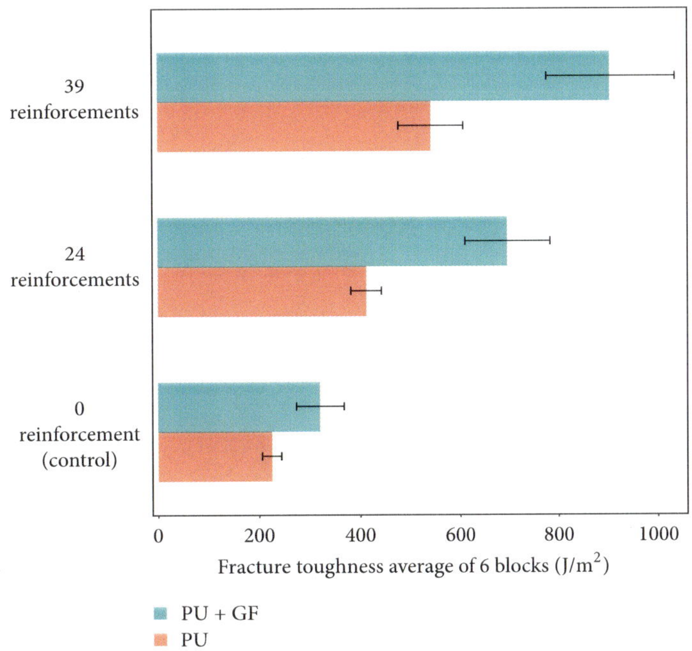

Figure 3: Fracture toughness values of specimens containing through-reinforcement and glass-fibre in adhesive bond lines versus fracture toughness of the controls.

values are dependent on the geometry of test specimens. The test results should therefore be regarded as "apparent fracture toughness" as suggested by fracture mechanics principles. Nevertheless, all the specimens (perforated and unperforated controls) consisted of wood veneers with the same grain pattern and similar geometry and they contained identical levels of adhesive. Therefore, the apparent fracture toughness values are comparable between specimens.

Fracture toughness results from the main experiment were subject to analysis of variance to determine the effects of the two experimental factors, through-reinforcement (39 reinforcements versus 24 reinforcements versus 0 reinforcement) and adhesive modification (PU adhesive versus glass-fibre reinforced PU adhesive) as well as the interaction of these two factors on fracture toughness. Statistical computation used GenStat (Release 17.1), and results were tested for statistical significance at the 5% level ($p < 0.05$). Results are presented in graphs plotted using the program R (version 3.2.3) and least significant difference bars derived from the ANOVA are included on each graph. These error bars can be used to determine if differences between individual means are statistically significant ($p < 0.05$).

2.4. Scanning Electron Microscopy of Fracture Surfaces. The microstructure of fractured surfaces in tested specimens containing 39 reinforcements was examined using scanning electron microscopy. Small wood specimens measuring 5 mm × 5 mm were carefully cut from the surface of fracture toughness specimens using a precision micro-table saw (Byrnes Model Machines Co.) equipped with a 10 cm diameter blade. Care was taken to ensure that fracture surfaces were not damaged during the preparation of specimens. Specimens were mounted on aluminium stubs using double-sided adhesive tape. They were then sputter coated with an 8 nm layer of gold and examined using a Zeiss UltraPlus analytical field emission scanning electron microscope (SEM) operating at 15 kV and a working distance of 13.1 to 14.0 mm. Selected secondary electron images were saved as TIFF files.

3. Results

3.1. Fracture Toughness Tests. The introduction of adhesive through-reinforcement in the birch wood composite significantly improved fracture toughness (Figure 3). The improvement of fracture toughness was significantly ($p < 0.05$) greater in samples containing higher levels of reinforcement (Figure 3). Fracture toughness was also significantly improved by reinforcing adhesive bond lines with glass-fibre (Figure 3). Analysis of variance indicated that there was no significant ($p > 0.05$) interaction of through-reinforcement and addition of glass-fibre to adhesive bond lines on fracture toughness. Nevertheless, the addition of glass-fibre to adhesive bond lines was more effective at improving the fracture toughness of the birch composite containing through-reinforcement than it was at improving the fracture toughness of the control. For example, the addition of glass-fibre to adhesive bond-lines improved the fracture toughness of the control by 42%, whereas comparable figures for composites containing through-reinforcement were 69 (24 reinforcements) or 67% (39 reinforcements).

The effects of through-reinforcement and addition of glass-fibre to adhesive bond lines on 5% maximum fracture toughness values of specimens, a reporting parameter for ASTM D5528-13, are shown in Figures 4 and 5, respectively. These figures also show other critical fracture characteristics of tested specimens, and demonstrates the effectiveness of

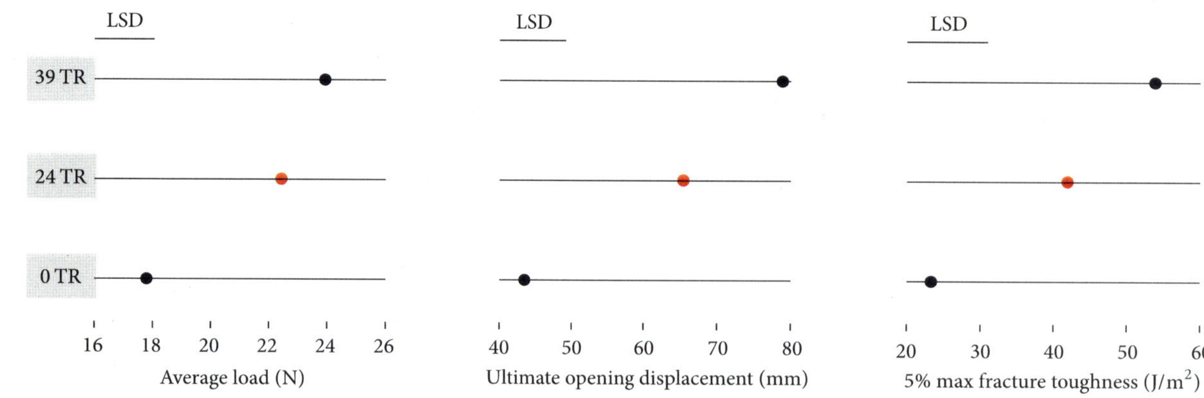

Figure 4: Effects of through-reinforcement (TR) (39 through-reinforcements versus 24 through-reinforcements versus 0 through-reinforcement) on some critical fracture characteristics of laminated birch wood specimens.

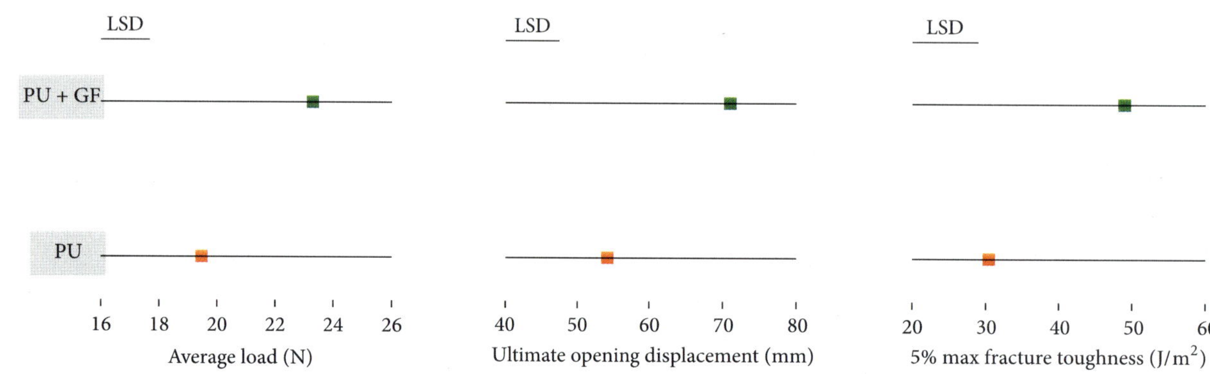

Figure 5: Effects of reinforcement of polyurethane adhesive (PU) with glass-fibre (GF) on some critical fracture characteristics of the laminated birch wood specimens.

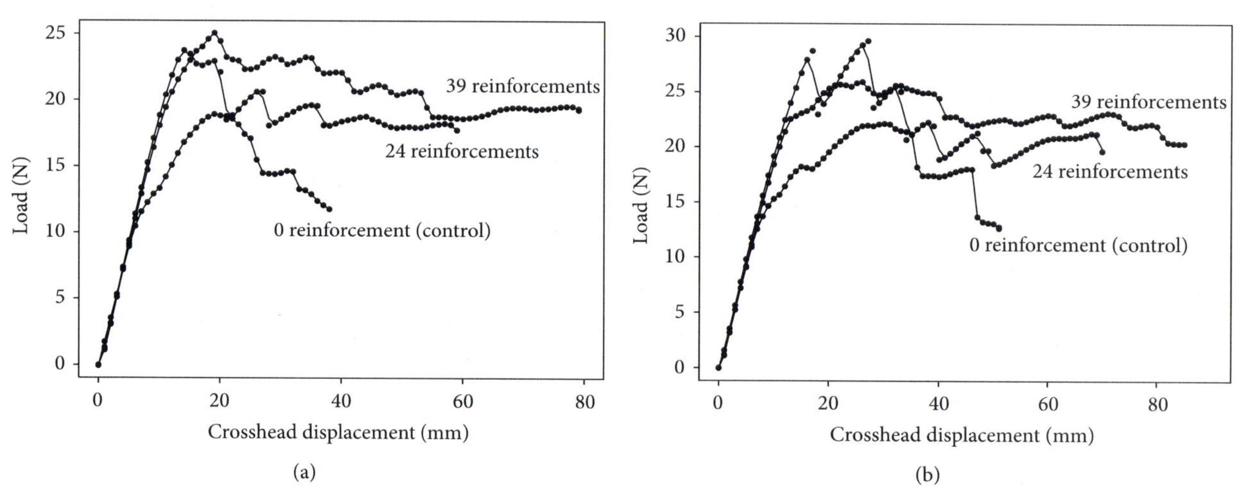

Figure 6: Load-displacement curves of specimens with different levels of through-reinforcement and bonded using unmodified polyurethane (PU) adhesive (a) or a glass-fibre reinforced PU (b).

through-reinforcement, and also adhesive bond-line reinforcement on the fracture toughness of the birch wood composite.

Through-reinforcement of the birch composite appeared to arrest propagation of the crack induced during fracture toughness testing. This effect is suggested by load-displacement curves of specimens during testing (Figure 6), which show that load increased abruptly and then increased more slowly. During the latter phase there were increases in load as the crack encountered through-reinforcement. Load increases are sinusoidal in specimens with through-reinforcement at the lower level, but higher levels of reinforcement appear to smooth the sinusoidal variation of load. The load carrying capability (y-axis) of the specimens with

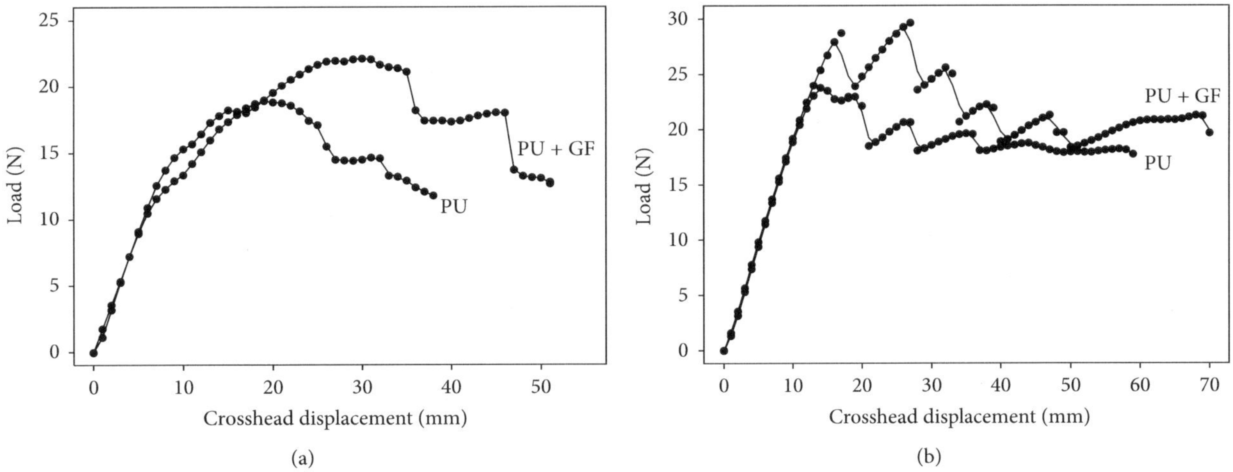

FIGURE 7: Load-displacement curves of specimens with unmodified polyurethane (PU) adhesive or adhesive modified by the addition of 5.5% w/w glass fibre (GF): (a) No through-reinforcement; (b) through-reinforcement (24 reinforcements).

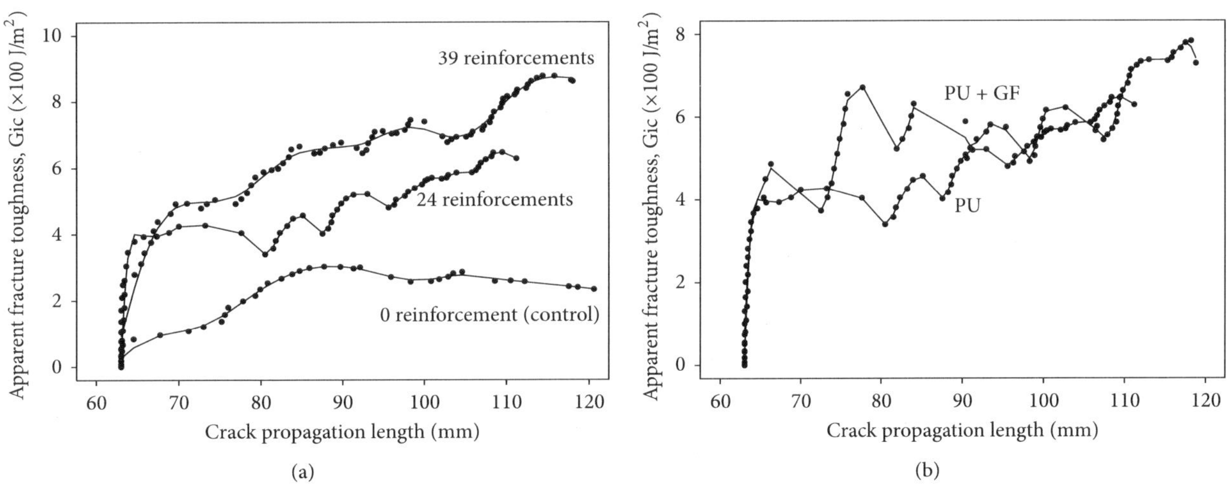

FIGURE 8: Mode I crack growth resistance curves (R-curves) of (a) polyurethane (PU) adhesive with different levels of through-reinforcement; and (b) different adhesive types (PU or PU + GF) with the same level of through-reinforcement (24).

through-reinforcement was generally higher than that of the control. Furthermore, specimens with through-reinforcement withstood greater mode I opening displacement (x-axis) until ultimate failure than the control (Figure 6).

Results in Figure 6 show the positive effects of through-reinforcement on load sustained by specimens bonded with an unmodified polyurethane adhesive (Figure 6(a)), and specimens bonded with a polyurethane adhesive reinforced with 5.5% glass-fibre (Figure 6(b)). The effect of adding glass-fibre to adhesive bond lines on load sustained by specimens can be clearly seen in Figure 7. This figure shows that adding glass-fibre to the PU adhesive increased the load carrying capacity of specimens and extended the maximum mode I opening displacement of specimens (Figure 7).

Crack resistance curves of specimens, also known as R-curves, accord with load displacement results and indicate the positive effects of through-reinforcement and adhesive modification on fracture toughness (Figure 8). They also support the suggestion that increases in toughness of specimens with through-reinforcement resulted from the ability of the reinforcement to arrest crack propagation.

Results in Figure 9 show the effects of different levels of adhesive through-reinforcement (0, 6, 12, 18, 51, and 66 reinforcements) on the load sustained by specimens bonded with unmodified PU adhesive. Table 1 shows the effects of different levels of adhesive through-reinforcement on the fracture toughness and mode I opening displacement of specimens. The results in Figure 9 and Table 1 confirm that through-reinforcement improves fracture toughness and indicate that improvements in fracture toughness depend on the level of reinforcement. The optimal level of reinforcement appeared to be 39 reinforcements per veneer strip (2.33 per cm^2).

3.2. Scanning Electron Microscopy of Fracture Surfaces. Scanning electron photomicrographs of fracture surfaces in laminated birch wood specimens after mechanical testing suggest how through-reinforcement and addition of glass-fibre to

Table 1: Effects of different levels of adhesive through-reinforcement on the fracture toughness and mode I opening displacement of laminated birch wood specimens bonded with an unmodified polyurethane adhesive.

No. of reinforcements	Density of reinforcements (N/cm^2)	Fracture toughness (J/m^2)	Displacement (mm)
0 (control)	0	225	41
6	0.36	287	40
12	0.72	287	47
18	1.08	327	48
24	1.43	412	56
39	2.33	542	66
51	3.04	340	57
66	3.94	353	52

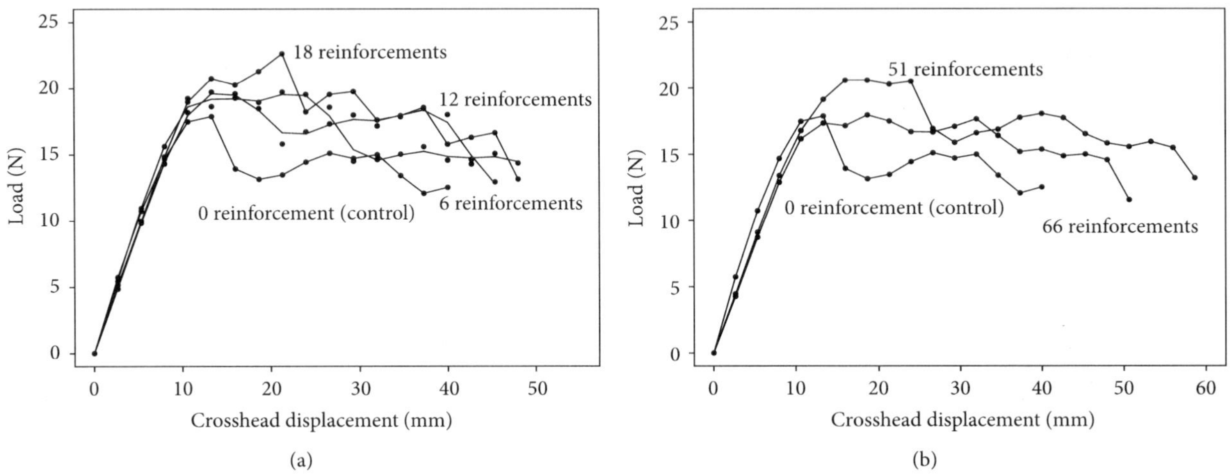

Figure 9: Load-displacement curves during double cantilever beam (DCB) mode I fracture toughness tests of specimens with 6 different levels of through-reinforcement (including the control): (a) reinforcement levels less than 24 (<1.43/cm^2); (b) reinforcement levels greater than 39 (>2.33/cm^2).

adhesive bond lines increased fracture toughness. They also cast some light on the bonding mechanism of polyurethane adhesives used with wood.

Polyurethane adhesive was clearly visible at the fracture surfaces of birch composites bonded with unmodified adhesive (Figure 10(a)). Residual adhesive at fracture surfaces exhibited tearing and pull-out, which may have contributed to adhesive bond strength (arrowed top left in Figure 10(a)). In addition, we observed pull-out of adhesive that had penetrated the cells in the rays of birch (arrowed in Figures 10(b) and 10(c)). The pillar-shaped structures that projected from the cells within rays had branches, some of which appeared to have fractured during testing (arrowed in Figure 10(d)). We also observed fracture, pull-out, and lateral displacement of adhesive columns that provided multiple through-reinforcement of the birch composite (arrowed in Figures 10(e) and 10(f)). The same patterns of failure were noted at fracture surfaces of composites bonded with polyurethane adhesive containing glass-fibre (Figures 11(a)–11(c)). In addition, we observed pull-out of fibre-bundles at fracture surfaces (arrowed in Figure 11(d)). Glass-fibre was clearly evident at fracture surfaces and there was evidence of pull-out of the glass-fibres in horizontal glue-lines (arrowed left in Figure 11(a)) and also within the adhesive columns that provided through-reinforcement (arrowed in Figures 11(e) and 11(f)).

4. Discussion

Our results support our hypothesis that introduction of through-reinforcement across veneers increases the fracture toughness of a laminated birch wood composite and show that improvements in fracture toughness depend on the level of reinforcement. Our results also show that glass-fibre reinforcement of adhesive bond lines significantly increases fracture toughness. They also suggest how the two different types of reinforcement increased fracture toughness. The addition of glass-fibre to adhesive bond lines appeared to provide reinforcement during crack propagation, hence consuming fracture energy. This suggestion accords with the results of many studies that have examined the use of glass-fibre to reinforce and improve the toughness of other composite materials, for example, vinyl-ester polymer composites [15], polypropylene composites [16], and dental particulate composites [17]. There have been no previous studies to our knowledge of the use of adhesive through-reinforcement to increase the fracture toughness of laminated wood composites. Therefore, we cannot compare our findings

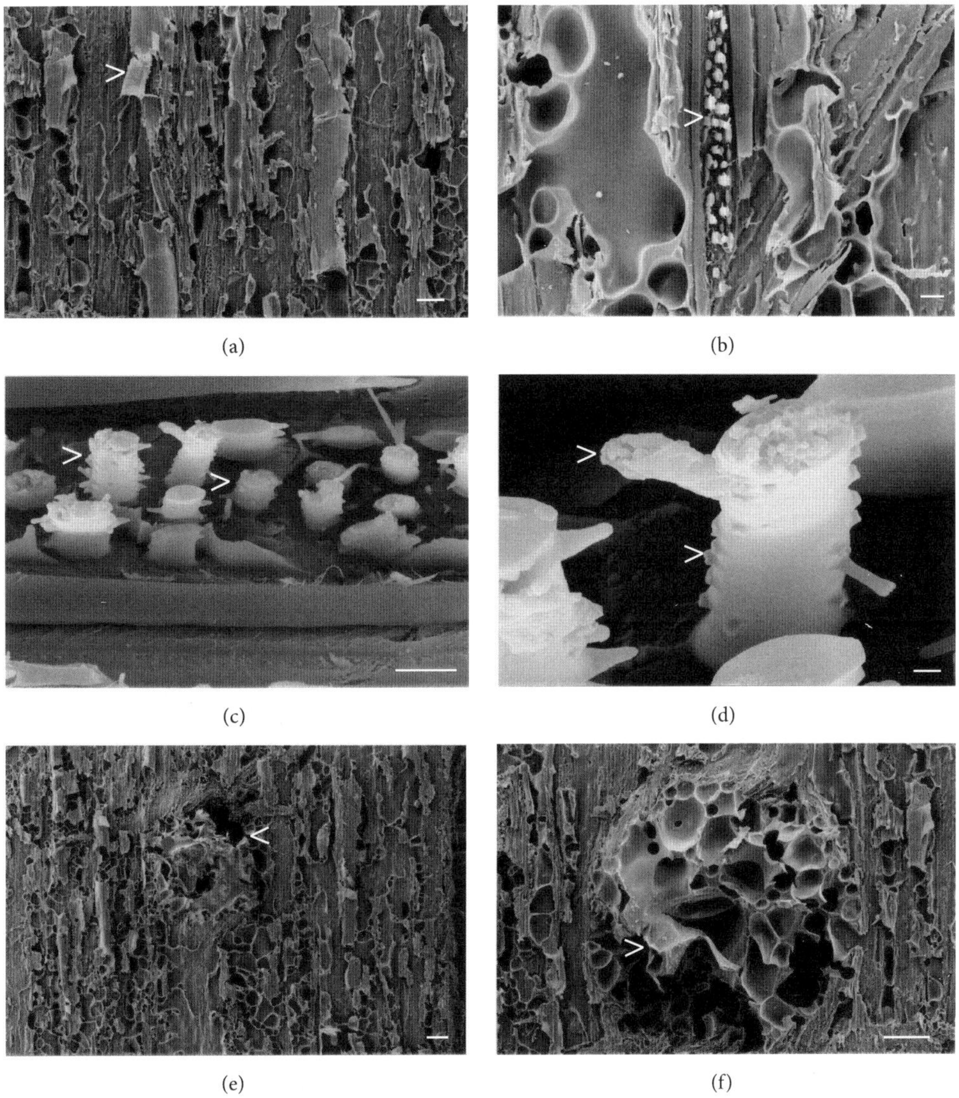

FIGURE 10: Fracture surfaces of a laminated birch wood composite bonded with unmodified single component polyurethane adhesive; (a) adhesive on fractured wood surface showing pull-out and microfracture of adhesive that partially covers the wood surface (arrowed left; scale bar = 100 μm); (b) fractured wood surface showing a multiseriate ray (centre) and pull-out of adhesive from the lumens of individual ray cells (arrowed centre; scale bar = 20 μm); (c, d) higher magnification photographs showing the pillars of adhesive that pulled out of ray cells shown in (b). Note the fractured ends of the pillars and the intact and fractured side-arms (branches) (arrowed left and centre; scale bars = 10 μm (c) and 1 μm (d)); (e) fractured column of adhesive that provided through-reinforcement to the composite (arrowed centre) and residual adhesive on the fractured wood surface (scale bar = 200 μm); (f) higher magnification photograph of the adhesive column shown in (e). Note the fracture and distortion of the adhesive column and the distortion of the adjacent wood (arrowed; scale bar = 200 μm).

with those of previous researchers. However, the multiple through-reinforcement provided by adhesive columns running radially in specimens has some similarities with that provided by rays (radial ribbons of woody tissue), which are aligned in the same direction as the adhesive columns engineered here. Our SEM images of fracture surfaces suggested that rays provided reinforcement, and Reiterer et al. [18] in their study of the fracture toughness of three hardwoods and the softwood, spruce, suggested that "the rays (in solid wood) can be considered as reinforcements in the radial directions." Furthermore, their load displacement graphs obtained during fracture testing of hardwoods have some similarities to those obtained here [18]. Hence, we suggest that adhesive through-reinforcement provided reinforcement behind the crack tip, hindering crack propagation and absorbing fracture energy. This suggestion is supported by the wave-shape of the crack propagation resistance curves (R-curves) of specimens containing adhesive through-reinforcement. The "wave peaks" indicate abrupt increases in fracture energy when the crack propagated through adhesive columns. This bridging effect occurred at the macroscale, but it complements that provided by adhesive bond reinforcement with glass-fibre. Such a complementary effect was more pronounced in specimens with through-reinforcement than in the controls, possibly because the glass-fibre reinforced the adhesive columns in the birch wood composite, in addition

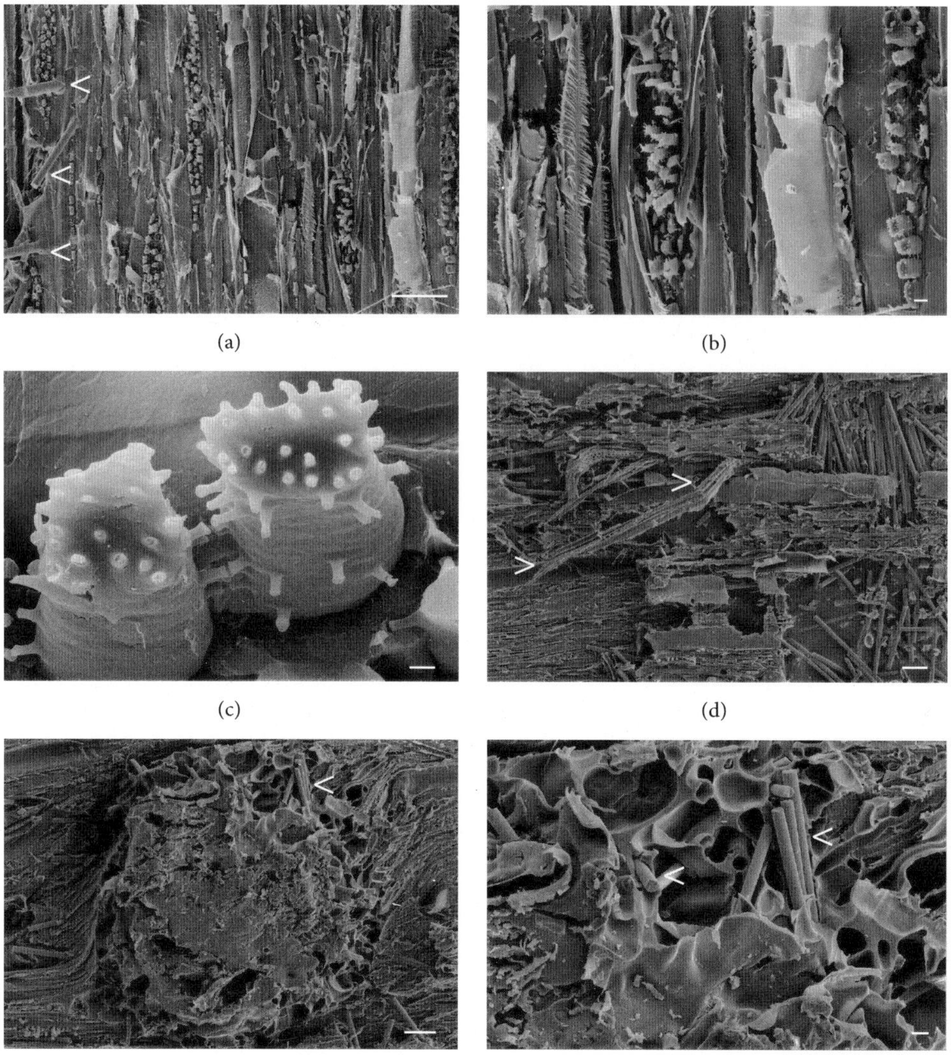

FIGURE 11: Fracture surfaces of a laminated birch wood composite bonded with a single component polyurethane adhesive containing 5.5% w/w milled (60 mesh) glass-fibre: (a) fractured wood surface showing pull-out and fracture of glass-fibre (arrowed left) and wood cells; pull-out of adhesive from ray cells; and adhesive that partially covers the wood surface (scale bar = 100 μm); (b-c) higher magnification photographs [from (a)] showing fracture of wood cells and pull-out of adhesive from the lumens of individual ray cells (scale bars are 10 μm (b) and 1 μm (c)); (d) fractured wood surface showing residual adhesive, pull-out of bundles of fibres (arrowed left and centre) and presence of numerous glass-fibre strands (right) (scale bar = 100 μm); (e-f) fractured column of adhesive containing embedded glass-fibre that provided through-reinforcement to the composite (arrowed top right of centre in (e), and centre and right in (f)). Note the fracture and distortion of the adhesive column and the distortion of the adjacent wood (scale bars are 100 μm (e) and 20 μm (f)).

to their ability to reinforce adhesive bond lines aligned in the x-y direction.

Birch plywood is a preferred material used for the construction of wooden aircraft, as mentioned above, and fracture toughness is an important property of composites used for this application. Nevertheless, as pointed out by Reiterer et al. [18] there are few studies of the fracture toughness properties of hardwoods such as birch. The fracture toughness properties of composites depend on good adhesion between adhesive and the matrix [19]. Adhesive bonding of wood involves a mix of physicochemical interactions including mechanical interlocking provided by penetration of adhesive into the porous microstructure of wood [20]. Penetration of adhesives into hardwoods occurs easily via the open pores (vessels) that are readily apparent in species such as oak. Hence, reviews of wood adhesion have focused on such an effect [20, 21]. The penetration of adhesives into rays was mentioned by Vick [20] as possibly having a detrimental effect on adhesion because "radially oriented rays can allow excessive flow and overpenetration" (of adhesive). Our SEM photomicrographs showed pull-out of adhesive from cells in rays suggesting that penetration of adhesive into the rays located at the tangential surface of rotary peeled birch veneer improved adhesion. The pillars of adhesive that pulled out from ray cells had side arms possibly produced by penetration of adhesive from ray cell lumens through pits (openings) into the lumens of adjacent cells. Some of these side arms were fractured, possibly indicating that they contributed to the fracture toughness of the composite. This observation suggests that penetration of adhesive into rays can have a positive effect on adhesion and underscores the need for further studies of the effects of wood microstructure on the adhesive properties and fracture toughness of composites made from hardwoods.

The fracture toughness of the model composite we tested was clearly improved by adhesive through-reinforcement and also the addition of glass-fibre to adhesive bond lines. Glass-fibre has been added to polymer composites to improve their mechanical properties, as mentioned above [16], and a study by Dorey and Cheng [22] showed that the mechanical properties of laminated spruce-pine-fir beams were significantly improved by reinforcing the phenol-resorcinol glue lines of the beams with pultruded glass-fibre straps. They concluded that "the application of glass-fibre reinforcing has a potential to play a significant role in glulam structures" [22]. Hence, the addition of glass-fibre to adhesives could be a practical means of improving the fracture toughness of veneer-based wood composites used in demanding applications. Clearly, our means of creating through-reinforcement, which also significantly improved the fracture toughness of our model composite, would be more difficult to implement in practice than adding glass-fibre to adhesive bond lines. However, the approach is attractive because significant increases in fracture toughness were achieved without increasing the level of adhesive in the composite. Therefore, it would be worthwhile to explore approaches that are more practical than precision drilling as a means of perforating veneer. One possible approach could involve passing veneer through a roller containing slitting knives, as has been done in the past to "reduce the tendency of rotary peeled veneer to distort when it is used to manufacture plywood" [23]. Further research would be needed to determine whether slit-like perforations and any resulting through-reinforcement would be as effective as the cylindrical connections used here to improve fracture toughness of composites made from birch wood veneer. An alternative related approach to improving fracture toughness would be to nail the composite in the Z-direction using thermoplastic polymer nails. Laminated wood composites that are bonded together with nails are available commercially [24], and such an approach might work with other types of wood composites. In support of this suggestion, the automated insertion of thin metal wires through laminates (Z-pinning) is highly effective at improving the fracture toughness of laminated synthetic composites and is used commercially to improve the properties of aerospace and automotive composites [3]. The Z-pins used in these applications are made from materials such as titanium alloy that are strong and stiff. This desirable property of Z-pins suggests that adhesives that are stronger than the polyurethane adhesive tested here might be better at increasing fracture toughness of laminated wood composites via adhesive through-reinforcement. Further research would be needed to test this hypothesis. We have demonstrated in recently published work that adhesive through-reinforcement has a positive effect at reducing the moisture-induced swelling of laminated wood composites [14], in addition to its beneficial effects on fracture toughness. Further research is needed to examine the effect of adhesive through-reinforcement on the elastic modulus, modulus of rupture, and internal bond strength of laminated wood composites. Significant improvements to the properties of such composites and also others derived from renewable materials, for example, bamboo, could allow them to compete more effectively in some applications with synthetic composites, with resulting benefits in terms of sustainability and the environment [25].

5. Conclusions

We hypothesised that introduction of through-reinforcement across veneers, and glass-fibre reinforcement of adhesive bond lines would significantly increase the fracture toughness of a laminated birch wood composite. Our results support this hypothesis, and we have shown that improvements in fracture toughness are related to the level of through-reinforcement. We have suggested how through-reinforcement and glass-fibre reinforcement of adhesive bond lines increase the fracture toughness of the composite. We have also suggested how the ability of through-reinforcement to increase fracture toughness could be improved further. Through-reinforcement is potentially an attractive, cost-effective, approach to improving fracture toughness, because increases in this property were achieved without the use of additives or increasing the level of adhesive in the composite. However, further research is needed to develop practical ways of creating perforations that facilitate adhesive through-reinforcement of laminated wood composites.

Competing Interests

The authors declare that they have no competing interests.

Acknowledgments

The authors thank Drs. Hua Chen and Frank Brink of the Centre for Advanced Microscopy at The Australian National University for expert assistance with scanning electron microscopy; Mr Vinicius Lube for assistance with macrophotography and image processing; Professor Dorin Ruse in the Department of Biomaterials at UBC for generous access to his Instron tensile testing equipment; Dr. Markus Haapasalo in the Department of Oral Biological & Medical Sciences at UBC for letting Wenchang He use his precision dental drilling equipment. Wenchang He thanks the China Scholarship Council for a scholarship. Philip D. Evans thanks Viance, Tolko, and FPInnovations for their support of his BC Leadership Chair at University of British Columbia, and The Australian National University for an honourary professorship in the Department of Applied Mathematics in the Research School of Physics and Engineering.

References

[1] N. Sela and O. Ishai, "Interlaminar fracture toughness and toughening of laminated composite materials: a review," *Composites*, vol. 20, no. 5, pp. 423–435, 1989.

[2] T. K. O'Brien, "Characterization of delamination onset and growth in a composite laminate," in *Damage in Composite Materials: Basic Mechanisms, Accumulation, Tolerance, and Characterization*, ASTM International, 1982.

[3] A. P. Mouritz, "Review of z-pinned composite laminates," *Composites Part A: Applied Science and Manufacturing*, vol. 38, no. 12, pp. 2383–2397, 2007.

[4] V. A. Franklin and T. Christopher, "Interlaminar fracture toughness estimation of aerospace composites by weighted residual approach," *Strength of Materials*, vol. 47, no. 6, pp. 789–796, 2015.

[5] Web source, http://europlywood.com/catalogue/aircraft-plywood/.

[6] A. D. Wood and T. G. Linn, *Plywoods: Their Development, Manufacture and Application*, W. & A. K. Johnston, 1950.

[7] V. Bucur and P. A. Martin, *Delamination in Wood, Wood Products and Wood-based Composites*, Springer Science + Business Media, Dordrecht, The Netherlands, 2011.

[8] J. F. Davalos, P. Qiao, and B. S. Trimble, "Fiber-reinforced composite and wood bonded interfaces: part 2. Fracture," *Journal of Composites Technology and Research*, vol. 22, no. 4, pp. 232–240, 2000.

[9] M. E. Launey and R. O. Ritchie, "On the fracture toughness of advanced materials," *Advanced Materials*, vol. 21, no. 20, pp. 2103–2110, 2009.

[10] U. G. K. Wegst, H. Bai, E. Saiz, A. P. Tomsia, and R. O. Ritchie, "Bioinspired structural materials," *Nature Materials*, vol. 14, no. 1, pp. 23–36, 2015.

[11] E. Schatzberg, "Ideology and technical choice: the decline of the wooden airplane in the United States, 1920–1945," *Technology and Culture*, vol. 35, no. 1, pp. 34–69, 1994.

[12] T. A. Bateman, J. D. Nelson, and B. M. Argrow, "A low cost, rapid construction unmanned aircraft design," in *Proceedings of the AIAA InfoTech at Aerospace Conference*, pp. 22–31, Rohnert Park, Calif, USA, May 2007.

[13] ASTM, "Standard test method for mode I interlaminar fracture toughness of unidirectional fiber-reinforced polymer matrix composites," ASTM D5528-13, ASTM International, New York, NY, USA, 2013.

[14] W. He and P. D. Evans, "Reducing the thickness swelling of a model wood composite by creating a three-dimensional adhesive network," *International Wood Products Journal*, vol. 7, no. 4, pp. 202–207, 2016.

[15] P. Compston, P.-Y. B. Jar, P. J. Burchill, and K. Takahashi, "The effect of matrix toughness and loading rate on the mode-II interlaminar fracture toughness of glass-fibre/vinyl-ester composites," *Composites Science and Technology*, vol. 61, no. 2, pp. 321–333, 2001.

[16] S.-Y. Fu, Y.-W. Mai, B. Lauke, and C.-Y. Yue, "Synergistic effect on the fracture toughness of hybrid short glass fiber and short carbon fiber reinforced polypropylene composites," *Materials Science and Engineering A*, vol. 323, no. 1-2, pp. 326–335, 2002.

[17] S. R. Dyer, L. V. J. Lassila, M. Jokinen, and P. K. Vallittu, "Effect of cross-sectional design on the modulus of elasticity and toughness of fiber-reinforced composite materials," *The Journal of Prosthetic Dentistry*, vol. 94, no. 3, pp. 219–226, 2005.

[18] A. Reiterer, G. Sinn, and S. E. Stanzl-Tschegg, "Fracture characteristics of different wood species under mode I loading perpendicular to the grain," *Materials Science & Engineering A*, vol. 332, no. 1-2, pp. 29–36, 2002.

[19] I. A. Ashcroft, D. J. Hughes, and S. J. Shaw, "Mode I fracture of epoxy bonded composite joints: 1. Quasi-static loading," *International Journal of Adhesion & Adhesives*, vol. 21, no. 2, pp. 87–99, 2001.

[20] C. B. Vick, "Adhesive bonding of wood materials," in *Wood Handbook-Wood as an Engineering Material, General Technical Report FPL-GTR-113*, USDA, Madison, Wis, USA, 1999.

[21] F. A. Kamke and J. N. Lee, "Adhesive penetration in wood—a review," *Wood and Fiber Science*, vol. 39, no. 2, pp. 205–220, 2007.

[22] A. B. Dorey and J. J. R. Cheng, *Glass Fiber Reinforced Glued Laminated Wood Beams*, A Joint Publication of the Canadian Forest Service and Land and Forest Services, Minister of Supply and Services Canada, 1996.

[23] T. D. Perry, *Modern Plywood*, Pitman Publishing Corporation, New York, NY, USA, 1948.

[24] I. Boyle and B. Gafner, Mass timber: Knowing your options, http://www.fastepp.com/images/attachments/Final_AIBC_AIA%20Conference%20for%20Upload%20ibbg_Oct24_2013%2010%2025.pdf.

[25] G. Koronis, A. Silva, and M. Fontul, "Green composites: a review of adequate materials for automotive applications," *Composites Part B: Engineering*, vol. 44, no. 1, pp. 120–127, 2013.

11

Mechanical Properties and Failure Behavior of Composite Samples

Hongwei Zhang, Zhijun Wan, Yuan Zhang, and Dong Wu

Key Laboratory of Deep Coal Resource Mining (CUMT), Ministry of Education of China, School of Mines, China University of Mining & Technology, Xuzhou 221116, China

Correspondence should be addressed to Zhijun Wan; zhjwan@cumt.edu.cn

Academic Editor: Dimitrios E. Manolakos

In underground coal mining systems, the occurrences of coal burst hazards and pillar failures relate not only to the condition of stress distribution but also the geometry of roof-coal-floor structures. To study the failure response of these structures, the rock-coal-rock (RCR) sample, in which a coal component is sandwiched between rocks, is always employed as the experimental subject. In this study, the effect of height ratio (a ratio represents the height percentage of coal component in an RCR sample) on the mechanical properties and deformation behavior of RCR samples was numerically investigated by using the distinct element model (DEM). The results reveal the following. (1) The uniaxial compression strength (UCS) of the RCR sample decreases with increasing height ratio as an inverse proportional function. (2) With increasing height ratio, the elastic modulus of the RCR sample decreases exponentially, while the postpeak modulus is strengthened in an inverse proportional manner. (3) Microcracking activity of the RCR sample is different from that of the pure sample during loading. Specifically, a reactive period always occurs after the quiet and active periods in the RCR sample. (4) The RCR sample fails in a progressive manner, in which cracking bands develop preferentially in coal and then extend to rocks. Expectably, the mechanical properties and failure behavior of RCR samples are height ratio dependent, which may contribute to predicting the hazard of coal bursts and estimating the failure of rock-coal-floor structures.

1. Introduction

Coal bursts occur frequently in underground coal mining systems, causing fatal injury and facility damage [1–3]. Therein, coal pillar bursts refer to failures of roof-pillar-floor systems that have been a significant safety concern in deep coal mines (Figures 1(a-1) and 1(b-1)) [4–7]. Since coal seam is sandwiched between the roof and floor strata, the destruction of pillar is influenced not only by the condition of stress distribution but also the geometry of roof and floor (Figures 1(a-2), 1(a-3), 1(b-2), and 1(b-3)) [8]. Therefore, understanding the mechanical properties and failure mechanisms of roof-pillar-floor structures under variable composite geometries is significantly important regarding safety mining.

In experimental scale modelling, pure coal and pure rock samples are always employed to conduct uniaxial compression tests [9–11]. However, the shortcoming is that the pure samples cannot accurately reflect the overall mechanical properties of roof-pillar-floor structures. Thus, two types of composite cylindrical samples were simplified to conduct laboratory and numerical investigations. The first is rock-coal (RC) type, in which roof and coal components are considered (Figure 1(c-1)) [12–14]; another is rock-coal-rock (RCR) type, where the coal is sandwiched between roof (rock) and floor (rock) components (Figure 1(c-2)) [15–17]. Geologically, coal seams and the surrounding strata may vary in thickness, which in turn indicates that the roof-pillar-floor structures may differ in physical geometry. In this study, the physical geometry refers to the height percentage (h_r) of the coal component in a composite sample.

To date, numerous attempts have been made to study the mechanical properties and deformation behavior of composite samples under uniaxial and triaxial compression conditions. In these studies, the most height ratios were between 40% and 60%, and the results show the following. (1) The uniaxial compression strength (UCS) of the composite sample depends mostly on the coal component [12]. (2) A higher UCS of rock component always implies a greater UCS, and the RCR sample with a lower height ratio tends to burst easier [18]. (3) Most failure modes of composite samples are conjugate X-shaped shearing failures [12, 13, 16]. Briefly, these foregoing investigations demonstrated that the UCS and failure behavior of composite samples are material properties dependent. However, these studies have been mainly concentrated on either the effect of material property or the damage of coal [19, 20].

In the aspect of geometrical effect, some studies investigated the failure response of composite rocks with different composite types (e.g., rock-coal, rock-rock-coal, and rock-coal-rock) [21], with different geometrical sizes [22] and with different inclined angles [23]. Interestingly, Chen et al. [14] found that the UCS of the RC sample decreases with increasing height ratio. Furthermore, a lower height ratio of the composite sample always implies a greater rock burst accident [24]. These studies contribute to our knowledge about the geometrical influence on the mechanical properties of composite samples. However, the relationship between height ratio and UCS of the RCR sample is not fully understood due to prior inadequate considerations of roof-coal-floor geometry. In addition, limited effort has been applied to determine the influence of the height ratio on the deformation evolution of RCR samples under uniaxial compressive loading. Therefore, in addition to the factors such as the physical properties of coal and rock components, it is also a significant issue to consider the height ratio of the RCR sample.

Apart from the mechanical properties and failure behavior of composite structures, acoustic emission (AE) is often a concern because related dynamical effects may lead to great failure and pillar damage in underground mining systems [25, 26]. Previous research has described the effect of the loading rate on the microcracking activities of RC and RCR samples [15, 27] but did not deal with the influence of the height ratio on the microcracking evolution of composite samples, which is a key issue to be investigated in this work.

This paper aims to study the effect of the height ratio on the mechanical properties and deformation evolution of RCR samples by conducting numerical simulation tests. Furthermore, by considering time-dependent cracks of the RCR sample, the microcracking characteristics in the deformation process of the RCR sample were simultaneously studied. In the following, the details of the methodology and results and discussion are explained.

2. Numerical Methods and Materials

The particle flow code in 2 or 3 dimensions (PFC2D (3D)) models the movement and interaction of circular (2D) or spherical (3D) particles using the DEM. It is possible for the blocks to break because they are composed of bonded particles [28, 29]. This software employs a DEM algorithm, and the DEM method iteratively solves Newton's second law of motion for an assembly of particles with a predefined contact bond model applied at the grain-grain contacts. The AE monitoring technology in discrete element modelling (DEM) is particularly suitable for studying the microcracking activities of samples.

2.1. Contact Bond Model in PFC. Two bonding behaviors are embodied in the contact-bond model (CBM) and parallel-bond model (PBM) (Figures 2(a) and 2(b)) [30], both of which can be envisioned as a kind of glue joining two neighboring particles. A contact bond also can be envisioned as a pair of elastic springs with constant normal and shear stiffness acting at a contact point. The two springs located between two adjacent particles have specified shear and normal strengths and control the micromechanical behavior of a contact bond.

The intrinsic difference between CBM and PBM is that the CBM (Figure 2(a)) can only transmit forces acting at the contact point, while the PBM (Figure 2(b)) can transmit both forces and moments between particles. Moreover, bond breakage in PBM immediately results in stiffness reduction since stiffness is contributed by both contact and bond stiffness (Figure 2(d)) [30]. In PFC, rock can be represented as a heterogeneous material composed of cemented grains. A PBM provides the mechanical behavior of a cement-like material substance between two contacting particles, and the PBM has been widely employed in estimating the mechanical behavior of rocks [31, 32]. Therefore, in this research, we chose the PBM to carry out the numerical simulations.

2.2. Numerical Samples and Calibrations. The RCR samples were composed of sandstone and coal (Figure 3(a)), and nine samples with height ratios ranging from 0% to 100% were generated for uniaxial compression tests (Figure 3(b)). It is clear that samples with height ratios of 0% and 100% are pure sandstone and pure coal samples, respectively.

According to the ISRM standard [33], the dimension of a sample that undergoes uniaxial compression is 50 mm in width and 100 mm in height. In this study, each RCR sample (φ50 mm × 100 mm) is discretized into 6673 particles, and the particle size (i.e., diameter) followed a uniform distribution ranging from 0.35 mm to 0.58 mm [34]. The interparticle friction for particle is 0.577, and the normal-to-shear stiffness ratio is set as 2.5 for both rock and coal particles. The average unit weights of sandstone and coal are 2600 kg/m^3 and 1800 kg/m^3, respectively. The parallel-bond tensile strengths of sandstone and coal are 24.0 MPa and 40.0 MPa, individually. The parallel-bond cohesive strength of coal is higher than that of sandstone: 15.0 MPa versus 8.4 MPa.

In PFC, numerical calibration requires back calculations using measured values of some commonly measured rock

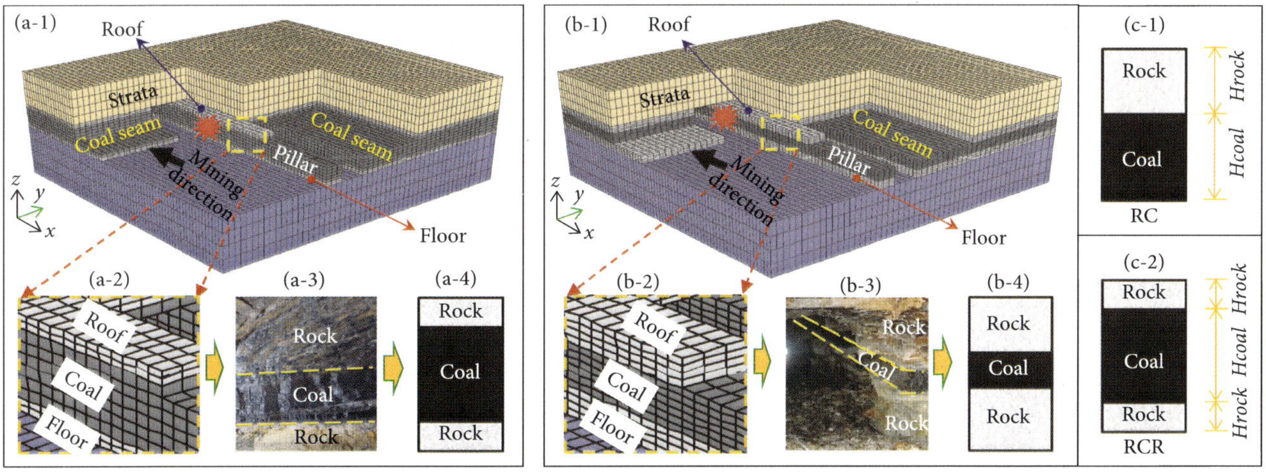

FIGURE 1: Coal pillars with two different geometries in longwall mining systems. The lower figures are in situ roof-coal-floor combinations and simplified RCR samples. (a-1) and (b-1) are the 3D schematics of longwall mining systems. (a-2) and (b-2) are the magnified figures of pillars. (a-3) and (b-3) are the two kinds of in situ roof-coal-floor structures with different height ratios. (a-4) and (b-4) are the schematics of RCR samples with different height ratios. (c-1) Rock-coal (RC) sample. (c-2) Rock-coal-rock (RCR) sample.

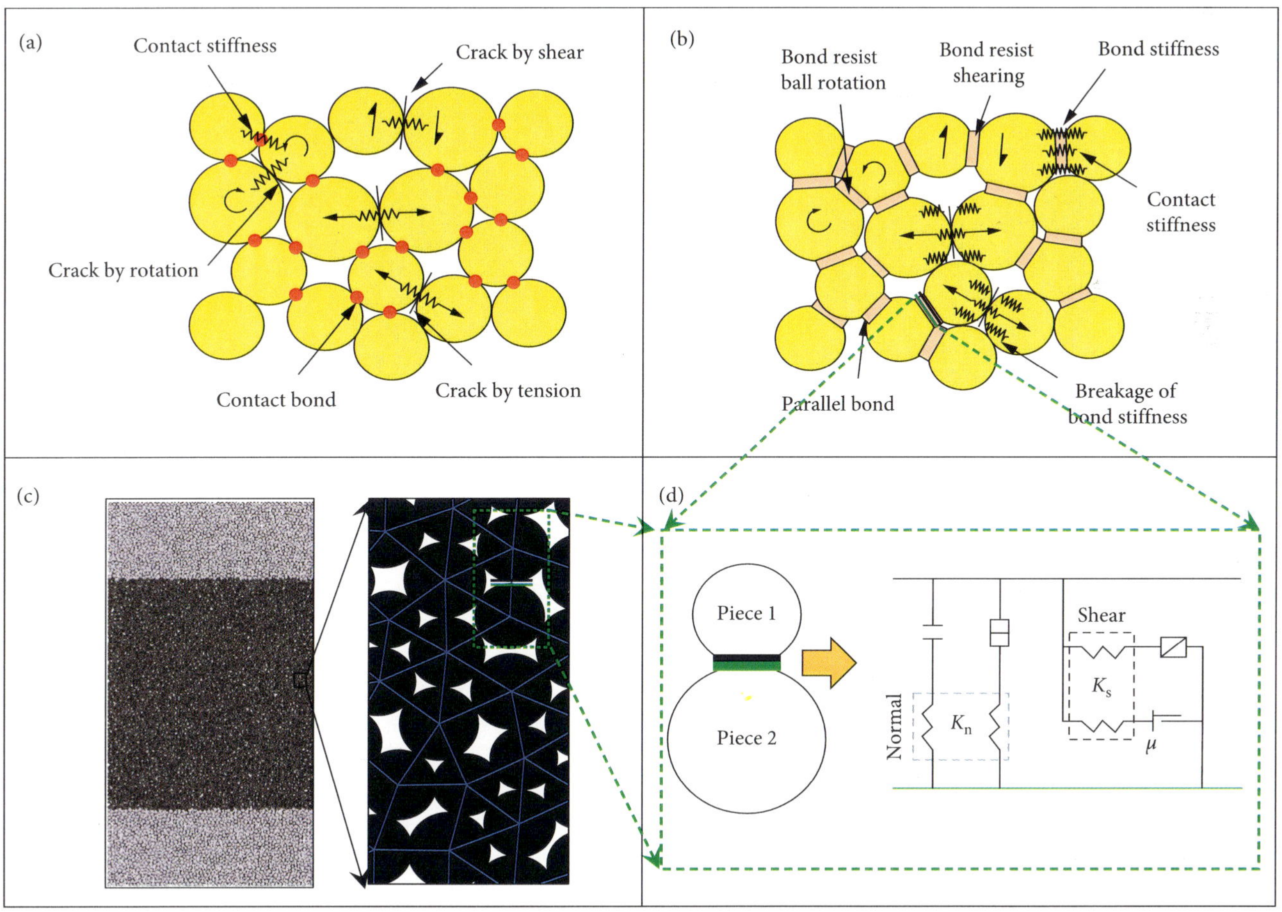

FIGURE 2: Illustration of bond models in PFC and numerical sample. (a) and (b) are contact-bond model (CBM) and parallel-bond model (PBM), respectively (modified from Cho et al. [30]). (c) A typical numerical RCR sample in this study. (d) Components of the linear parallel-bond model (modified from Itasca [28])), where K_n, K_s, and μ represent normal stiffness, shear stiffness, and friction coefficient at a contact point, respectively.

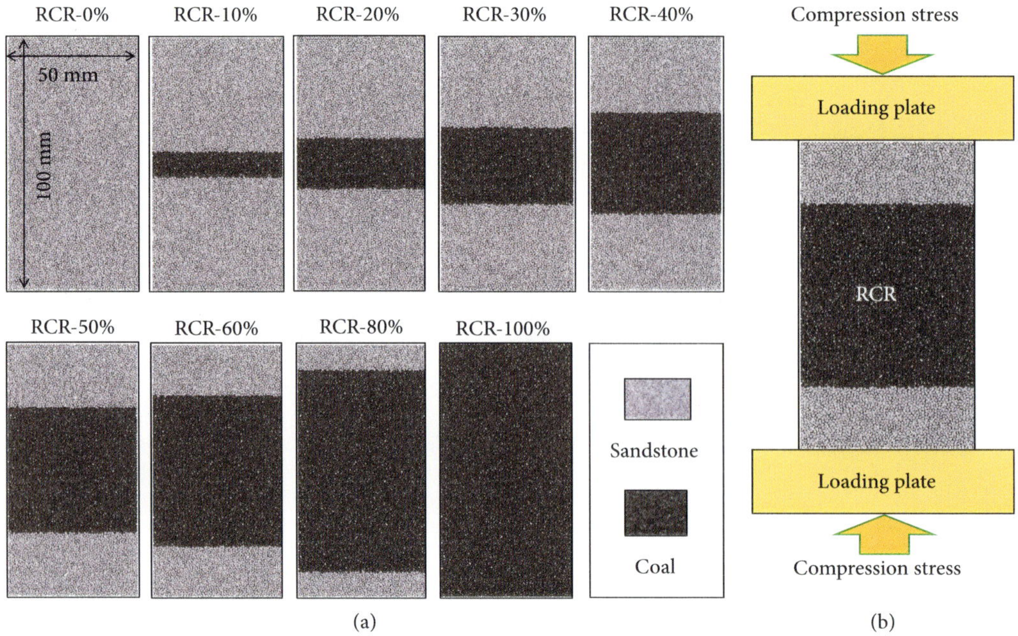

FIGURE 3: Numerical RCR samples with height ratios ranging from 0% to 100% (a), and the framework of uniaxial compression test (b).

properties, such as elastic modulus and UCS. Table 1 presents the calibrated results, which are comparisons between experimental and numerical data of sandstone and coal samples under uniaxial compression tests, and the coal and sandstone were taken from Yunhe coal mine (Jining, Shandong Province, China) [14].

TABLE 1: Calibrated results [14].

Rock type	Peak strength (MPa)		Elastic modulus (GPa)	
	Experimental	Numerical	Experimental	Numerical
Sandstone	87.30	89.30	8.13	8.37
Coal	29.24	30.60	3.06	3.11

2.3. Microcracking Activity Measurement and Testing Procedures

2.3.1. Microcracking Activity Measurement Method. In order to understand the failure mechanisms in rock, monitoring AE events caused by microcracking activities is always employed [35–37]. In PFC, bond breakages can be simultaneously monitored using the program's internal language (FISH). Furthermore, the initiation and propagation of microcrack can be represented as a progressive breakage of contact bonds [28, 35]. Therefore, the microcracking activities were historically recorded by monitoring the bond breakage events.

2.3.2. Uniaxial Compression Method. During the test, an external load was applied on the top of the RCR sample in the axial direction (Figure 3(b)). As recommended by the user manual of PFC [28], the loading rate for compression tests was set as 0.02 m/s. It should be noted that the numerical loading rate is related to time step, which is different from that employed in laboratory tests. In this word, the time step in each calculation cycle was 1.16×10^{-7} s. Thus, the loading rate of 0.02 m/s can be translated to 2.32×10^{-6} mm/step, which means that it requires more than 431,000 steps to move the loading plate 1.0 mm. A brief review of the loading rate setting in PFC can be found in the research of Zhang and Wong [38].

The simulated samples were compacted and tested as follows. (1) Generating an RCR sample; (2) removing floating particles; (3) identifying the coal component; (4) applying linear parallel contact bond to particles and setting the mechanical properties of the RCR sample; (5) applying a constant velocity of 0.02 m/s on the top wall in all numerical runs; and (6) ceasing the compression process until 40% of the UCS is reached. During the loading process, mechanical and physical parameters (e.g., axial compression stress, axial strain and bond breakage events) were recorded in real time.

3. Results and Discussion

Axial stress-strain curves for RCR samples under uniaxial compression tests are shown in Figure 4, in which the stress-strain curves of the pure sandstone sample (height ratio equals to 0%) and pure coal sample (height ratio equals to 100%) are also plotted. The curves of the RCR samples are located between those of pure sandstone and pure coal samples. From the shapes of these curves, it is clear that the height ratio has a key effect on the mechanical properties and deformation behavior of RCR samples, which will be analyzed in detail.

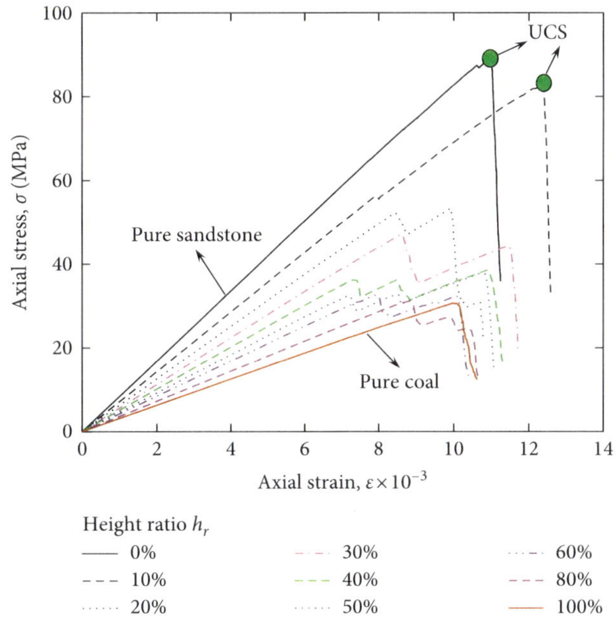

FIGURE 4: Stress-strain curves of the samples.

3.1. Effect of Height Ratio on the UCS of RCR Samples. In this section, the influence of the height ratio on the UCS of the RCR sample is investigated. In accordance with the stress-strain curves shown in Figure 4, the UCS values (σ_c) for all samples are given in Table 2 and presented graphically in Figure 5. Generally, the UCS decreases with increasing height ratio. Specifically, the UCS values of RCR samples are all lower than that of pure sandstone (89.36 MPa) and higher than that of the pure coal sample (30.64 MPa), which is consistent with previous studies [14, 16]. Figure 5 demonstrates the relation between height ratio (h_r) and UCS (σ_c) of samples (namely, $h_r \sim \sigma_c$ relationship), and the relationship can be illustrated by an inversed proportional function:

$$\sigma_c = 24.7 + \frac{5.85}{h_r}, \quad (1)$$

where σ_c and h_r are UCS (MPa) and height ratio of the RCR sample, respectively.

The decreasing trend of the $h_r \sim \sigma_c$ curve is most pronounced for the RCR and pure coal samples, and the correlation coefficient of the regressive curve is 0.992. According to the UCS values, the $h_r \sim \sigma_c$ curve can be divided into three stages. In stage I (i.e., height ratios between 10% and 40%), the UCS decreases sharply from 82.89 MPa to 38.57 MPa. In stage II (i.e., 40% to 60%), the UCS experiences a slight decrease from 38.57 MPa to 32.72 MPa, while a proximate stable trend can be identified afterward in stage III (from 32.72 MPa to 31.14 MPa). Therefore, the critical height ratios of 40% and 60% may have key effects on both the mechanical and deformation behavior of RCR samples.

In quasistatic compression of the RCR sample, the load is equally applied on the tops of coal and rock components. Once the external load reaches the UCS of the coal component, failure of coal would occur and the whole RCR sample may lose load-bearing capacity. Thus, the UCS of the RCR is determined by the strength of coal [12]. However, it is not always the same for composite samples with variable height ratios [14]. From Figure 5, it can be concluded that the UCS of the RCR sample is strongly influenced by the height ratio. Thus, a question may be posed: does the UCS of the RCR sample determined by the size effect of coal or rock component?

In this part, the size effects on the UCS of individual rock component and RCR sample were studied. Tuncay and Hasancebi [39] reviewed the size effect of length to diameter ratio on the UCS of the sample, indicating that the height-diameter ratio is highly correlated with USC of the rock sample. Hoek and Brown [40] suggested that the UCS decreases with increasing sample diameter (Figure 6(a)), and they suggested the following equation to convert the UCS value of a different diameter sample to that of a 50 mm sample:

$$\frac{\text{UCS}}{\text{UCS}_{50}} = \left(\frac{50}{\varphi}\right)^{0.18}, \quad (2)$$

where UCS_{50} is the UCS of the sample with a diameter of 50 mm (ASTM standard). φ is the diameter of a test sample. Notice, the height of all samples employed in this method is 100 mm (Figure 6(a)).

According to the definition of the height-diameter ratio, the height ratio (h_r) of an RCR sample can be converted into the height-diameter ratio. As shown in Figure 6, an individual coal or rock component can be represented by a pure sample with a height of 100 mm. However, the diameter of the pure sample should be zoomed out or zoomed in according to the height-diameter ratio.

TABLE 2: Uniaxial compressive strength, elastic modulus, and postpeak modulus of RCR samples.

Sample	Height ratio (h_r)	Size (mm)	σ_c (MPa)	E (GPa)	E_r (GPa)	Note
RCR-0%	0%	$\varphi 50 \times 100$	89.36	8.37	−225.19	Pure sandstone
RCR-10%	10%	$\varphi 50 \times 100$	82.89	7.10	−270.09	RCR
RCR-20%	20%	$\varphi 50 \times 100$	53.30	6.23	−235.00	RCR
RCR-30%	30%	$\varphi 50 \times 100$	47.05	5.52	−114.50	RCR
RCR-40%	40%	$\varphi 50 \times 100$	38.57	4.95	−65.03	RCR
RCR-50%	50%	$\varphi 50 \times 100$	37.94	4.54	−104.16	RCR
RCR-60%	60%	$\varphi 50 \times 100$	32.72	4.16	−75.39	RCR
RCR-80%	80%	$\varphi 50 \times 100$	31.14	3.58	−78.76	RCR
RCR-100%	100%	$\varphi 50 \times 100$	30.64	3.11	−38.98	Pure coal

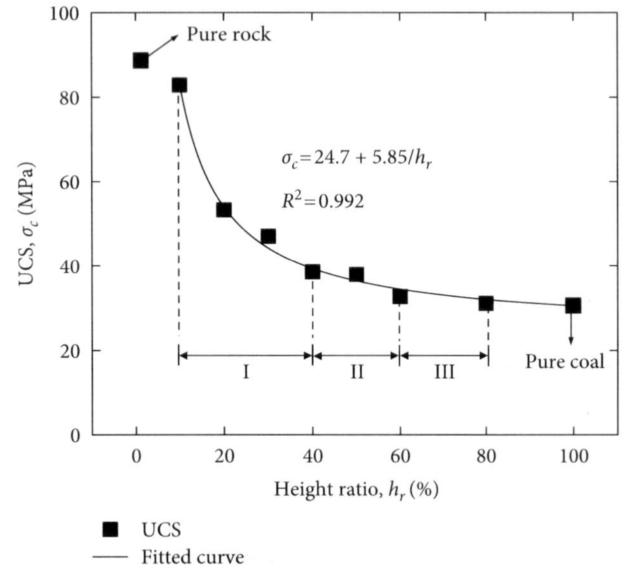

FIGURE 5: Relation between height ratio and UCS.

Based on Equation (2), the theoretical UCS/UCS$_{50}$ values of sandstone and coal components in an RCR sample can be calculated and plotted in Figure 7. In accordance with the UCS values of RCR samples shown in Table 2, the numerical UCS/UCS$_{50}$ values of coal and rock components in RCR samples are also plotted. Interestingly, the theoretical curve is between the numerical curves of rock and coal samples. Furthermore, the numerical UCS/UCS$_{50}$ curve of the sandstone component has a similar trend with the theoretical curve, while the numerical curve of the coal component experiences a reversed shape. Therefore, the $h_r \sim \sigma_c$ relationship cannot be simply illustrated as a result of size effect but may also be caused by the effect of coal-rock interface [20]. The strength of weak coal is enhanced near the interface, while the strength of sandstone is weakened, thus the global strength may be modified.

3.2. *Effects of Height Ratio on the Elastic and Postpeak Modulus of RCR Samples*

3.2.1. Relation between Height Ratio and Elastic Modulus. The elastic modulus (E) is a mechanical property of an object which represents resistance to being deformed elastically when loading. In a stress-strain curve, the elastic modulus is defined as the slope of the elastic deformation region (Figure 8). At the stage of elastic deformation, the compression stress increases linearly with axial strain, and the elastic deformation dominates a stress-strain curve in the prepeak stage. According to the stress-strain curves shown in Figure 4, the elastic modulus values of RCR samples which are listed in Table 2 and presented in Figure 9 can be regressively derived. The relation between height ratio and elastic modulus (namely, $h_r \sim E$ relationship) is also presented.

As shown in Figure 9, the height ratio has a key effect on the value of the elastic modulus. The elastic modulus values of RCR samples are between the values of pure sandstone (7.10 GPa) and pure coal (3.11 GPa) samples. On the whole, the elastic modulus of the sample nonlinearly decreases with increasing height ratio. According to the regressive curve, the $h_r \sim E$ relationship of the RCR sample is expressed as

$$E = 2.58 - 5.58 \times 0.1196^{h_r}, \quad (3)$$

where E and h_r are the elastic modulus (GPa) and the height ratio of the RCR sample, respectively.

As mentioned in Section 3.1, the effect of rock-coal interface should be taken into consideration when analyzing the UCS of RCR samples. In an RCR, the frictional force exists on the rock-coal interface to restrict the deformation of rock and coal, which is different from the no-frictional interface in the idealized sample (Figure 10(a)). Thus, a mechanical model was established (Figure 10) and discussed in the following parts.

Let E_{rock}, E_{coal}, and K_{rock}, K_{rock} denote, respectively, the elastic modulus and stiffness where the subscripts "coal" and "rock" identify the coal and rock components, and suppose $E_{\text{rock}} > E_{\text{coal}}$ and $K_{\text{rock}} < K_{\text{coal}}$. It is known that the stiffness of a rock is a measure of the resistance offered by an elastic body to deformation. For an elastic rock, the stiffness is defined as

$$K = \frac{F}{\delta}, \quad (4)$$

where F and δ are external load on the top of an elastic body and elastic deformation, respectively.

For the series-connected sample (Figure 10(a)), the rock-coal interaction at a connecting layer can be neglected due to the eliminated friction caused by rollers. The stiffness of such series-connected sample, K_w, can be expressed as

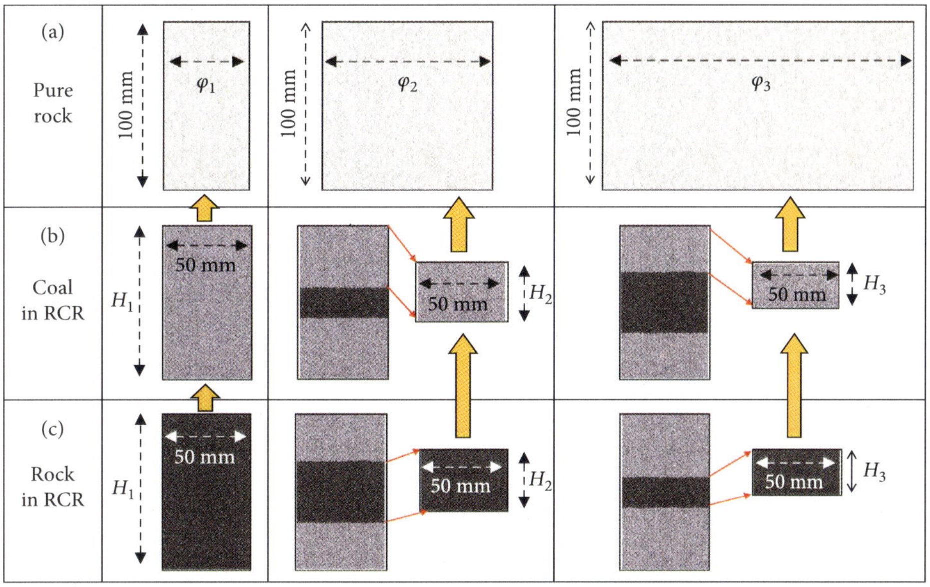

FIGURE 6: Size effect of samples. (a) Samples with a constant height (100 mm) employed in Equation (2). (b) Rock components with a constant diameter of 50 mm and different heights ($H_1 \sim H_3$) in RCR samples. (c) Coal bodies with a constant diameter of 50 mm and different heights ($H_1 \sim H_3$) in RCR samples.

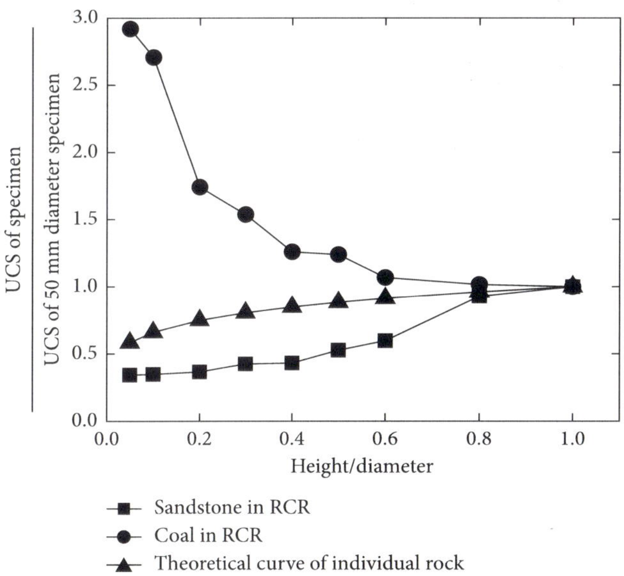

FIGURE 7: Size effect on the UCS of the samples.

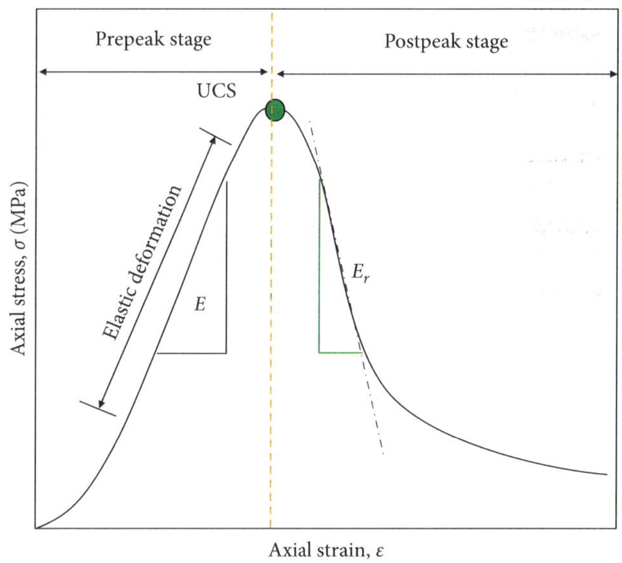

FIGURE 8: The USC, elastic modulus, and postpeak modulus in a typical stress-strain curve [41, 42].

$$K_w = \frac{1}{\sum_{i=1}^{n} (1/K_i)}, \quad (5)$$

where K_w is the stiffness of the composite sample, N/m; K_i is the stiffness of an individual component; and n refers to the number of components.

The elastic modulus of a material is not the same as the stiffness of a component made from that material [43]. Specifically, stiffness is an extensive property of the elastic body that is dependent on the material properties, shape, and boundary conditions. For a rock in compression, the axial stiffness is

$$K = \frac{EA}{H}, \quad (6)$$

where E is the elastic modulus of samples. A and H are the cross-sectional area and height of a rock, respectively.

By arranging Equations (5) and (6), the elastic modulus of a composite sample without interface friction can be obtained as follows:

$$E_w = \frac{H}{A \sum_{i=1}^{n} (H_i / E_i A_i)}, \quad (7)$$

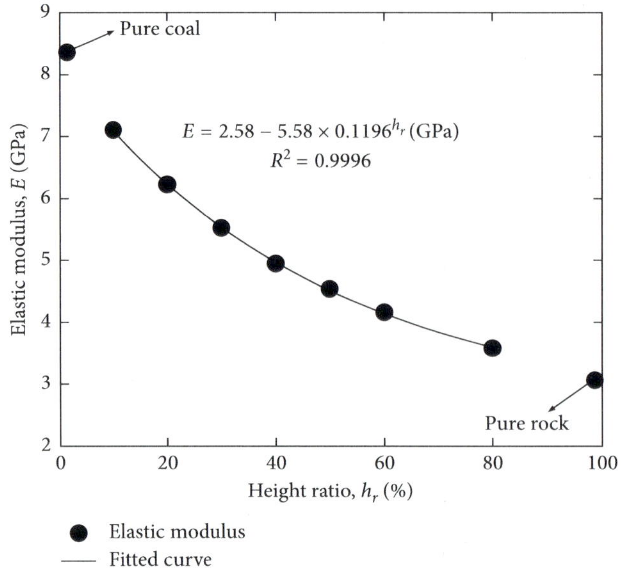

FIGURE 9: Relationship between height ratio and elastic modulus.

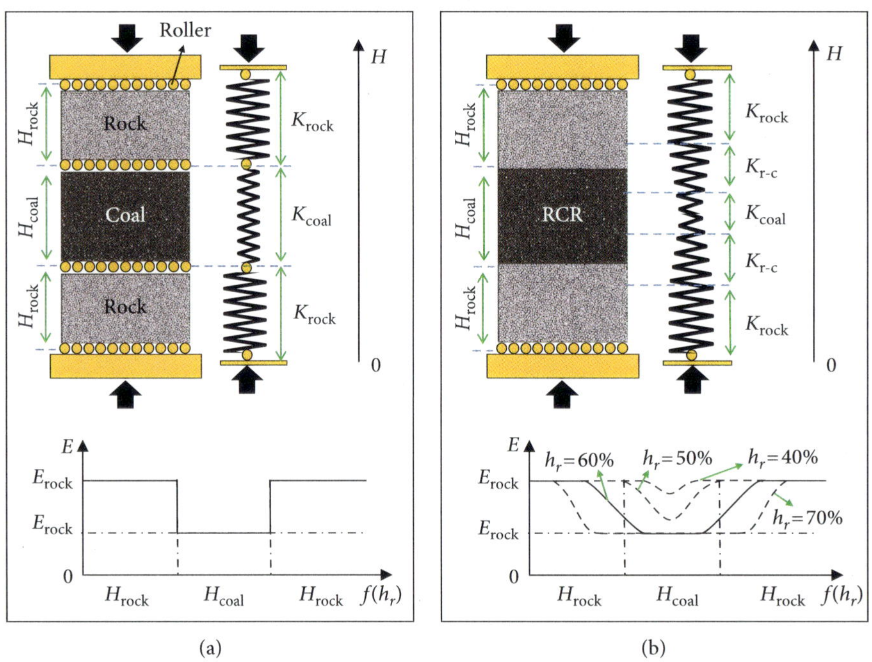

FIGURE 10: Mechanical models of the RCR samples. (a) Series-connected sample without rock-coal surface interaction, in which rollers are installed to reduce the friction. The lower represents the elastic modulus versus height of the series-connected sample. (b) RCR sample with rock-coal surface friction. The lower shows the elastic modulus versus height of the RCR sample. In the rock-coal interface area, the elastic modulus gradually changes from E_{rock} to E_{coal}.

where E_w is the elastic modulus of the composite sample without interface friction. A_i, H_i, and E_i are the cross-sectional area, elastic modulus, and the height of individual component, respectively.

In the RCR (Figure 10(b)), the mechanical properties of a composite sample not only are influenced by the size effect but may also by the effect of the rock-coal interface. Considering the elastic modulus of the area of rock-coal interface (K_{r-c}) is nonconstant, but it varies with the height of this area, which is different from that of the series-connected sample shown in Figure 10(a). By taking Saint-Venant's principle (i.e., boundary conditions) into consideration [44], the elastic modulus of an RCR can be expressed as

$$E_w = \frac{H}{\sum_{i=1}^{n}\left(f^{i}{}_{(h_r)}/f_{(E_i)}\right)}. \quad (8)$$

With

$$f_{(E_i)} = \begin{cases} E_{\text{rock}} f_{(h_r)}, & \text{rock component,} \\ f_{(E_{\text{rock}}, E_{\text{coal}})} f_{(h_r)}, & \text{rock} \sim \text{coal component,} \\ E_{\text{coal}} f_{(h_r)}, & \text{coal component,} \\ f_{(E_{\text{coal}}, E_{\text{rock}})} f_{(h_r)}, & \text{rock} \sim \text{coal component,} \\ E_{\text{rock}} f_{(h_r)}, & \text{rock component,} \end{cases} \tag{9}$$

where $f_{(E_i)}$ is the function of the elastic modulus of a component, $f_{(h_r)}$ is a shape function that determines the heights of components, and $f_{(E_{\text{coal}}, E_{\text{rock}})}$ is the elastic modulus in the area of rock-coal interface.

The elastic modulus along the height of the RCR sample can be primarily depicted in the below parts of Figure 10. According to Equation (5), the elastic modulus of the series-composite sample (Figure 10(a)) is a constant value and is unrelated to the height ratio. However, as can be seen easily in Figure 10(b), the elastic modulus along the height of the RCR sample gradually changes from E_{rock} to E_{coal}, which is different from that of the series-connected sample (Figure 10(a)). Further, according to Equation (7), the height ratio has a key effect on the elastic modulus along the height, because the height ratio determines the values of the elastic modulus in rock-coal areas. Therefore, the coupled effects of the height ratio and elastic modulus of each component may result in the nonlinear phenomena in $h_r \sim \sigma_c$ and $h_r \sim E$ curves.

3.2.2. Relation between Height Ratio and Postpeak Modulus. The postpeak stage of a typical strain-strain curve shows that the strength decreases with increasing strain (Figure 8). The postpeak modulus (E_r) [42], determined based on the progress of a postpeak failure stage, is a mechanical parameter of great importance in solving mining geomechanics problems. The value of postpeak modulus is determined by the tangent of angle of the inclination of a line being a linear approximation of the stress-strain curve. More detailed descriptions on the postpeak modulus can be found in the research of Bukowska [41].

The approximate linear portions of postpeak stages are plotted in Figure 11, in which linear fitted curves are also presented. The correlation coefficients of these linear fitted curves are between 0.953 and 0.993. The values of the postpeak modulus are listed in detail in Table 2. It can be seen that the postpeak modulus of pure sandstone and pure coal samples are −225.19 GPa and −38.98 GPa, respectively. In Figure 12, $h_r \sim E_r$ relationship is graphically presented. It is clear that the postpeak modulus grows up with increasing height ratio as an inversely proportional function which can be illustrated as

$$E_r = -43.8 - 24.7/h_r, \tag{10}$$

where E_r and h_r are postpeak modulus (GPa) and height ratio of RCR sample, respectively.

The fitted $h_r \sim E_r$ curve is primarily divided into two stages by the critical height ratio of 40%. In stage A, the postpeak modulus increases sharply with height ratio (10%–40%), which means the postpeak curve tends to fall more rapidly with a steeper slope in the postpeak stage. However, in stage B, the postpeak modulus keeps stable with increasing height ratio (above 40%).

3.3. Effect of Height Ratio on the Microcracking Activity of Samples. The microcracking activity can be presented and counted by monitoring bond breakage events in PFC [28]. Figure 13 shows the stress-strain curves and microcracking evolutions of pure sandstone and pure coal samples during the whole loading history, while Figure 14 shows those of RCR samples. In accordance with the microcracking activity and stress-strain results, a detailed investigation for the microcrack characteristics of pure and composite samples was carried out.

In Figure 13, the microcracking activity of pure samples is divided into two periods, namely, the quiet period and the active period. During the quiet period, there is no bond breakage, and the stress linearly grows up with increasing strain. During the active period, stress-strain curves of pure sandstone and coal samples undergo two different stages, i.e., the stage of elastic deformation and the stage of crack growth and propagation. Moreover, microcracking activities of the pure coal sample are much more intense than those of the pure rock sample. This is because the sandstone with a higher brittleness experienced a sharp stress drop after failure, while the coal sample failed in a progressive manner.

The microcrack characteristics of RCR samples are different from those of pure samples. As shown in Figure 14, the reactive period occurs after the quiet and active periods. The microcracking behavior of RCR samples during the quiet period is not dependent on the height ratio. In the active period, the stress increases linearly at the stage of elastic deformation. However, when it increases to small peak stress, the stress-strain curve occurs to depart from the linear curve and shows distinctly stress drop behavior. After the active period, another active period (i.e., reactive period) occurs, because the propagation of microcrack causes the redistribution of the contact force. As shown in the elliptical zones in Figure 14, the RCR samples progressively fails. It can be seen that before the failure of the whole sample, the component of coal fails with several stress drops.

3.4. Discussion on the Failure Mode of RCR Samples

3.4.1. Validation of the Simulation Results

(1) Comparison between the Simulation Results and Laboratory Test Results. In this section, we conducted a comparison between the simulation results and the laboratory test results. These final failure modes of RCR samples (RCR-20% and RCR-40%) are shown in Figure 15, wherein the left are simulation results and the right are laboratory test results.

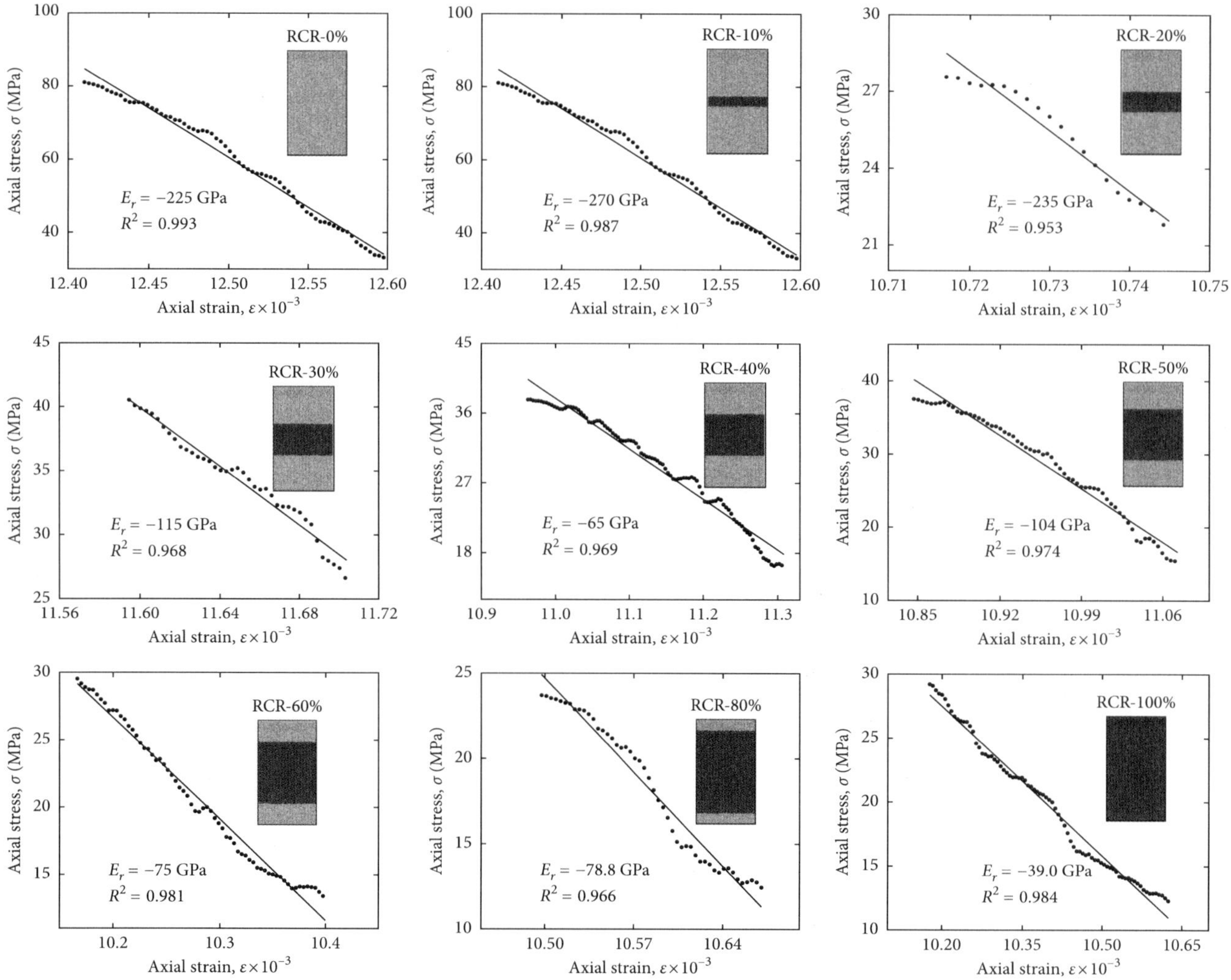

FIGURE 11: Postpeak stages and linear fitted curves.

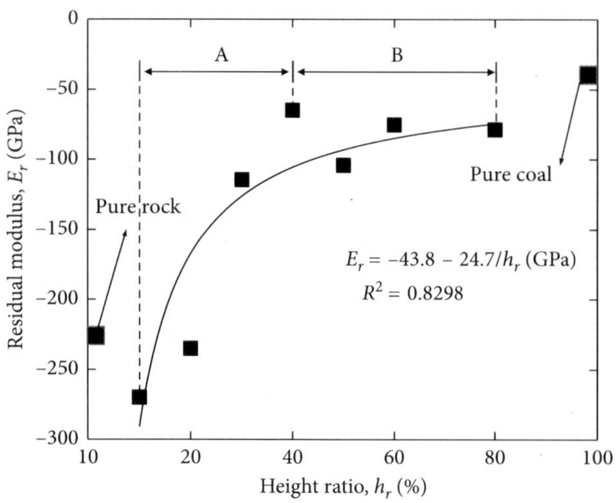

FIGURE 12: Relation between height ratio and the postpeak modulus.

The failure modes of laboratory test results are consistent with those of simulation results. As can be seen in Figure 15(a), with an increase of uniaxial compression stress, the deformation of rocks is not apparent and lateral deformation of the coal is marked and much greater than that of rocks; the coal ruptures vertically first. With further increase of compressive stress, the coal section splits and smashes into a columnar form; at the time of coal failure, the rock generates splitting failure rapidly. As can be seen from both the simulation and laboratory test results (Figure 15(b)), the coal bodies are broken more thoroughly than the rock bodies. The lower part of the coal body is tapered, suggesting that the X-type shearing fracture surface must have been formed inside the coal before failure. Shear fracture mainly occurs before splitting, and the final destruction of the coal is dominantly splitting [16].

(2) Comparison between the Simulation Results and the in Situ Test Results. Figure 16 presents the comparison between the simulation results and the in situ test results, wherein Figure 16(a) is the simulation failure mode of the RCR sample, and Figures 16(b) and 16(c) are the in situ failure modes. The in situ test pillar (Figure 16(b)) is about 2 m in width [45, 46].

In the coal component of the RCR sample, the conjugated shear bands can be easily identified (Figure 16(a)). Similarly, such conjugated shear bands can be identified in the in situ

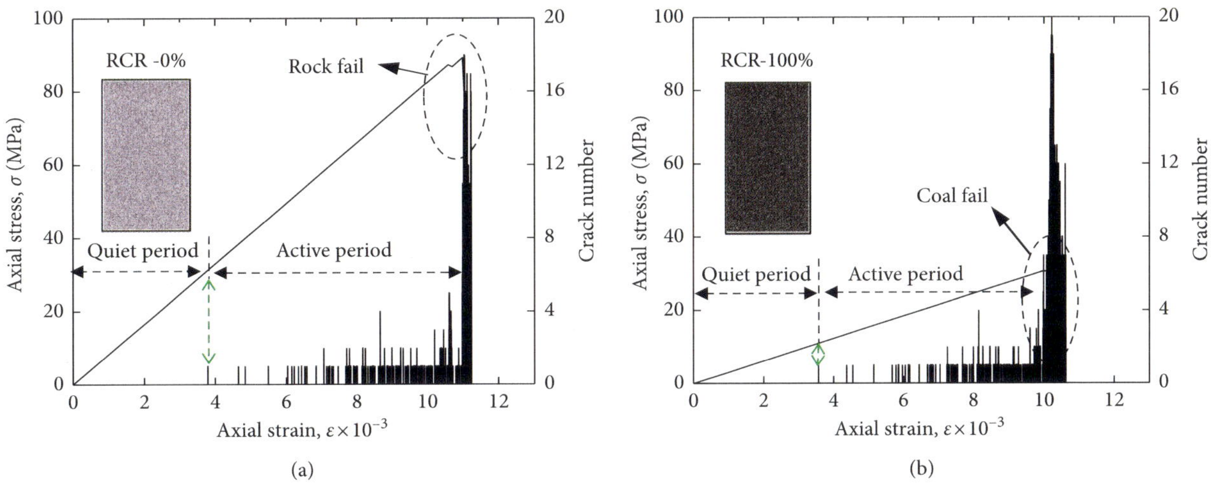

FIGURE 13: Microcracking evolution of pure sandstone (a) and pure coal (b) samples under uniaxial compression tests.

rock/coal pillars (Figure 16(b) and 16(c)). This failure mode was also called double-pyramid type of failure [46]. Generally, the influence on the pillar strength of the lateral confinement generated at the pillar ends decreases with distance from the end interfaces. As shown in Figure 16, this is evidenced by the hourglass shape of the stressed pillars. It is one reason why the strength of a pillar with a given width decreases with increasing pillar height. This is consistent with the effect of the height ratio on the failure mode of the RCR sample.

3.4.2. Effect of Height Ratio on the Failure Mode of RCR Sample. Figure 17 illustrates the influence of the height ratio on the ultimate failure modes of the RCR samples under uniaxial compression tests. The crack distributions are depicted as short colored lines, and the red line occurred later than the yellow and green lines.

In pure sandstone and pure coal samples, the conjugated shear bands can be easily identified. However, in the RCR samples, fractures initiated in coal components and then propagated to the whole RCR samples, resulting in multi-stress drops in the stress-strain curves (Figure 14). In accordance with the time-dependent microcracking activities shown in Figure 17, it can be concluded that the failure of coal component causes the first stress drop in the stress-strain curve, and the last stress drop results from the collapse of rock component or the whole sample.

3.5. Progressive Application of the Results. The failure of rock samples in laboratory testing is similar to the dynamic disasters in the field [47]. The pure coal sample experiences splitting and comminuting failure. However, the failure of the RCR sample is progressive. The cracking bands develop preferentially in the component of coal, and then the development of cracks gradually extends to the rock portion. This is because the constraint force exerted on the interface between the coal and the rock in the direction perpendicular to the axis is small, and the tensile stress caused by the radial expansion of the sample exceeds the tensile strength of coal first, thus resulting in a columnar splitting failure of the coal section [15].

The determiner, controlling the failure of the whole RCR sample, changes from rock to coal component with increasing height ratio. According to Figures 14 and 17, the critical height ratio can be identified as 60%. Prior to the critical height ratio, the UCS of the RCR sample is determined by the failure of rock, while the UCS of the RCR sample is controlled by the failure of coal component afterward. We can wonder that the in situ roof-coal-floor structure experiences such progressive failure. Thus, the coal failure in the roof-coal-floor structure with a height ratio lower than 60% would not result in systematical failure, and the systematical failure often takes place after one or two dense AE events. However, if the height ratio of a roof-pillar-floor structure is higher than 60%, the failure of coal would lead to systematical failure of the whole structure. These results may contribute to predicting the hazards of coal bursts and estimating the failure of rock-coal-floor structures in underground coal mining systems.

4. Conclusions

In this paper, we studied the mechanical properties and deformation behavior of RCR samples with variable height ratios ranging from 0 to 100%. The effects of the height ratio on the UCS, elastic modulus, postpeak modulus, microcracking activity, and failure mode of RCR samples were systematically investigated. The following can be concluded:

(1) The UCS of the RCR sample is strongly affected by the height ratio, and the increase of the height ratio decreases the UCS as an inverse proportional function. In addition, the UCS versus height ratio curve is divided into three stages, namely, dramatical decrease stage (height ratio: 0 to 40%), slight decrease stage (height ratio: 40% to 60%), and stable stage (height ratio: 60% to 100%).

(2) The elastic modulus tends to decrease exponentially with increasing height ratio, while the

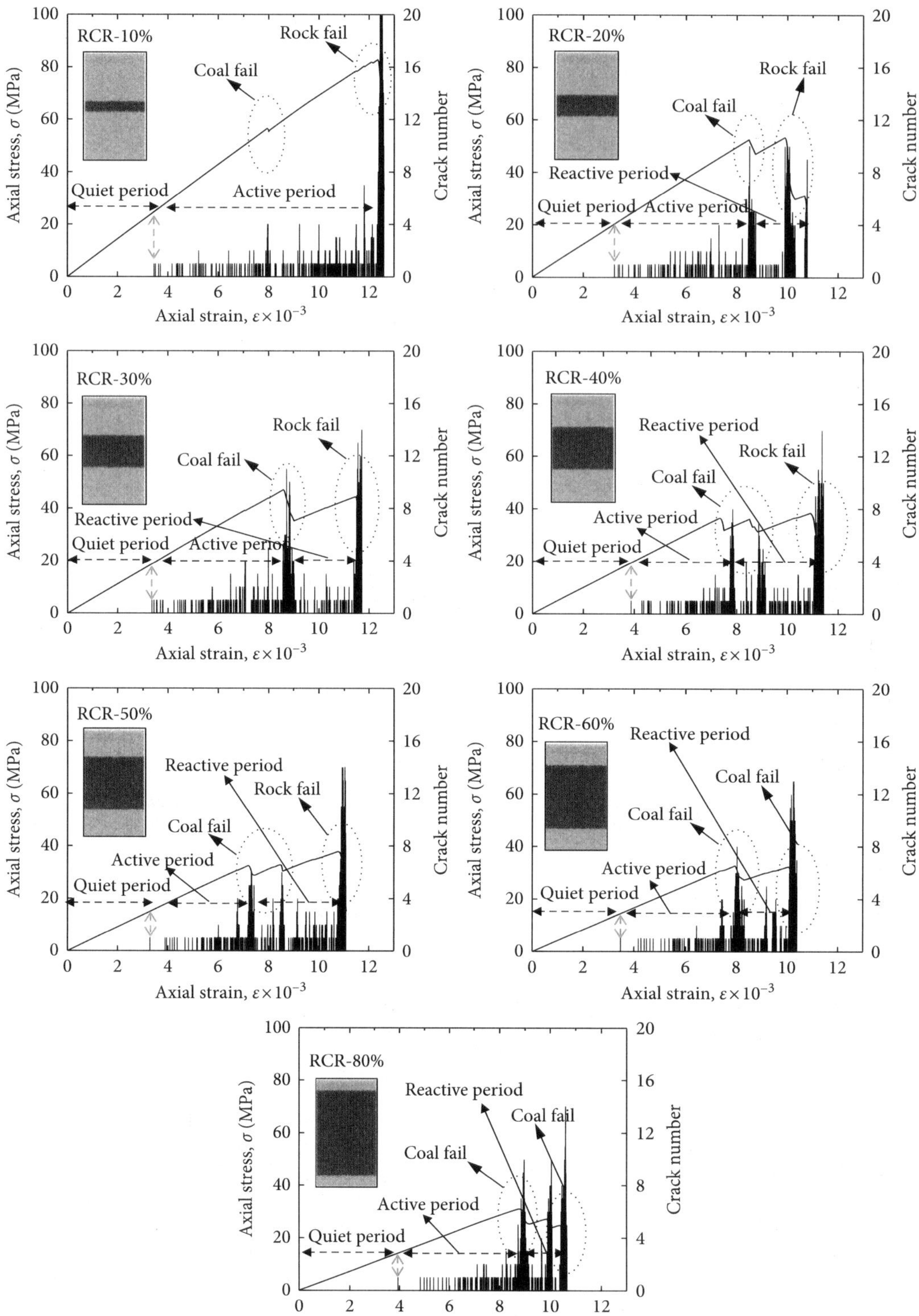

FIGURE 14: Relation between stress and microcracking evolution of RCR samples.

(a)

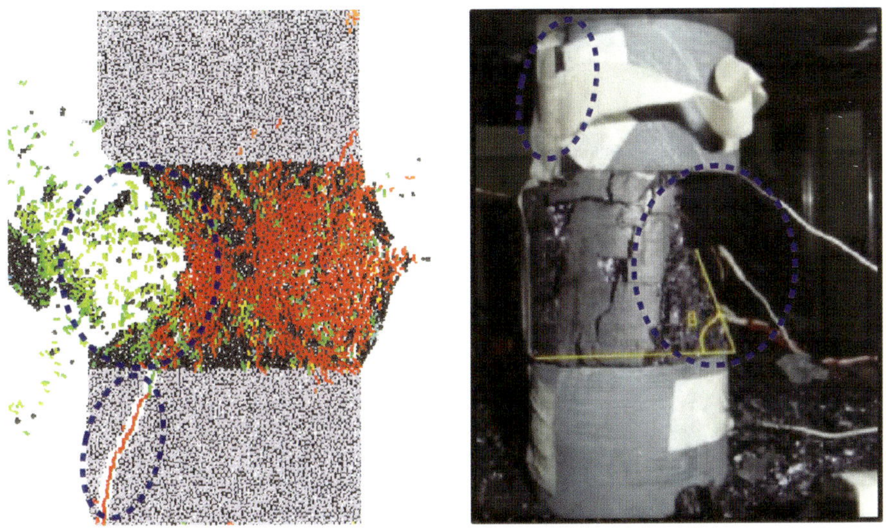

(b)

Figure 15: Comparison between the simulation results (left) and the laboratory test results (right). (a) Failure mode of the RCR sample with a height ratio of ~20%. The laboratory test result is derived from Huang and Liu [15]. (b) Failure mode of the RCR sample with a height ratio of 40%. The laboratory test result is derived from Liu et al. [16].

(a) (b) (c)

Figure 16: Comparison between the simulation results (left) and the in situ test results (right). (a) Failure mode of an RCR sample; (b) and (c) are the failure modes of the in situ pillars [45, 46].

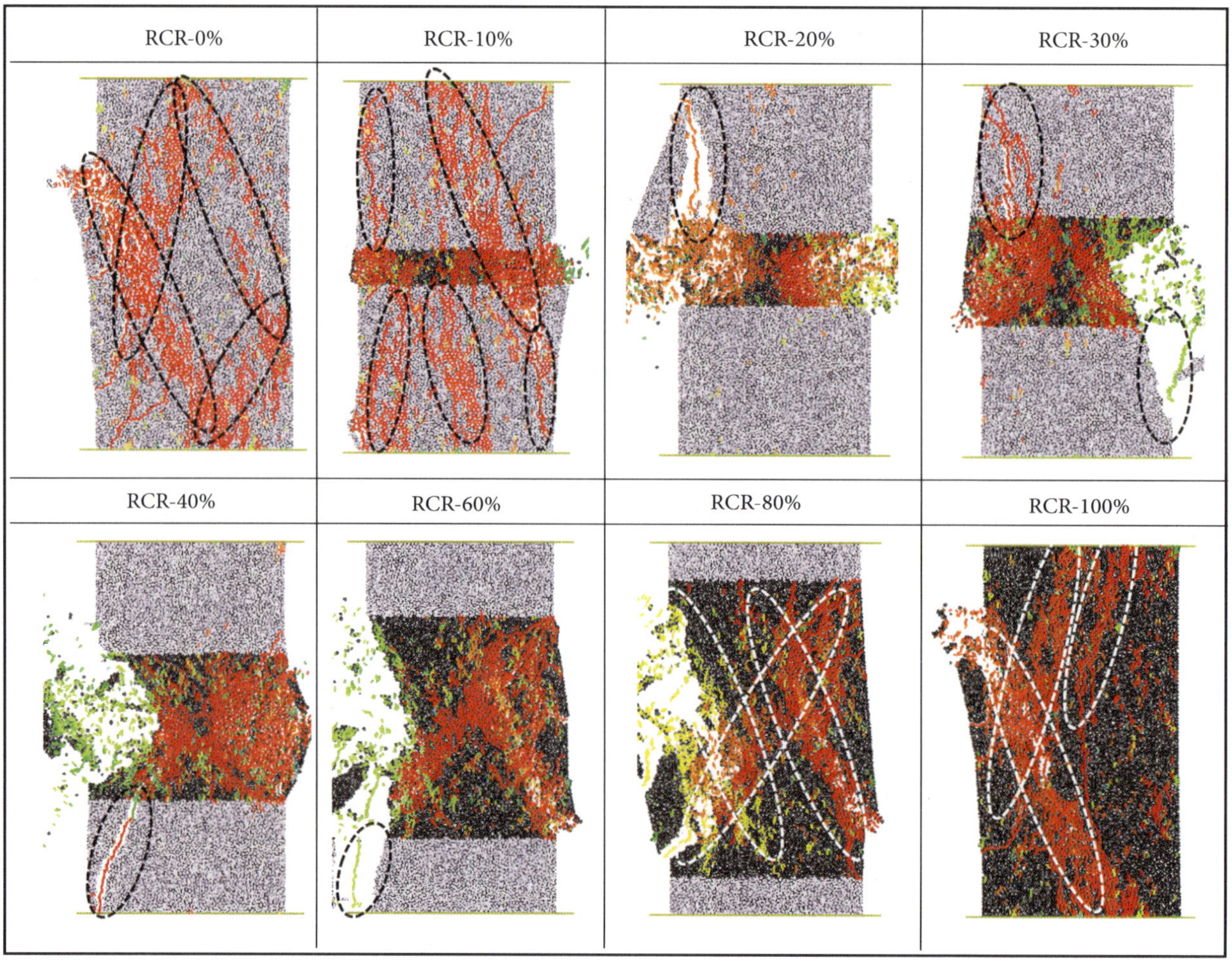

FIGURE 17: Failure modes of the RCR samples.

postpeak modulus value prefers to rise with the increasing height ratio in an inverse proportional manner.

(3) The microcracking activity of RCR samples is different from that of pure samples. The reactive period always occurs in RCR samples after the quiet and active periods. The active period in an RCR sample comes up with the failure of coal component, while the reactive period always implies the failure of the whole RCR sample.

(4) The RCR sample fails in a progressive manner. The cracking bands develop preferentially in the component of coal, and then the development of cracks gradually extends to rocks. However, the determiner of the UCS of the RCR sample varies in the height ratio. Prior to the critical height ratio of 60%, the UCS is determined by the failure of rock, while the UCS is controlled by the failure of coal component afterward.

Conflicts of Interest

The authors declare no conflicts of interest.

Acknowledgments

This work was supported by the National Basic Research Program of China (2014CB046300) and supported by the Priority Academic Program Development of Jiangsu Higher Education Institutions (PAPD).

References

[1] N. G. W. Cook, "Seismicity associated with mining," *Engineering Geology*, vol. 10, no. 2-4, pp. 99–122, 1976.

[2] L. Driad-Lebeau, F. Lahaie, M. A. Heib, J. P. Josien, P. Bigarré, and J. F. Noirel, "Seismic and geotechnical investigations following a rockburst in a complex french mining district," *International Journal of Coal Geology*, vol. 64, no. 1-2, pp. 66–78, 2005.

[3] L. Dou, C.P. Lu, Z. Mu, and M. Gao, "Prevention and forecasting of rock burst hazards in coal mines," *International Journal of Mining Science and Technology*, vol. 19, no. 5, pp. 585–591, 2009.

[4] Y. Jiang, H. Wang, S. Xue, Y. Zhao, J. Zhu, and X. Pang, "Assessment and mitigation of coal bump risk during extraction of an island longwall panel," *International Journal of Coal Geology*, vol. 95, pp. 20–33, 2012.

[5] Z. Li, L. Dou, W. Cai et al., "Failure-pillar induced rock burst mechanism of thick coal seam in deep mining," *Chinese*

Journal of Rock Mechanics and Engineering, vol. 32, pp. 333–342, 2013.

[6] Y. Pan, Z. Li, and M. Zhang, "Distribution, type, mechanism and prevention of rockburst in China," *Chinese Journal of Rock Mechanics and Engineering*, vol. 22, pp. 1844–1851, 2003.

[7] C. Wang, F. Jiang, P. Wang et al., "Microseismic events distribution characteristics and mechanical mechanisms of rock bursting induced by a coal pillar," *Journal of China Coal Society*, vol. 34, pp. 1169–1173, 2009.

[8] C. P. Lu, L. M. Dou, B. Liu, Y. S. Xie, and H. S. Liu, "Microseismic low-frequency precursor effect of bursting failure of coal and rock," *Journal of Applied Geophysics*, vol. 79, pp. 55–63, 2012.

[9] E. Hoek and E. T. Brown, "Practical estimates of rock mass strength," *International Journal of Rock Mechanics & Mining Sciences*, vol. 34, no. 8, pp. 1165–1186, 1997.

[10] T. P. Medhurst and E. T. Brown, "A study of the mechanical behaviour of coal for pillar design," *International Journal of Rock Mechanics & Mining Sciences*, vol. 35, no. 8, pp. 1087–1105, 1998.

[11] H. Zhang, Z. Wan, D. Ma, Y. Zhang, J. Cheng, and Q. Zhang, "Experimental investigation on the strength and failure behavior of coal and synthetic materials under plane-strain biaxial compression," *Energies*, vol. 10, p. 500, 2017.

[12] Z. Zhao, X. Lv, W. Wang, and Y. Tan, "Damage evolution of bi-body model composed of weakly cemented soft rock and coal considering different interface effect," *Springerplus*, vol. 5, p. 292, 2016.

[13] J. Zuo, H. Xie, W. U. Aimin, and J. Liu, "Investigation on failure mechanisms and mechanical behaviors of deep coal-rock single body and combined body," *Chinese Journal of Rock Mechanics & Engineering*, vol. 30, pp. 84–92, 2011.

[14] S. Chen, D. Yin, B. Zhang, H. Ma, and X. Liu, "Study on mechanical characteristics and progressive failure mechanism of roof-coal-pillar structure body Chinese," *Journal of Rock Mechanics and Engineering*, vol. 36, pp. 1–11, 2017.

[15] B. Huang and J. Liu, "The effect of loading rate on the behavior of samples composed of coal and rock," *International Journal of Rock Mechanics & Mining Sciences*, vol. 61, pp. 23–30, 2013.

[16] J. Liu, E. Wang, D. Song, S. Wang, and Y. Niu, "Effect of rock strength on failure mode and mechanical behavior of composite samples," *Arabian Journal of Geosciences*, vol. 8, no. 7, pp. 4527–4539, 2015.

[17] Z. H. Zhao, W. M. Wang, and J. X. Yan, "Strain localization and failure evolution analysis of soft rock-coal-soft rock combination model," *Journal of Applied Sciences*, vol. 13, no. 7, pp. 1094–1099, 2013.

[18] L. Dou, C. Lu, and Z. Mu, "Rock burst tendency of coal-rock combinations sample," *Journal of Mining & Safety Engineering*, vol. 23, pp. 43–46, 2006.

[19] Z. Zhao, W. Wang, L. Wang, and C. Dai, "Compression–shear strength criterion of coal–rock combination model considering interface effect," *Tunnelling & Underground Space Technology*, vol. 47, pp. 193–199, 2015.

[20] Z. H. Zhao, W. M. Wang, C. Q. Dai, and J. X. Yan, "Failure characteristics of three-body model composed of rock and coal with different strength and stiffness," *Transactions of Nonferrous Metals Society of China*, vol. 24, pp. 1538–1546, 2014.

[21] J. Liu, C. Tang, W. Zhu, and T. Yang, "Effects of combination mode on mechanical properties and failure characteristics of the coal-rock combinations," *Journal of China Coal Society*, vol. 26, pp. 276–280, 2004.

[22] X. B. Deng, H. J. Hu, G. Xu, X. T. Li, and F. Y. Chen, "Numerical simulation for burst failure of two-body rock structure," *Journal of Mining & Safety Engineering*, vol. 29, pp. 833–839, 2012.

[23] D. M. Guo, J. P. Zuo, Y. Zhang, and R. S. Yang, "Research on strength and failure mechanism of deep coal-rock combination bodies of different inclined angles," *Rock & Soil Mechanics*, vol. 32, pp. 1333–1339, 2011.

[24] J. Li, Q. Qi, D. Mao, and Y. Wang, "Discussion on evaluation method of bursting liability wit composite model of coal and rock," *Chinese Journal of Rock Mechanics & Engineering*, vol. 24, p. 4805, 2005.

[25] P. K. Kaiser and C. A. Tang, "Numerical simulation of damage accumulation and seismic energy release during brittle rock failure—part II: rib pillar collapse," *International Journal of Rock Mechanics & Mining Sciences*, vol. 35, no. 2, pp. 123–134, 1998.

[26] V. A. Mansurov, "Prediction of rockbursts by analysis of induced seismicity data," *International Journal of Rock Mechanics and Mining Sciences*, vol. 38, no. 6, pp. 893–901, 2001.

[27] X. Wang, C. Lu, J. Xue et al., "Experimental research on rules of acoustic emission and microseismic effects of burst failure of compound coal-rock samples," *Rock and Soil Mechanics*, vol. 34, pp. 2569–2575, 2013.

[28] C. G Itasca, *PFC (Particle Flow Code in 2 and 3 Dimensions), Version 5.0 (User's Manual)*, Minneapolis, 2014.

[29] D. O. Potyondy and P. A. Cundall, "A bonded-particle model for rock," *International Journal of Rock Mechanics and Mining Sciences*, vol. 41, no. 8, pp. 1329–1364, 2004.

[30] N. Cho, C. D. Martin, and D. C. Sego, "A clumped particle model for rock," *International Journal of Rock Mechanics and Mining Sciences*, vol. 44, no. 7, pp. 997–1010, 2007.

[31] P. A. Cundall, *Numerical Experiments on Rough Joints in Shear Using a Bonded Particle Model*, Springer, Heidelberg, Germany, 2000.

[32] D. O. Potyondy, "Simulating stress corrosion with a bonded-particle model for rock," *International Journal of Rock Mechanics & Mining Sciences*, vol. 44, no. 2, pp. 677–691, 2007.

[33] Z. T. Bieniawski and M. J. Bernede, "Suggested methods for determining the uniaxial compressive strength and deformability of rock materials: part 1. Suggested method for determining deformability of rock materials in uniaxial compression," *International Journal of Rock Mechanics and Mining Sciences and Geomechanics Abstracts*, vol. 16, no. 2, pp. 138–140, 1979.

[34] H. Zhang, D. Elsworth, and Z. Wan, "Failure response of composite rock-coal samples," *Geomechanics and Geophysics for Geo-Energy and Geo-Resources*, vol. 4, no. 2, pp. 175–192, 2018.

[35] H. Shimizu, S. Murata, and T. Ishida, "Distinct element analysis for rock failure considering ae events generated by the slip at crack surfaces," *Journal of Acoustic Emission*, vol. 27, pp. 194–211, 2009.

[36] H. Horii and S. Nemat-Nasser, "Compression—induced microcrack growth in brittle solids: axial splitting and shear failure," *Journal of Geophysical Research Solid Earth*, vol. 90, no. B4, pp. 3105–3125, 1985.

[37] D. Lockner, "The role of acoustic emission in the study of rock fracture," *International Journal of Rock Mechanics and Mining Science and Geomechanics Abstracts*, vol. 30, pp. 883–899, 1993.

[38] X. P. Zhang and L. N. Y. Wong, "Choosing a proper loading rate for bonded-particle model of intact rock," *International*

Journal of Fracture, vol. 189, no. 2, pp. 163–179, 2014.

[39] E. Tuncay and N. Hasancebi, "The effect of length to diameter ratio of test specimens on the uniaxial compressive strength of rock," *Bulletin of Engineering Geology and the Environment*, vol. 68, no. 4, pp. 491–497, 2009.

[40] E. Hoek and E. T. Brown, *Underground Excavations in Rock*, Stephen Austin and Sons Ltd., London, UK, 1980.

[41] M. Bukowska, "Post-peak failure modulus in problems of mining geo-mechanics," *Journal of Mining Science*, vol. 49, no. 5, pp. 731–740, 2013.

[42] W. R. Wawersik and C. Fairhurst, "A study of brittle rock fracture in laboratory compression experiments," *International Journal of Rock Mechanics and Mining Sciences and Geomechanics Abstracts*, vol. 7, pp. 561–564, 1970.

[43] G. Pharr, W. Oliver, and F. Brotzen, "On the generality of the relationship among contact stiffness, contact area, and elastic modulus during indentation," *Journal of Materials Research*, vol. 7, no. 3, pp. 613–617, 1992.

[44] R. A. Toupin, "Saint-venant's principle," *Archive for Rational Mechanics and Analysis*, vol. 18, no. 2, pp. 83–96, 1965.

[45] Z. T. Bieniawski, "Mechanism of brittle fracture of rock: part II—experimental studies," *International Journal of Rock Mechanics and Mining Sciences and Geomechanics Abstracts*, vol. 4, no. 4, pp. 407–423, 1967.

[46] Z. T. Bieniawski, "Note on in situ testing of the strength of coal pillars," *Journal of the South African Institute of Mining and Metallurgy*, vol. May, pp. 455–465, 1968.

[47] N. G. W. Cook, "The failure of rock," *International Journal of Rock Mechanics and Mining Sciences and Geomechanics Abstracts*, vol. 2, no. 4, pp. 389–403, 1965.

12

Electrical Resistivity of Concrete for Durability Evaluation

Pejman Azarsa and Rishi Gupta

Department of Civil Engineering, University of Victoria, Victoria, BC, Canada

Correspondence should be addressed to Rishi Gupta; guptar@uvic.ca

Academic Editor: Gianfranco Palumbo

Degradation processes in reinforced concrete structures that affect durability are partially controlled by transport of aggressive ions through the concrete microstructure. Ions are charged and the ability of concrete to hold out against transfer of ions greatly relies on its electrical resistivity. Hence, a connection could be expected between electrical resistivity of concrete and the deterioration processes such as increase in permeability and corrosion of embedded steel. Through this paper, an extensive literature review has been done to address relationship between concrete electrical resistivity and its certain durability characteristics. These durability characteristics include chloride diffusivity and corrosion of reinforcement as these have major influence on concrete degradation process. Overall, there exists an inverse or direct proportional correlation between these parameters. Evaluated results, from measuring the concrete electrical resistivity, can also be used as a great indicator to identify early age characteristics of fresh concrete and for evaluation of its properties, determination of moisture content, connectivity of the micropores, and even condition assessment of in-service structures. This paper also reviews and assesses research concerning the influential parameters such as environmental conditions and presence of steel rebar and cracks on measuring electrical resistivity of concrete. Moreover, concrete resistivity concept, application, and its various measurement techniques are introduced.

1. Introduction

The durability of concrete is defined as its ability to resist weathering action, chemical attack, abrasion, or any other deterioration process to retain its original form, quality, and serviceability when exposed to harsh environment [1]. To a large extent, it is commonly accepted that concrete durability is governed by concrete's resistance to the penetration of aggressive media. This media may be present in a liquid or gaseous state and that may be transported by various mechanisms such as permeation, diffusion, absorption, capillary suction, and combinations of the items just mentioned. Hence, for concrete in service, a combined action of various media may prevail and mixed modes of transport processes occur. Moreover, there are correlations between transport parameters of concrete and the following durability characteristics: carbonation, sulphate attack, alkali-aggregate reaction, frost resistance, leaching, soft water attack, acid attack, abrasion, chloride ingress, and reinforcement corrosion. Consequently, the transport of ions through microstructure of concrete plays an important role in the control of concrete durability. When ions are charged, then it is the concrete's ability to withstand transfer of charged ions which is highly dependent upon its electrical resistivity. In this study, since chloride ingress and reinforcement corrosion are reported as major concrete deterioration processes, one of the main concentration areas is on these durability characteristics and their relationship with concrete electrical resistivity.

Over the last few decades, a great deal of attention has been paid to research and development of electrical resistivity measurement techniques as a nondestructive technique (NDT) to evaluate the durability of concrete structures. This method is becoming more popular especially for field evaluations due to its simplicity, rapidness, and cost during test conduction. However, the inclusion of these methods into the standards and guidelines is quite slow. Electrical resistivity has been standardized in 2012 by ASTM C1760 [2] to measure the concrete bulk resistivity and also by AASHTO

TP 95-11 [3] to quantify the surface resistivity of concrete. However, there is a gap that still exists between the current knowledge and industry practice.

Electrical resistivity is a material property that can be used for various purposes, one of which is to identify early age characteristics of fresh concrete. When the fresh concrete sets and hardens, depercolation (discontinuity) of the capillary pore space leads to an increase in its electrical resistivity. Since electrical current is conveyed by dissolved charged ions flowing into the concrete pore solution, it is a good indicator of concrete pore structures [4]. This pore structure formation at early-ages can define the long-term durability of concrete. In addition, the tensile strength of cementitious materials at early-ages is low and the material is prone to cracking. This initial cracking also serves as a pathway for deleterious materials to ingress into the matrix. This cracking can also be captured by resistivity measurements and thus helps predict the long-term durability of concrete. In addition, electrical resistivity can be used as an index to determine the moisture content and the connectivity of the micropores in the concrete [5].

Several researchers attempted to characterize the effects of various parameters on electrical resistivity measurements. One of the important factors affecting the measurements is environmental conditions such as temperature, rainfall, and relative humidity. During testing, good electrical connection between concrete and electrodes as well as specimen geometry plays a key role in having a reliable measurement. The electrical resistivity measurements are highly influenced by the moisture content of concrete. For instance, when the moisture content is reduced, the resistivity is increased significantly. Therefore, considering all these influencing parameters for on-site resistivity measurements and to make meaningful conclusions is not a simple task.

In this paper, the correlation between electrical resistivity and certain durability characteristics of concrete is discussed. These concrete characteristics include chloride permeability, corrosion rate, and compressive strength. Also, different approaches in the measurement of concrete resistivity including bulk and surface resistivity measurements are presented. This paper reviews the effect of several influencing parameters such as external environment (e.g., temperature) and concrete mixture on the electrical resistivity. In addition, some of bulk and surface resistivity test setups (both of laboratory and field tests) conducted by authors are also presented.

2. Theoretical Background

2.1. Concept. Electrical resistivity (ρ) of a material is defined as its capability to withstand the transfer of ions subjected to an electrical field. It is largely dependent on the microstructure properties of concrete such as pore size and shape of the interconnections (i.e., tortuosity) [6]. Specimens with similar degree of water saturation and temperature should be used as both of these factors affect resistivity. Lower permeability results from a finer pore network with less connectivity and eventually leads to higher electrical resistivity. The range spanned by resistivity is one of the greatest of any material property [12]. For concrete, it varies from 10^6 Ω·m for oven dried samples to 10 Ω·m for saturated concrete [13]. Electrical resistivity is the ratio between applied voltage (V) and resulting current (I) multiplied by a cell constant and the electrical current is carried by ions dissolved in the pore liquid [7, 14]. Thus, it is a geometry independent property and an inherent characteristic of a material, as described in the following [6, 14]:

$$\rho = k \cdot R = k \cdot \left(\frac{V}{I}\right), \tag{1}$$

where R is the resistance of concrete; k is a geometrical factor which depends on the size and shape of the sample as well as the distance between the probes on the testing device [6]. There are several factors that may affect electrical resistivity of concrete, and they can be divided into two groups: (1) intrinsic factors affecting the electrical resistivity of concrete, such as w/c ratio, aging, and pore structure; (2) factors affecting the resistivity measurements, including specimen geometry, moisture content, temperature, electrode spacing, and presence of rebar. For instance, more pore water as well as wider pores results in lower concrete resistivity and environmental factors such as higher temperature decreases the resistivity values [7]. Furthermore, adding reactive supplementary cementitious materials such as blast furnace slag and fly ash leads to lower permeability and higher electrical resistivity due to reduction in capillary porosity and hydroxyl ions (OH^-). Both carbonation and chloride penetration also individually cause an increase in concrete resistivity in particular in Portland cement concrete but penetrated chloride impact is relatively small [7]. The effects of the above-mentioned parameters will be discussed in detail later in this paper.

2.2. Measurement Techniques. Electrical resistivity measurements can be performed in several ways nondestructively: using electrodes positioned on a specimen surface, or placing an electrode-disc or linear array or a four-probe square array on the concrete's surface. Types of device techniques that can be used typically to measure resistivity physically include (1) bulk electrical resistivity test, (2) surface disc test, (3) Wenner four-point line array test, and (4) four-probe square array test.

2.2.1. Bulk Electrical Resistivity Test. In the bulk resistivity method (or uniaxial method), two electrodes are placed on the concrete surface (usually two parallel metal plates) with moist sponge in between (Figure 1(a)). Generally, only standard cylinders/prismatic specimens or cores taken from existing structures are used in this method. The geometrical factor in this method can be obtained by the following equation:

$$k = \frac{A}{L}, \tag{2}$$

where A is the cross-sectional area perpendicular to the current and L is the height of sample. Although this non-destructive test takes only a few seconds, its application is limited for field evaluation because electrodes access to

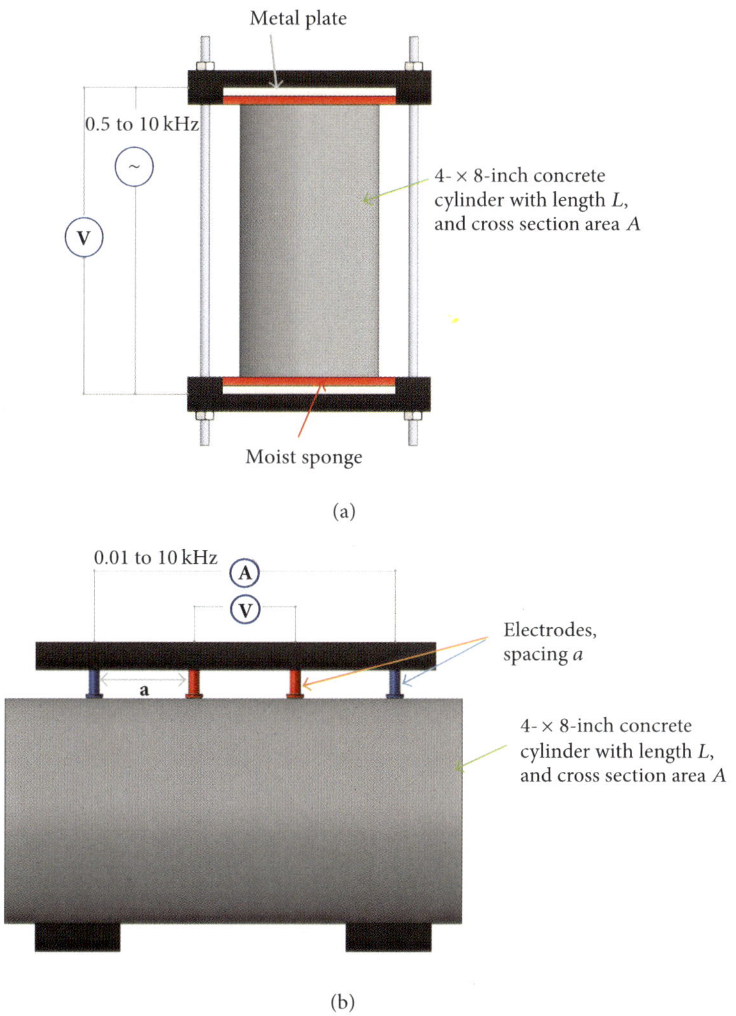

FIGURE 1: Electrical resistivity measuring techniques: (a) two-point uniaxial method and (b) four-point (Wenner probe) method (reproduced from [6]).

opposite sides of the concrete element is not possible all the time; while other above-mentioned resistivity measurement (surface disc test, Wenner four-point line array test, and four-probe square array test) methods may use probes placed on only one side surface of specimen.

2.2.2. Surface Disc Test. The electrode-disc test method includes an electrode (disc) placed over a rebar and measuring the resistance between the disc and the rebar, as shown in Figure 2 [7]. One disadvantage of this method is a connection requirement to the steel reinforcement and full rebar continuity. In this technique, a cell constant is dependent on cover depth (which varies over the surface) and the rebar diameter whose precise measurements are impossible due to lack of exact current flow prediction [7]. For cover depth, disc and bar diameters being 10–50 mm, the cell constant is approximately 0.1 m. Hence, the resistivity can be derived using

$$\rho\,(\text{disc}) = 0.1 \times R\,(\text{disc} - \text{bar})\,. \tag{3}$$

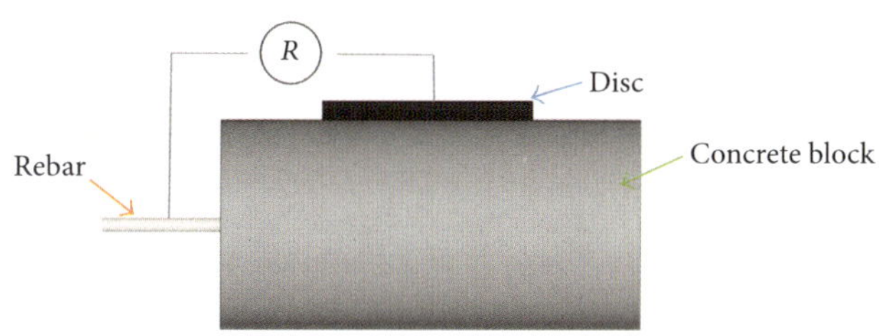

FIGURE 2: Setup of one electrode-disc: measurement of concrete resistivity (reproduced from [7]).

2.2.3. Wenner Four-Point Line Array Test. The Wenner probe technique was first introduced for the geologist's field in order to determine soil strata by Wenner at the National Bureau of Standards in the 1910s and then modified through time for concrete application [15]. In this technique, four equally spaced linear electrodes are used to measure the surface electrical resistivity of concrete (Figure 1(b)). The two exterior electrodes apply an AC current to the concrete surface while the electrical potential is measured from the interior probes. It should be noted that DC current is not desirable as it may result in inaccurate readings because of polarization effect. The effect of current frequencies on measurements is discussed in Section 5.2.6 (studies on electrical signal shape and frequency). The constant cell is defined as (4) for semi-infinite homogenous material [6]:

$$k = \gamma \cdot a, \tag{4}$$

where a is the distance between the equally spaced electrodes and γ is the dimensionless geometry factor which is equal to 2π for semi-infinite concrete elements such as concrete slabs [6]. However, the geometry factor is different for tests conducted in a laboratory condition on small cylinders or cubic specimens. To measure the surface electrical resistivity, AASTHO TP 95-11 is the only specified standard which requires an electrode spacing of 1.5 inch (or 38 mm) with an AC frequency of 13 Hz [3]. Due to its configuration, this method is reliable for on-site measurement; however many factors that will be discussed in Section 5 can affect the results such as rebar and cracks presence, surface conditions, concrete mixture, and environmental conditions.

2.2.4. Four-Probe Square Array Test. The four-probe square array consists of the four probes that are arranged in square position with spacing of 50 to 100 mm [10].

2.3. Applications. Electrical resistivity can be related to certain performance characteristics of concrete and can be used as a promising quality assurance tool for fresh or hardened concrete [6]. Some of these correlations will be discussed in the following sections. It can be used as a measure of concrete resistance to chloride ingress as well as corrosion initiation and rate measurements. The concrete diffusion coefficient as an important factor in the service life estimation of structures also can be obtained by electrical resistivity technique. In addition, it is a reliable test method to detect and monitor the initiation and propagation of cracks in concrete since they change the connectivity of concrete pore structure, and thus its electrical conductivity [16]. Cement mortars and concrete setting time can be determined through the concept of electrical resistivity. However, the correlation between setting time and concrete durability is not fully understood. Another potential application of the electrical resistivity method is to compute the moisture content of concrete, although reliability of this method is still under question [5]. However, electrical resistivity method is a simple and reliable nondestructive test method; the application and reliability of this method in determining certain characteristics of concrete has yet to be widely evaluated. This is more due to the limited knowledge in this area especially for on-site evaluation.

3. Objective and Methodology

The primary objective of this paper is to review the existing state of practice on the electrical resistivity measurements technique. This paper also identifies the applicability and limitation of electrical resistivity method and reviews the correlation between resistivity and certain durability properties of concrete. Correlation between surface and bulk electrical resistivity and their applications is also discussed. Finally, key parameters affecting the electrical resistivity readings are identified for future research in the area.

An extensive literature search was undertaken from most relevant publications in the area. A comparison was made of the experimental setup (Section 4), and the way in which the correlated data was obtained between resistivity and durability properties of concrete (Sections 6–8). Several parameters influencing the concrete resistivity were identified and compared (Section 5). The information observed from the literatures was based on experimental and numerical studies. The reviewed data was compiled in tables and later compared. Detailed information on the experimental setups is presented in Abbreviations section and Tables 1–4. The literature search covered both laboratory and field investigation.

4. Comparison of the Experimental Investigations

In this section, experimental setups developed by other researchers have been summarized in Tables 1–4. These tables consist of specimens' configuration, materials type, resistivity measurement techniques, and specimen curing/exposure conditions. The data in the tables is arranged in the order in which the citations appear in Sections 5–8. An additional row that contains authors' data on measuring electrical resistivity of simulated field circular hollow-section columns is also included. The extent to which differences in the setups can influence electrical resistivity measurements are discussed later in Sections 5–8 using data presented in this section. A comparison of the experimental setups is given in Sections 4.1–4.3. Abbreviations and symbols are defined in Abbreviations section.

4.1. Specimen Geometry and Setup. Frequently, in the electrical resistivity studies, samples with dimensions between 100 and 400 mm were used (Table 1). Specimens with dimensions over 1000 mm to simulate real-world condition were more seldom used. According to Table 2, the steel rebar diameter varied from 4 to 25 mm. In most cases, no detailed information was provided about the type of steel embedded; both smooth and ribbed steel was used. Cover depth ranging from 10 to 80 mm was considered for steel reinforcement bars in the majority of the experimental investigations. For those studies investigating the relationship between steel reinforcement corrosion and concrete resistivity, chloride ingress was the major cause of corrosion. However, no information was provided on the size of the anode and the ratio between anode and cathode in the reviewed articles. Only one study concentrated solely on carbonation-induced corrosion.

4.2. Materials and Exposure Conditions. According to data in Table 1, concrete or mortar samples were casted with a w/b ratio between 0.4 and 0.65 by mass in most reviewed

Table 1: Details of the specimen geometry (in terms of specimen size), material type, and number of specimens.

Refs	Specimen configuration				Materials type						w/b ratio	Rebar presence
	Type	Size (mm)	TNVM	Mix type	OPC	SLG	FA	MK	SF	Others		
[4]	Disk	Φ100 × 50	44	CON/CEM	×	×	×	—	× (5% & 10%)	NP (30% and 50%)	0.35, 0.4, 0.45	N
[16]	Slab	254 × 76 × 12.7	18	CON	× Type I	—	×	—	—	—	0.26	N
[22]	Cylinder	Φ47 × 95	30	CEM	× Type I	—	× (classes C & F)	—	—	$CaCO_3$ (10% & 15%)	0.35	N
[23]	Slab	250 × 200 × 120	1	CON	×	—	—	—	—	—	0.5	Y
[8]	Slab, cube, cylinder	300 × 300 × 135, 300 × 300 × 200, 100 × 100 × 100, Φ100 × 200	6	CON	× (blended)	—	× (20% blended)	—	—	—	0.4	Y
[24]	Prism	300 × 300 × 150	7	CON	× Type I	—	×	—	—	—	0.42	Y
[25]	Cylinder, prism	Φ100 × 200, Φ150 × 300, 200 × 200 × 175, 160 × 160 × 140, 120 × 120 × 110	NR	CON	× Type I	—	—	—	—	—	0.4, 0.6	Y
[26]	Cylinder, prism	Φ150 × 300, 40 × 40 × 160	NR	MOR	×	—	—	—	—	—	0.5	Y
[27]	Prism	400 × 400 × 100	6	MOR	×	—	—	—	—	—	0.6	Y
[28]	Cylinder	Φ150 × 220	4	CON	×	—	—	—	—	—	0.4, 0.6	Y
[29]	Slab, cube	650 × 650 × 100, 150 × 150 × 150	1	CON	×	—	—	—	—	—	0.5	N
[30]	Slab	600 × 600 × 120	9	CON	×	—	—	—	—	—	0.36, 0.48, 0.61	Y
[11]	Beam	1500 × 200 × 100	2	CON	×	—	—	—	—	×	0.7	N
[31]	Cylinder	Φ100 × 200	21	CON	× Type I/II	× (grades 100 & 120, 50%)	× (classes C & F, 20%)	—	× (5% & 10%)	—	0.35, 0.5, 0.65	N
[32]	Cylinder	Φ100 × 200	33	CON	× Type I/II	—	×	—	×	—	0.41	N
[33]	Cube	150 × 150 × 150	47	CON	×	×	—	—	—	—	0.4–0.55	N
[34]	Prism	100 × 100 × 170	12	CON	×	× (50% and 70%)	—	—	—	WPC	0.3, 0.42, 0.55	N
[35]	Cylinder	Φ150 × 300	3	CON	× Type I	—	—	—	—	—	0.45, 0.55, 0.65	N
[36]	Cylinder	Φ75 × 150	NR	CON	× Type V	—	—	—	—	—	0.45	N
[37]	Cube, cylinder, block	100 × 100 × 100, Φ100 × 200, Φ150 × 300, 300 × 300 × 200	NR	CON	× (blended)	—	× (18% blended)	—	—	—	0.4	N
[38]	Cube	150 × 150 × 150	NR	CON	×	×	—	—	×	PFA	0.59–0.7	N
[39]	Cylinder	Φ100 × 200	12	CON	×	×	×	×	×	Micro-FA	0.3–0.4	N
[40]	Cube	150 × 150 × 150	33	CON	× Type I & Type V	×	—	—	—	—	NR	N
[41]	Cylinder	Φ100 × 200	19	CON	×	×	×	—	×	—	0.41	N
[42]	Slab, cylinder	280 × 280 × 102, Φ102 × 204	NR	CON	×	×	×	×	—	MS	0.35–0.45	N
[43]	NR	NR	NR	MOR/CEM	× Type I	—	—	—	—	—	0.42	N
[44]	Cylinder	Φ100 × 200	12	CON	×	×	×	—	—	—	0.41	N

Table 1: Continued.

Refs	Specimen configuration				Materials type						w/b ratio	Rebar presence
	Type	Size (mm)	TNVM	Mix type	OPC	SLG	FA	MK	SF	Others		
[45]	Block	300 × 300 × 200	3	CON	×	×	—	—	—	PFA	0.39, 0.4, 0.44	Y
[46]	Slab, cylinder	610 × 610 × 152, Φ100 × 200, Φ150 × 300	10	CON	×	×	× (class F)	—	—	MS	0.43	Y
[47]	Prism	1000 × 1000 × 300, 150 × 270 × 150	NR	CON	×	—	×	—	—	—	0.35–0.65	Y
[48]	Cylinder	Φ100 × 200	NR	CON	× Type I/II	—	× (20%)	—	× (8%)	—	0.4	N
[49]	Cylinder	Φ100 × 200	10	CON	×	× (SC)	—	—	—	—	0.45, 0.65	Y
[50]	Prism	100 × 100 × 300	12	CON	×	×	×	—	—	—	0.4, 0.45, 0.5	Y
[51]	Cylinder, cube	Φ100 × 200, 100 × 100 × 100	12	CON	×	×	—	—	×	—	0.25, 0.28, 0.35	N
[52]	Cylinder	Φ100 × 200	12	CON	×	×	×	×	×	(Super fine fly ash)	0.28–0.49	N
[53]	Cylinder	Φ100 × 200	11	CON	×	×	×	—	—	—	0.37–0.45	N
[54]	Cylinder	Φ100 × 200	343	CON	× Type I/II	—	× (class C, 25%)	—	—	—	0.42	N
[55]	Cylinder	Φ100 × 200	514	CON	×	×	× (class F)	—	—	—	0.41	N
[56]	Cylinder	Φ100 × 200	57	CON	× Type I	—	—	×	×	RHA, NP	0.4–0.6	N
[57]	Cube, slab	100 × 100 × 100, 250 × 250 × 100	10	CON	×	×	—	(5%, 10%, 20%)	—	MS (5% and 10%) & PFA (30%)	0.52	N
[58]	Disk	Φ100 × 50	6	CON	×	× (30%)	—	—	× (10%)	—	0.25, 0.28, 0.35	N
[59]	Cylinder	Φ150 × 300	24	CON	×	× (50%)	—	—	8%	NP (12%, 25%)	0.28–0.6	N
[60]	Cylinder	Φ150 × 200	4	CON	×	—	—	—	—	—	0.4, 0.6	Y
[21]	Cylinder, prism	Φ102 × 178, 406 × 76 × 102	NR	MOR	×	—	—	—	—	WPC	0.42	N
[61]	Cylinder	Φ100 × 200	33	CON	× Type II–V	(grades 100 & 120)	(classes C & F)	×	×	—	0.44	N
[62]	Cylinder	Φ100 × 200, Φ150 × 300	23	CON	× Type II–V	× (grade 120)	(class F, 20%)	×	×	—	0.44	N
Authors	Hollow section, cylinder	Inner dia. 152 and external dia. 304, Φ100 × 200	18	CON	×	×	—	×	—	CA (2%), PLC	0.5	Y

TABLE 2: Details of the reinforcements and measurement methods used to record corrosion rate.

Refs	Reinforcement			Cause of corrosion	Corrosion rate		
	Φ (mm)	Length (mm)	Cover depth (mm)		Technique	Details	Correction for ohmic drop
[23]	10	200	1, 10, 20	Carbonation	—	—	—
[8]	10	300	50 & 75	Not studied (NS) (only effect of rebar presence on resistivity measurement was considered)	—	—	—
[24]	16	300	50	Cyclic ponding with sea water	—	—	—
[25]	4	110, 160, 200	Various (53.5–100)	NS	—	—	—
[26]	8	40	80	NS	—	—	—
[27]	13, 19, 25	410	20, 30, 40	NS	—	—	—
[28]	10	250	150	NaCl solution/marine exposure	LPR	Embedded steel rebar	N
[30]	10	—	25	NS	—	—	—
[45]	16	200	50	NaCl solution/marine exposure	—	—	—
[46]	NR	NR	NR	NaCl solution	LPR	NR	NR
[47]	NR	NR	70	NaCl solution	—	—	—
[49]	16	NR	42	NaCl solution	NR	Embedded steel rebar	NR
[50]	8	150	10 or 30	NaCl solution	LPR	Embedded CE & RE on the surface	N
[60]	10	200	15	NaCl solution	—	—	—
Authors	10	914	19–38	NaCl solution	LPR	Embedded steel rebar	Y

experimental programs. The mixture proportions and cement content varied and blended cements such as fly ash or slag cements were used in parts of studies. In a couple of studies, no detailed information was provided about the cement type. However, ASTM Type I and CEM I/II cements were used in most of the articles. Only one reported study used White Portland Cement (WPC) [34]. Also, Rice Husk Ash (RHA) as a cementitious supplementary material was only studied by Gastaldini et al. [63]. Work done by authors of this paper seems to be the only one that considered crystalline admixture as a healing agent to investigate its effect on electrical resistivity of concrete.

The specimens were cured and exposed to various and/or changing conditions over the testing period (Table 3). In most studies, samples were cured in the lime-saturated water tank with controlled temperature to eliminate the temperature effect on resistivity measurements. The temperature was kept constant between 20°C and 25°C in most experiments. To achieve a wide range of concrete resistivity, drier climates were considered occasionally. In most experiments, specimens were kept in a water tank during resistivity measurements or exposed to a high relative humidity (RH). For those studies focused on accelerating corrosion process, RH between 90% and 95% was chosen as an exposure regime. In parts of studies, samples were exposed to outside climates, in particular marine conditions (similar to authors' experimental setup). In general, laboratory experiments were undertaken over a period between 28 and 365 days. Only a few studies measured electrical resistivity for a period over one year [42, 50, 64, 65].

4.3. Measurement Methods. Either two-electrode or four-point electrode (Wenner probe setup) techniques were employed to record concrete electrical resistance, which is then converted into resistivity by multiplying it with an appropriate geometrical factor. The limitations of 2-electrode method resulted in using Wenner probe configurations in most studies specially for field investigations. In experimental studies that attempted to find correlation between concrete electrical resistivity and its durability parameters, other destructive and nondestructive testing techniques from standardized measuring protocol including Rapid Chloride Permeability (RCP) test, Rapid Chloride Migration (RCM) test, Bulk Diffusion (BD) test, and Ultrasonic Pulse Velocity (UPV) were employed. Authors ongoing work also employs

TABLE 3: Details of the curing conditions, exposure conditions, and measurement period.

Refs	Curing conditions	Temperature (°C)	Exposure conditions	Measurement period (days)
[4]	Lime-saturated water tank	20 (except one type mixture kept in water having 5, 20, 35 temperature)	Lime-saturated water tank/lab condition and oven dry state (only one type mixture)	90 (except ten of the mixtures tested at age of 365 days)
[16]	Lime-saturated water tank	23 ± 1	Lime-saturated water tank	28
[22]	NR	15, 26, 40	Plastic wrapped	1
[23]	NR	20	Laboratory (air dry)	1000
[8]	Water tank	20	Water tank (except two slabs were kept in air after 7 days)	30
[24]	NR	NR	Various	120
[25]	Lime-saturated water tank	23	Various	28
[26]	100% relative humidity in a chamber	20 ± 2	100% relative humidity in a chamber	28
[27]	Water tank	20	Water tank	45
[28]	NR	NR	Seashore Exposure/immersed in saline solution	1000
[29]	Plastic wrap	20	Room temperature	90
[30]	NR	NR	NR	28
[11]	Room temperature	25 ± 2	Room temperature	28
[31]	Lime-saturated water tank	Laboratory	Lime-saturated water tank	56
[32]	NR	10–45	Room condition with RH > 95%	2,190
[33]	Water tank	20	Water tank	181
[34]	Wet chamber with RH > 95%	23 ± 2	Wet chamber with RH > 95%	91
[35]	Water tank	23 ± 0.5, 105 ± 2	Various	28
[36]	Wet burlap	20	Oven dried and then water bath	31
[37]	Water tank	20 & 5	Water tank (after 7 days, some cylinders subjected to air condition)	30
[38]	Water tank	20 ± 2	Water bath	720
[39]	Lime-saturated water tank	23 ± 2	Lime-saturated water tank	91
[40]	Water tank	NR	Water tank	365
[41]	Various	21–45	Various	1,100–2,200
[42]	Lime-saturated water tank/wet burlap	NR	Lime-saturated water tank/wet burlap for 3 or 7 days	91
[43]	Lime-saturated water tank	Various (10–45)	Various	65
[44]	Various	Various	Various	500
[45]	Wrapped in damp Hessian and stored under polythene tentage	15–20	Maine exposure	140
[46]	Various	18–32	Salt ponding	90
[47]	Water tank and laboratory air	20 ± 2	Actual tidal zone, wet and dry cycle	NR
[48]	Five various curing regimes (tap water, NaCl solution, rog room)	Room	Various	1500
[49]	Various	20 ± 2/uncontrolled	Various	90

TABLE 3: Continued.

Refs	Curing conditions	Temperature (°C)	Exposure conditions	Measurement period (days)
[50]	Six days in fog room/three weeks in room condition	20	26 weekly cycles of 24 h 3% NaCl solution penetration and drying for 6 days/after 30 weeks, various conditions	52
[51]	Water tank	25	Water tank	90
[52]	Moist room with a sustained 100% humidity	23	NaCl solution	1,092
[53]	Lime-saturated water tank	22	Lime-saturated water tank	56
[54]	Changing exposure	23 ± 2	Changing exposure	90
[55]	Lime-saturated water tank	21 or 36	Various	1000
[56]	Lime-saturated water tank	23 ± 2	Lime-saturated water tank	28
[57]	Water tank for 14 days/14 days in drying cabinet at 40°C	20 ± 1 & 40 ± 1	Salt ponding (wet & dry cycle)	270
[58]	Lime-saturated water tank	25	Lime-saturated water tank	90
[59]	Chamber with RH > 95%	21	Chamber with RH > 95%	28
[60]	Laboratory environment (kept in plastic bag)	14–27, 3–13	Seashore condition	1,000
[21]	Water tank	21 ± 3	Water tank	7
[61]	Lime-saturated water tank	NR	Lime-saturated water tank	1
[62]	Lime-saturated water tank	NR	Lime-saturated water tank	730
Authors	14 days wet burlap and 14 days air dry	Uncontrolled	Natural environment and simulated seashore condition	720

use of both 2-electrode and 4-electrode techniques as well as UPV technique.

In summary, the experimental setup can have a significant effect on the electrical resistivity measurements. Specifically, the measurement methods and environmental conditions comprise a variety of parameters affecting the obtained data. The geometry of specimen and the general setup have a minor influence on the recorded resistivity values. To investigate electrical resistivity, the material, curing condition, and exposure condition should be carefully selected. Simulating the real-world conditions is in any case desirable since the recorded data can be used later as input in prediction models. As field survey data is rarely reported in the reviewed studies, it seems critical to identify possible deviations between laboratory investigations and field conditions. Also, authors' focus currently is to find these deviations to fill this knowledge gap by measuring concrete resistivity on both laboratory-size and field-size specimens.

5. Influencing Parameters on Electrical Resistivity Measurements

In the following sections, investigated parameters in published literature that influences the electrical resistivity readings have been discussed. For simplicity, they have been divided into two subgroups: (1) factors affecting the intrinsic electrical resistivity of concrete and (2) factors affecting the electrical resistivity measurements.

5.1. Intrinsic Factors Affecting Electrical Resistivity of Concrete

5.1.1. Effect of Water to Cement (w/c) Ratio. Generally, water to cement (w/c) ratio is one of the main factors contributing in permeability of concrete and its properties. Higher w/c ratio results in a high percentage of porosity (more voids) and leads to a lower electrical resistivity value indicating a more permeable concrete [31]. However, concrete containing supplementary cementitious materials such as slag showed an irregular behaviour for various w/c ratios [31]. For instance, an increase in w/c ratio from 0.35 to 0.65 caused an increase in electrical resistivity values, which means a less permeable concrete. Additionally, the electrical resistivity measurements are affected by the degree of hydration as further hydration typically reduces the concrete porosity and how these pores are interconnected [32]. The results from experimental study conducted by Van Noort et al. [33] on different concrete compositions with various w/c ratio ranges also indicated that concrete's electrical conductivity increased as w/c ratio increased. Within a hardened concrete matrix, electrical conduction flows through the fluid contained in the pores; therefore the relative volume of interconnected pores controls the concrete's electrical resistivity. Increasing the w/c ratio (at fixed cement content) leads to a higher volume fraction of hydrated cement paste in the concrete mix and results in higher concrete electrical conductivity. Similar tendency has been observed [34] even for concrete containing White Portland Cement. When w/c ratio was reduced, conductivity

TABLE 4: Details of the different measurement methods used in the literature.

Refs	Measurement technique: Concrete resistivity — Two-electrode	Concrete resistivity — Four-point method	Rapid chloride permeability test	Others
[4]	×	—	—	Rapid chloride migration test (NT Build 492) & ASTM C1760
[16]	×	—	—	—
[22]	×	—	—	—
[23]	×	×	—	Multiring electrodes
[8]	×	×	—	NT Build 492
[24]	—	×	—	—
[25]	—	×	—	—
[26]	×	×	—	—
[27]	—	×	—	—
[28]	—	×	—	Steel potential
[29]	×	×	—	—
[30]	—	×	—	Ultrasonic Pulse Velocity
[11]	—	×	—	Electric imaging
[31]	—	×	×	—
[32]	—	×	—	—
[33]	×	—	—	NT Build 492
[34]	—	×	—	—
[35]	—	×	—	—
[36]	×	—	—	—
[37]	×	×	—	NT Build 492
[38]	×	—	—	—
[39]	×	×	—	—
[40]	×	×	—	—
[41]	—	×	—	—
[42]	×	×	× (ASTM C1202)	—
[43]	×	×	—	—
[44]	—	×	—	—
[45]	×	—	—	—
[46]	×	×	× (AASHTO T227)	Half-cell potential
[47]	×	×	—	—
[48]	—	×	—	—
[49]	—	×	—	—
[50]	×	—	—	Steel corrosion potential
[51]	×	—	—	NT BUILD 492
[52]	—	×	—	ASTM C1556-04
[53]	×	×	× (ASTM C1202)	—
[54]	—	×	×	KDOT Boil Testing
[55]	—	×	—	Bulk diffusion test (NT Build 443), NT Build 492
[56]	—	×	×	Water Penetration Depth
[57]	—	—	—	Resistivity using disc method (one external electrode)
[58]	×	×	—	NT Build 492
[59]	—	×	—	Natural diffusion test (90 days)
[60]	—	×	—	Half-cell potential method

TABLE 4: Continued.

Refs	Measurement technique			
	Concrete resistivity		Rapid chloride permeability test	Others
	Two-electrode	Four-point method		
[21]	×	×	—	—
[61]	×	×	—	—
[62]	×	×	—	—
Authors	×	×	×	UPV, half-cell potential, infrared camera

of pore solution was increased due to the greater ionic concentration of the solution. Su et al. [35] studied the effect of moisture content on concrete resistivity measurements. It was found that the electrical resistivity difference for mixes with 0.55 and 0.65 w/c ratios was not noticeable, although it became significant for specimens with a w/c ratio of 0.45. It seems that both the capillary pore size effect and interconnectivity effect improve resistivity for saturated concrete with a higher w/c ratio. Long-term experimental study also showed the reduction in concrete resistivity with the increase of w/c ratio until approximately 500 days. However, after 500 days, the resistivity results revealed a contrary behaviour because the concrete specimens kept in unsaturated condition (in a laboratory environment) with higher w/c ratio favored the carbonation process that led to larger resistivity values for more advanced ages. Saleem et al. [36] also found similar trend for concrete samples contaminated with sulphate/chloride. About 15–20% reduction in electrical resistivity values was reported when the water/cement ratio increased from 0.4 to 0.6 [25].

5.1.2. Effect of Aggregate Size and Type. In general, aggregates depending on their location and size have a higher electrical resistivity compared to hardened cementitious paste because they have less porosity; thus electrical current can easily flow through the pore system of the paste. Hence, a number of researchers attempted to investigate aggregates' effect on electrical resistivity measurements. The experimental study performed by Sengul [4] indicated that increasing aggregate content resulted in higher electrical resistivity. He also observed that the resistivity of the mixture containing 60% aggregate with the size of 16–32 mm was approximately 3 times higher than that of the hardened cement paste [4]. Increase in aggregate content and reduction in cement paste for a given volume resulted in higher resistivity values because of replacing the porous hardened cement paste with denser aggregates. The investigation on comparing effect of two different aggregate sizes (0–4 mm and 16–32 mm) on electrical resistivity showed that larger aggregate size resulted in higher electrical resistivity values. Morris et al. [18] also reported that the variability was greater on the specimens with larger maximum aggregate size. Two possible causes of this variability originate from the tortuosity effect and formation of more interfacial transition zone (ITZ) (more porous structure compared to bulk cement paste) for smaller aggregate/particle size. Therefore, variation in aggregate content and size should be taken into account when comparing the resistivity values of different concretes.

As reported in the Sengul study [4], aggregate type also affected the electrical resistivity of concrete. For electrical resistivity measurements, comparison between the crushed limestone aggregate and gravel showed higher values when crushed limestone was used [4]. Gravel was rounded shaped aggregates with smooth surface whereas the limestone aggregates have rough surface texture. Therefore, using rounded aggregates such as gravel results in poor bonding between gravel and cement paste which may also be a reason behind the variations in resistivity readings. In addition, tortuosity can be higher for crushed stone aggregates due to the rough surface texture and irregular particle shape, which, in turn, may reduce the rate of electrical flow and affect resistivity [4]. Comparable standard deviation values were also observed when different aggregate type was used with the same maximum aggregate size [18]. Furthermore, using granite as coarse aggregate with fly ash also resulted in higher resistivity measurements than the mixture containing limestone aggregate type [44]. Hence, the effects of aggregate type should not be ignored during resistivity measurements.

5.1.3. Effect of Curing Conditions. The resistivity evolution of concrete is affected by the curing regimes [48]. Two key elements influence this variation in resistivity: the degree of hydration of the cementitious material and the degree of saturation of the specimen. The numerical study performed by Weiss et al. [70] attempted to simulate a mortar with a water to cement ratio of 0.42 with three curing conditions: (a) sealed during curing and testing, (b) sealed during curing and saturated during testing, and (c) saturated during curing and testing. It was concluded that the specimen that was sealed during both curing and testing had the highest resistivity whereas the sample that was sealed during curing and saturated at the time of testing had the lower most resistivity [70]. This difference can be explained by the saturation degree of the sample. The results recommend that storing a sample underwater in the lab may cause a remarkably different degree of hydration than what may occur in a field structure. The sample that was continually saturated and the sample that was sealed and saturated at the time of testing had a similar resistivity for the same degree of hydration; however, the continually saturated sample had a higher degree of hydration at the same age [70]. For specimens cured under saturated lime water, it has also been hypothesized that the volume of solution in which samples are stored can affect resistivity measurements due to possible pore solution concentration or dilution via leaching [43, 70]. According to AASHTO TP 95-11 [3], for the samples cured in a lime-water tank, the average resistivity value needs to be multiplied by 1.1 while this coefficient is 1 for the specimens stored in a 100% relative humidity moist room. For a given water-cement ratio, it was observed that better curing procedure yielded higher electrical resistivity [49]. For both wet and dry curing conditions experimented in the study [8], the resistivity distinctly increased with increasing age. Sample storage and curing conditions are important, as they can influence the degree of hydration, the degree of saturation, and the pore structure and solution through leaching [43]. Differences in resistivity can develop as a result of sample storage conditions and wetting the specimens prior to the resistivity measurements is recommended.

5.2. Factors Affecting the Electrical Resistivity Measurements

5.2.1. Presence of Rebar. A number of researchers have been exploring the effect of embedded rebar presence on concrete electrical resistivity through experimental and numerical investigation. Theoretically, electrical current fluxes take pathways having the least amount of resistivity and when there is embedded rebar in concrete, the current field is distorted. However, the alternation in current field is dependent on many factors such as orientation of rebar with respect to the probe, rebar diameter and spacing, and depth at which it is located [17, 23, 71].

Millard [71] and Gowers and Millard [17] utilized an experimental setup with steel rebar in the tanks filled with conductive medium solution and its finite element modeling in order to study the effects of concrete cover thickness as well as rebar diameter and spacing on concrete resistivity using four-point Wenner probe. According to this study, distance between the probe and embedded rebar was found to be the main influential parameter when measurements were taken on top of the bar. It was also reported that rebar diameter is not impactful in its disturbance. Moreover, it was found that measurement errors were increased by reducing rebar spacing while measurements are taken between two parallel rebars. However, it should be noted that results were obtained from measurements on the conductive solution tank and not from real concrete block. Similar study of resistivity measurements utilizing Wenner probe on concrete block with embedded steel reinforcement showed that orienting the probe perpendicular to reinforcements significantly reduced their influence on resistivity measurements [23]. It is more common in reinforced concrete structures that rebars are available in both directions and electrically linked together but, in the tested concrete block, no lateral rebars were present which then may have different effects on measured resistivity.

Practical general guidelines were developed by Polder's work [7] from summarizing literature for the RILEM TC 154 [14] technical recommendation for taking resistivity measurement on concrete. It was identified that placing all four electrodes over an embedded rebar at 10 or 20 mm depth can result in errors by a magnitude of two to six times that

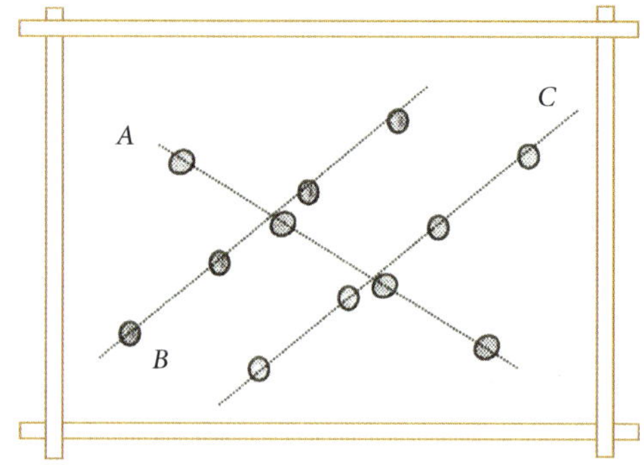

FIGURE 3: Resistivity using four electrodes at various spots in the same area to minimize influence of rebars [7].

of true resistivity and even if one of the four electrodes was near a rebar, results will lead to errors. Because of lack of research on resistivity measurements over rebar meshes, it was recommended that resistivity measurement with Wenner probe are taken in diagonal alignments on the concrete surface (Figure 3). Five measurements, each a few millimetres in distance from one another, and taking the median of them, are also suggested for collecting the resistivity value of the interested area. However, no recommendations were made for the case where it is not possible to fit all four electrodes inside the mesh unit created by the rebars. In addition, the recommended scheme in this study was not supported by any experimental and numerical works.

Another similar experimental investigation, done by Sengul and Gjorv [8], studied the effects of different parameters on concrete resistivity measurement using Wenner probe setup due to rebar presence. The study included the effective parameters: cover thickness, probe measurement directions relative to embedded rebar, electrode spacing, and probe measurement distance away from the embedded rebar. In total, five different probe positions with respect to the location of embedded steel reinforcement were considered, where four of these configurations were parallel to the rebar and the last one was perpendicular (Figure 4). Similar to Weydert and Gehlen's study [23], only a single rebar was positioned in the slab and two different thicknesses of 50 mm and 70 mm were studied. Their findings similar to the previous works stated that placing the probe orthogonally to the rebar did not influence the resistivity measurements, although significant errors were obtained once measurements were taken directly above and parallel to the rebar. It was also suggested that all measurements should be captured as far away as possible from embedded steel to reduce errors and if it is not possible due to dense reinforcement configuration, then space between electrodes should be kept relatively small. As only one rebar was placed in the tested concrete block as well as a small slab size was being used, this possibly contributed to errors due to edge effect. Hence, it may be difficult to extrapolate these conclusions to real cases.

Presuel-Moreno et al. [24, 72] have recently attempted to numerically and experimentally understand the influences of the number and configuration of embedded steel reinforcement along with the location and angle of the Wenner

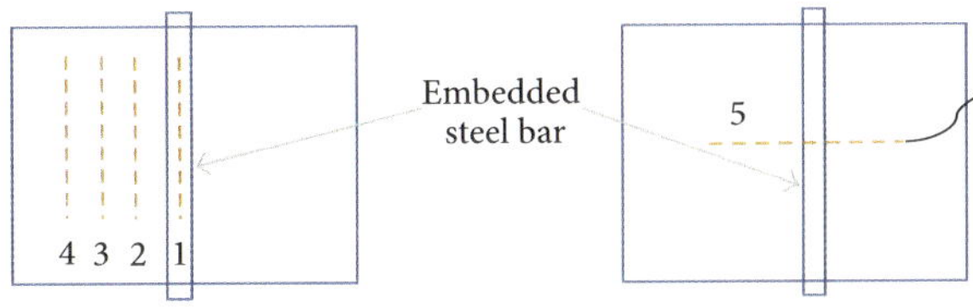

FIGURE 4: Five Wenner probe configurations with respect to embedded rebar tested by Sengul and Gjorv [8].

probe with respect to rebar alignment on concrete resistivity measurements. One of the few studies considered the effect of rebar mesh as well as orientation of the Wenner probe and demonstrated the difference between data achieved once there is a rebar mesh rather than a single rebar. Overall, six and five different orientations were investigated for the presence of a single rebar and rebar mesh, respectively. Like previous studies, it is also recommended to take the measurements perpendicular to the rebar location. However, the performance of Wenner probe due to variation in rebar spacing, cover thickness, or location and orientation of the probe with respect to the rebar mesh was not considered in this study.

Salehi et al. [9, 73] numerically characterized the effects of different concrete and slab thicknesses, rebar diameter, and probe arrangements with respect to the rebar mesh and rebar mesh densities on concrete electrical resistivity measurements with rebar presence using the four-point Wenner probe technique. It was concluded that the smallest error will result while setting up the probe parallel to the top rebar within the rebar mesh and perpendicular to bottommost rebar during measurements taken, as illustrated in Figure 5. It was also found that the observed resistivity decreased once the rebar mesh densities increased and the rebar diameter effect on concrete resistivity measurements can be neglected, although the numerical study results were not validated by experimental investigation.

For cylindrical concrete specimens with a single embedded steel rebar, study conducted by Chen et al. [25] suggests a correction factor to be applied to resistivity measurements corresponding to the ratio of specimen length to electrode spacing as well as the ratio of specimen diameter to electrode spacing. It was stated that no correction factor for prismatic specimens was necessary with the possibility that the applied current did not pass through the reinforcement. This research also lacked a discussion on the use of a multiple rebar and consideration of larger concrete specimens.

The effect of rebar presence on mortar electrical resistivity conducted by the four-point Wenner method was also investigated numerically and experimentally by Garzon et al. [26] and Lim et al. [27]. In Garzon et al.'s experimental study, small scaled cylindrical and prismatic specimens were casted. As polarization will happen due to double layer at the steel and concrete interface acting as a resistance-capacitor, resistivity measurements taken directly above rebar will result in errors. Hence, a rebar factor was suggested to be applied to the obtained resistivity results. In addition, modified Wenner equations are recommended for various geometric parameters [26]. Only reinforced cylindrical and

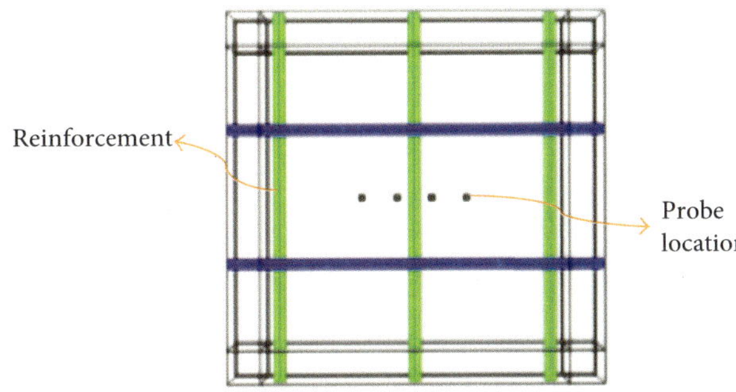

FIGURE 5: Probe configuration with respect to rebar mesh suggested to reduce electrical resistivity measurement error [9].

prismatic specimens were included in the experimental setup without considering a reinforced slab. However, in a numerical study, a slab with embedded rebar was considered [26]. The experimental investigation lacked in using concrete mixture instead of a mortar mixture which is not exactly representative of real-world cases and may lead to more errors. Furthermore, the proposed rebar factor may not be applicable to a large concrete slab with multiple rebar because their experimental conclusions are based on laboratory testing. Lim et al. [27] also studied the effects of cover depth, electrode spacing, rebar diameter, and the resistivity of concrete and reinforcement in the numerical model. However, only one probe configuration taken right above and parallel to rebar was considered. It was suggested to apply a geometric effect rate that ranges from 0 to 1 in order to estimate the reinforcement geometry impact and this rate is derived utilizing a resistivity estimation model. The geometric effect rate was also validated through the experimental investigation for on-site measurements. Based on experimental findings, it was stated that the geometric effect rate decreased with increasing concrete cover thickness and increased with increasing rebar diameter and increasing electrode spacing. Again, using mortar mixture, only one single rebar, and a single probe configuration with respect to rebar is not completely representative of real-world conditions. An error to resistivity measurements may also be introduced while the epoxy coating on the mortar specimens was used in this study due to a barrier between the electrodes and mortar surface.

The last and recent study in this category belongs to Sanchez et al. [74] who numerically proposed a modified 4-point Wenner method based on the experimental data, deployed on a bridge over the River Danube in Romania.

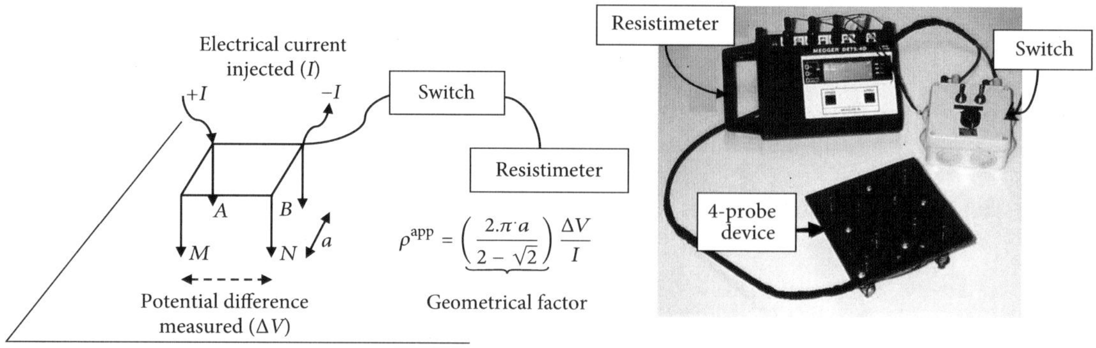

FIGURE 6: Four-probe square array principle [10].

In this study, a "rebar factor," f_b, was introduced through the modified version of Wenner method to measure the resistivity in reinforced concrete structures with thin concrete covers. Effect of rebar presence on concrete electrical resistivity measurements also can be found in detail in similar studies by [17, 28, 29, 65, 75].

5.2.2. Presence of Cracks in the Concrete Cover. Due to the presence of cracks, apart from embedded rebar, the electrical resistivity measurements may vary using Wenner probe technique because it is initially assumed that concrete is homogeneous and isotropic with semi-infinite geometry. In this section, some researchers' investigations in order to characterize cracks in concrete conducted by electrical resistivity measurements are summarized.

Lataste et al. [10] attempt to identify and locate cracks and spalling in concrete by means of electrical resistivity. For this a utilized instrument was built to measure the electrical resistivity through the use of four electrodes in a square configuration, rather than linearly conventional Wenner probe arrangement (Figure 6). The specified built probe allows taking the measurements in two orthogonal directions without having to rotate the probe between measurements. To change the function of electrodes from current imposing to potential measurement, the use of an electrical switch was considered. Both on-site measurements on a reinforced concrete slab and laboratory measurements on a cracked reinforced concrete beam were experimentally studied. To observe the effects of crack characteristics, such as crack opening and bridging degree between crack lips and depth of crack, a numerical model was developed as well and validated experimentally. It was reported that when a conductive crack was present, depending on the direction of imposed electrical current, resistivity readings could overestimate or underestimate the true resistivity. Once current was orthogonally imposed to the crack, no impact was detected; however, reduction in electrical resistivity measurements was observed while the crack was parallel to the imposed electrical current. For an insulated crack, perpendicular readings overestimated resistivity, and parallel measurements underestimated it. It was also concluded that crack depth has a direct relation to the electrical resistivity measurements (increase in crack depth leads to increase in the resistivity measurements).

A couple of assumptions and conclusions in Lataste et al.'s work [10] may not be adaptable with reality. First, it was assumed that the rebar effect on the resistivity measurements is independent of crack depth or type which might not be true for the cases that rebar mesh or conductive crack present. Second, simulations on limited size concrete block can possibly exaggerate the crack impact. Finally, the four-electrode square configuration, which is a less common electrode setup compared to the Wenner probe array, may generate different measurement errors.

Goueygou et al. [30] used the same square probe configuration proposed in Lataste et al.'s work [10] to compare electrical resistivity measurements with transmission of ultrasonic waves for characterizing, detecting, and localizing the surface cracks. For taking measurements in both directions (parallel and perpendicular), concrete beam specimens bent via three-point loading to induce one main crack were constructed. Both nondestructive techniques were capable of identifying the main simple crack inside the concrete specimens; however, complexity rose when the number of cracks increased and in most cases became impossible to detect cracks depth and patterns. Similarly, experimental and numerical investigations in detecting and characterizing cracks using electrical resistivity measurements with a square probe were done by Shah and Ribakov [76]. In their study, two experimental setups including a set of five cubic laboratory concrete specimens and a small area on a 40-year-old reinforced concrete slab as well as numerical model to identify the crack depth and differentiating insulated and conductive cracks were involved. Overall, it was observed that higher resistivity values were obtained from insulative cracks while lower values resulted from conductive ones. From field data and numerical work, variations on electrical resistivity measurements were observed for different crack depth and opening [76].

Experimental study conducted by Wiwattanachang and Giao [11] also investigated the capability of concrete electrical resistivity measurements with Wenner probe method in detecting a crack development in the concrete beams. Artificial cracks made up of plastic sheet inside a concrete beam to simulate insulated cracks as well as cracks being induced in a beam by a four-step loading test on its tension face were studied in this work. After correcting resistivity

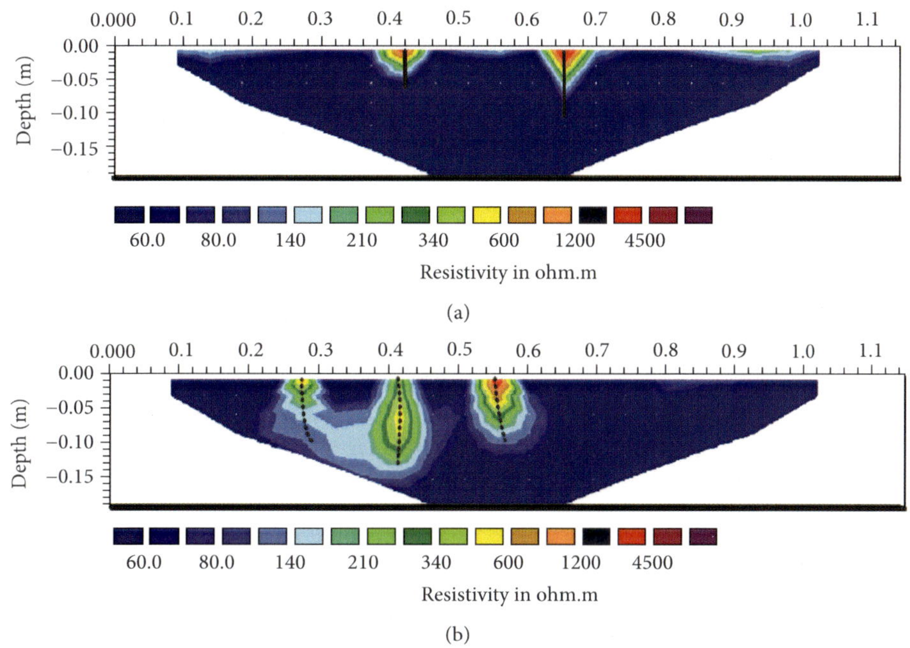

FIGURE 7: Electrical resistivity image of a concrete beam with cracks [11]: (a) concrete beam with artificial plastic sheets as crack; (b) concrete with cracks being developed from a four-point loading.

readings for the specimens' geometry, it was concluded that obtained data were increased for both crack types; although only insulated cracks were examined, conductive cracks were not. Simulated images of electrical resistivity were plotted from obtained results, as demonstrated in Figure 7. Although using electrical resistivity measurements for detecting cracks inside concrete was explored in the study, investigations on different type of cracks and their orientations toward the probe were ignored.

It was also numerically found that when cracking and delamination were present in reinforced concrete structures, electrical resistivity measurements were different from when they were not present [77]. Similar to previous studies, this finding gave rise to the conclusion that delamination at early stages can be detected using resistivity measurements with 21 linearly aligned electrodes instead of four-point Wenner method. However, proposed model in Chouteau and Beaulieu [77] only identified the effect on resistivity measurements as being different when cracking was present. In addition, to evaluate and detect cracks and discontinuities, such as joints, in massive concrete structures within preexisting boreholes, electrical resistivity measurements with a DC current was stated to be a good quality assessment indicator [78]; however, using a DC current rather than AC current may result in unfavorable results due to polarization effect which was not considered in Taillet et al. [78].

A comprehensive numerical study by Salehi et al. [79] on the effect of different cracks types, depths, and widths incorporating both the presence of cracks and rebar mesh indicated that measurements on conductive cracks result in lower electrical resistivity values. For conductive cracks, numerical results showed that decreasing crack depth did not significantly disturb the electrical resistivity measurements. Furthermore, for an insulated crack between two inner electrodes, electrical resistivity readings led to a maximum error of about 200% higher than actual concrete resistivity. Conductive crack in here represents as a crack filled with water and insulated one denotes a crack without bridging and filled with air. It was also concluded that once the crack depth decreased, lower errors were observed. Also, the rebar and crack were found to act independently of one another while rebar mesh was present. Salehi et al.'s work lacked validating numerical investigation with an experimental study. Following this study, Morales [75] experimentally investigated the effect of rebar, chloride ingress, corrosion, and various crack types on concrete electrical resistivity measurements. For all moisture conditions, it was suggested that discrete cracks of all depth and conductivity properties should be avoided in order to minimize potential errors when performing resistivity measurements. Although it was observed that electrical resistivity measurements were not significantly affected by surficial microcracking, it may be able to identify delamination as the difference observed between measured resistivity of in-tact and delaminated concrete covers. Still, investigations on the effects of cracking induced by corrosion and insulated and conductive cracks with bridging conducting electrical resistivity measurements lacked in this study.

5.2.3. Probe Spacing. Concrete is considered to be a heterogenous material and this is one of the assumptions behind the Wenner probe method. However, this assumption appears to be a likely source of error since aggregates inside concrete typically have greater resistivity and they propagated widely in different locations with various sizes. Hence, this inconsistency in the initial assumption of concrete homogeneity may affect resistivity measurements. To mitigate this issue, some researchers recommended considering enough wide space between electrodes (usually between 20 mm and 70 mm) in order to reduce the influence of nonhomogeneity due to the aggregates presence [8, 18, 71]. It was suggested to take several readings at various locations and then these measurements should be averaged. Many commercial instruments are also equipped with a variable probe spacing as well to allow the device to measure concrete resistivity involving larger aggregate size.

One recommendation to help reduce variance in resistivity measurements is to consider probe spacing 1.5 times higher than the maximum aggregate size [17]. It was observed that when probe spacing became smaller than the maximum aggregate size, standard deviation in the measurements increased to around 10% (Figure 8) [17]. For various probe spacing (16 mm, 25 mm, and 50 mm), Millard [71] experimentally found that as the maximum aggregate size became closer to the probe spacing, the scattering in the observed results increased. Therefore, to compensate for the local effect of aggregates, larger electrode spacing should be considered for practical purposes. For concrete cubes (100 × 100 × 100 mm), while the electrode spacing was changed in different steps from 20 to 35 mm, the relative resistivity measurements increased by approximately 70% [37]. Increasing electrode spacing also resulted in increasing resistivity values to even a greater extent than that in the cubes for concrete cylinders [37]. Increase in resistivity observed due to wider spacing is in part also due to finite geometry and not just the aggregate size. The results of the resistivity measurements preformed on 28-day water-cured concrete slabs with and without embedded steel rebar indicated only a small difference for probe spacing less than 30 mm [8]. For larger electrode spacing, however, both the steel rebar and the probe spacing showed significant impact on the electrical resistivity measurements, and the larger the electrode spacing, the larger the effect of the steel rebar. For instance, increase in electrode spacing from 20 to 70 mm led to increase in resistivity by approximately 26% for the slab without any steel reinforcement whereas the resistivity values either increased by 33% or decreased by 25% depending on the orientation of taken measurements (perpendicular or parallel) for the slab with rebar [8].

According to Polder [7], the electrical current may travel through the concrete volume with approximately the same depth as that of the electrode spacing. Hence, as the probe spacing increases, the current flows deeper inside the concrete volume and when the electrical current reaches the surface of the rebar, the current is transported through the reinforcement and, thus, results in lower resistivity observation [8]. For prismatic specimens, it was also suggested by Chen et al. [25] that the effects by the probe spacing can be ignored when the spacing is larger than 40 mm; however, the resistivity values increased with less electrode spacing. For application of Wenner probe method, the important role of electrode spacing should be definitely considered during electrical resistivity measurements as it will affect the obtained results.

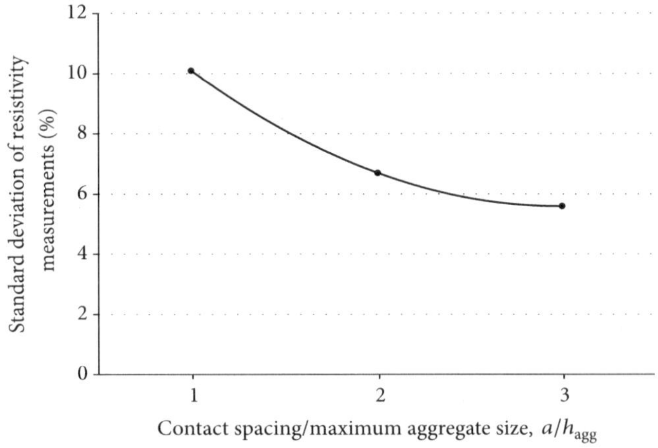

FIGURE 8: Effect of contact spacing on resistivity measurement [17].

5.2.4. Electrode Contact. The contact area between the electrodes and concrete surface may alter the electrical resistivity measurements using four-point Wenner probe. The experimental investigation in electrolytic tanks and finite element modeling resulted in the fact that the contact between the concrete surface and the probes has no significant influence on Wenner probe resistivity measurements [17, 71]. It was also reported a maximum error of 6% in resistivity (without stating lower or higher readings) when the diameter of electrode contact area varied from 1 mm to 40 mm [71]. According to Gowers and Millard work [17], misleading values can be lessened by use of a fairly low frequency and alternating current (AC). Practically, electrode contact area becomes vital for both external current imposing and internal potential measurement electrodes. It was observed by Ewins [80] that contact resistance between electrodes and concrete surface can lead to significant increase in electrical resistivity values. Both symmetry of the system and the probe performance as it was also confirmed in Gowers and Millard's work [17] may be affected by the contact resistance [80].

In two-electrode method (bulk resistivity measurements), poor contact between the plate electrode and the test cylinder surface is mainly responsible for electrode resistance. One solution to minimize the contact resistance effect is to use flexible electrodes [18]. Also, to mitigate this issue in laboratory testing, using an aid that allows for a good electrical contact such as an electrically conductive jelly was recommended [81, 82]. Other alternative solutions included the use of soft conductive medium, saturated sponges, chamois cloth, and paper towels [38, 83]. By using a saturated sponge, only an average of 2% difference in resistance was reported for contact resistance between the sponge and concrete surface [39]. The sponge resistance is largely dependent on the moisture content of the sponges and the conductivity of the solution in which they are saturated. It was shown that, in the two-point measurement procedure, a sponge contacting system can give a higher resistivity than those obtained using four-point techniques, and careful consideration must be given to the electrode-sample contacting system when trying to evaluate concrete resistivity [40]. According to McCarter et al.'s work [40], the sponge contacting method introduced a misleading resistance originating from the sponge-sample interface that was in series with the bulk resistance of the sample. Therefore, an operator using an electrical resistivity device needs to ensure proper contact between the electrodes and concrete surface as poor contact may affect the electrical resistivity readings. The influence of electrode contact is less governing in the Wenner probe method than in the uniaxial method and, hence, measurements can be performed in a wider frequency range (10 Hz to 10 kHz) [6].

5.2.5. Specimen Geometry. In the four-point Wenner method, the electrical resistivity measurements are initially presumed to be performed on the domain of semi-infinite medium which is not a practically accurate assumption. This assumption leads to deviation from the ideal condition of having infinitely large geometry which can possibly occur in different electrode orientations. For relatively small size concrete elements (e.g., cylinder or prism specimens), constriction of current to flow into a different field pattern is one of the major reasons for this deviation. Even though several researchers have realized the effect of specimens' geometry, only very limited information is available on this topic.

To account for interference between current flow and coarse aggregates in a small size sample, a suggested correction coefficient factor (or geometry correction factor) has been established in Spragg et al.'s work [84], using the simulations developed by Morris et al. [18]. The correction coefficient factor proposed formula is outlined below (see (5)) and only needs to be used when $d/a \leq 6$ or $L/a \leq 6$ (where a and d are parameters related to electrode spacing and specimen diameter, resp.) [84]:

$$k = 1.10 - \frac{0.730}{d/a} + \frac{7.34}{(d/a)^2}. \quad (5)$$

For a standard cylinder size (φ100 mm × 200 mm), the correction coefficient value ranges from 1.8 to 1.9 while using the probe spacing of 38 mm [41, 84]. Also, for the correction factor (k), Morris et al. [18] plotted a graph based on the finite element modeling data to study a wide range of geometrical variations in concrete cylinders (Figure 9). However, this study provides wide-ranging values for correction factor; different electrode spacing and configuration as well as various geometrical concrete element types were not investigated (only 25.4 mm electrode spacing for concrete cylinders was considered).

According to Millard [71] and Gowers and Millard [17] through experimental findings, electrode spacing should be 1/4 times smaller than concrete section dimensions and half the distance of the contact area from any element edge due to the three-dimensional current flow restriction closer to the edge. It was also observed that when the domain of the medium becomes smaller than the ideal semi-infinite condition, overestimated resistivity values resulted. Study performed by Bryant et al. [42] found an average of 24% higher electrical resistivity values for cylindrical samples in comparison with concrete slab specimens for various ages even when the geometry correction factor was used; however, Shahroodi [85] reported an average 25.8% lower resistivity

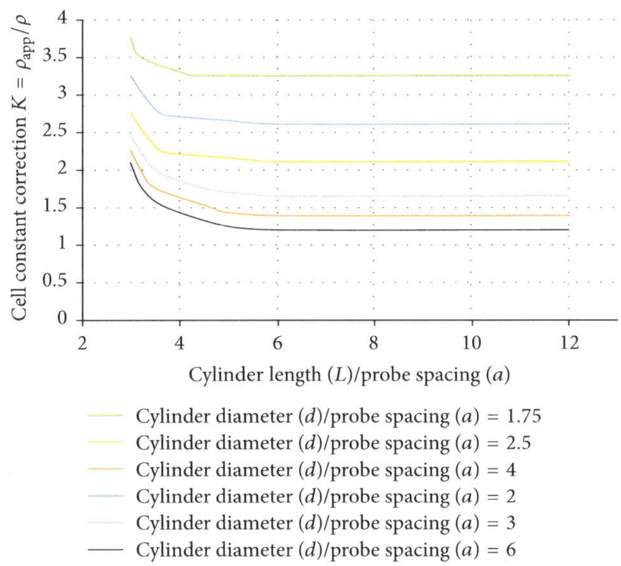

FIGURE 9: Cell constant correction to determine the concrete resistivity [18].

value. The differences in the mentioned values could possibly originate from variation in geometry correction coefficient, surface texture, device, operator, material, production, and curing process [86]. Measured resistivity values can also change by the geometry of the measuring plane. The variation in resistivity values observed experimentally by Chen et al. [25] on the curved surface and the cutting flat surface showed an increase with the volume of cutting portion. Therefore, correction factor should be applied accordingly, especially for large size specimen. They also confirmed that the resistivity of the cylindrical samples varied with the specimens' size even though the electrode spacing remains the same [25]. Furthermore, for the two-electrode method, since electrical current passes through the entire specimen volume, this measurement method is independent of specimen geometry while, for the four-point method, the depth of the electrical current passing through the concrete volume is related to both the geometry of the sample and the distance between the electrodes. This independency was also reported in Sengul and Gjorv's work [37] when the electrical resistivity values using both two- and four-electrode methods for concrete block of size 300 mm × 300 mm × 200 mm (semi-infinite geometry) and standard cylinder samples were almost the same.

5.2.6. Electrical Signal Shape and Frequency. Due to resistor-capacitor circuits' behaviour in saturated cementitious system, it introduces a phase difference between electrical current imposing and the measured potential (impedance) [43]. At different frequencies, there is a significant difference in impedance and it follows that the real component of the impedance at zero phase angle is the true uniaxial resistance. Since the phase is almost never zero, the meters record the total impedance: the real and imaginary components added in quadrature. The impedance spectrum consisting of

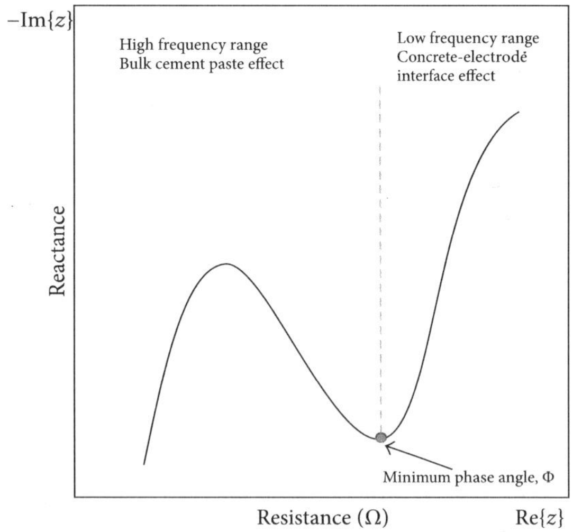

FIGURE 10: Schematic representation of the AC Impedance response of concrete [6].

two arcs in high and low frequency ranges is illustrated in Figure 10 [6]. At higher frequencies, the impedance spectrum characteristics are featured by the concrete microstructure but these characteristics are mainly attributed to the conditions at the electrode-concrete interface for low frequency domain. Generally, the frequency range in between 0.5 and 10 kHz is used for uniaxial electrical resistivity measurements method; however, no general statement can be found on the optimum frequency since it varies with moisture contents and mixture proportions. At lower frequencies (below 500 Hz), it was stated that electrical resistivity measurement can result in overestimated data because of electrode-concrete contact interface impact [6]. As it is described in Section 5.2.4, this effect is less dominant in the four-point probe method, and a wide-ranging frequency (10 Hz to 10 kHz) can be conducted [6].

Two signal shapes used usually for electrical resistivity measurements are sine-wave and square-wave AC current. Ewins [80] reported 800% error in results when high frequency (600 Hz) sinusoidal signals with poor contact were used for taking the resistivity measurements. Presence of parasitic capacitors and uneven contact resistance were the two major error sources that affected the accuracy of the measurements. However, impact from parasitic capacitors was larger. To avoid unwanted effects, using low frequency square-wave signal, instead of sine-wave, was recommended in Ewins work [80]. Similarly, it was found that sine-wave Wenner probe with 300 Hz frequency was adversely affected by the addition of external resistors compared to the case for which square-wave Wenner probe was used or frequency of the sine-wave signal was reduced [71]. The external resistors were used to mimic the unbalanced contact area resistance. Having unbalanced contact resistor in the voltage measurement circuit has the most significant effect on the accuracy of four-point probe method while high frequency sinusoidal AC signal is being used.

5.2.7. Effect of Temperature. Temperature variation has been reported to have a significant influence on electrical resistivity of concrete, and an increase in temperature results in a decrease in resistivity. Electrical current generally flows through the ions dissolved in the pore solution and can be affected by temperature which causes changing the ion (Na^+, K^+, Ca^{2+}, $SO_4^{\ 2-}$, and OH^-) mobility and ion-ion and ion-solid interactions, as well as the ion concentration in porous media [7, 87, 88]. It was reported that the effect of temperature on bulk pore solution resistivity was noticeably different from that of cement paste or mortar with the same ion concentration, which was most likely because of strong ion-solid interactions in cement paste or mortar and less internal ionic frictions [17]. Increase in temperature has an impact on concrete isotherm curves and also tends to reduce the moisture content of concrete which indirectly affects the resistivity readings [89]. So, temperature plays an important role in concrete electrical resistivity and a number of researchers attempted to correlate its effect with true concrete resistivity values. It should be noted that not all studies performed the resistivity measurements for the different temperatures at the same moisture content. If not saturated (i.e., immersed), the concrete needs to be left for a long enough time at a given temperature and moisture. Castellote et al. [87] as well as Liu and Presuel-Moreno [41] have attempted to produce equations to convert from one temperature to a normalized temperature.

Studies conducted by Millard et al. [90] and Gowers and Millard [17] found that decrease in ambient temperature led to increase in resistivity values and suggested a correction factor of 0.33 KΩ-cm/°C in order to compensate for the variation in temperature; however, the range of temperatures used in their studies was limited. For simplicity, a change of 3% to 5% on resistivity measurements per °C has been recommended for temperature range of 0°C to 40°C [7, 67]. A linear relationship has been developed on temperature dependency of resistivity [45, 88, 91]:

$$\rho = \rho_0 (1 + \alpha \cdot \Delta T), \quad (6)$$

where ρ is resistivity at temperature T (°C), ρ_0 is resistivity at reference temperature T_0, ΔT is temperature difference between T and T_0 and α is temperature coefficient (between 0.022 and 0.035/°C).

It was found that the above equation is only applicable over a limited interval of ±5°C to the reference temperature [45, 88, 91]. A model developed in DuraCrete [92] also describes the correlation between ρ_T at temperature T (°C) and resistivity ρ_{20} at 20°C:

$$\rho_T = K_T \cdot \rho_{20}, \quad (7)$$

where K_T is characteristic value of the temperature factor for resistivity and K_T is defined as

$$K_T = \frac{1}{1 + K(T - 20)}, \quad (8)$$

where $K = 0.025$ when $T < 20$°C and $K = 0.073$ when $T > 20$°C.

Several researchers also defined a wide agreement for this correlation using Arrhenius law [41, 45, 87, 88, 91, 93]:

$$\rho_T = \rho_0 \cdot \exp\left[\frac{E_{a,\rho}}{R}\left(\frac{1}{T} - \frac{1}{T_0}\right)\right], \quad (9)$$

where ρ is resistivity at temperature T (K), ρ_0 is resistivity at reference temperature T_0 (K), R is gas constant (8.314 kJ^{-1}mol^{-1}), and $E_{a,\rho}$ is activation energy for resistivity (J/mol) (ranges from 16.9 J/mol to 42.77 J/mol).

The activation energy of conduction ($E_{a,\rho}$) can be determined using the slope of a plot of the natural logarithm of resistivity and the inverse of temperature [43]. The slope of the best-fit line is multiplied by the negative of the universal gas constant [8.314 kJ^{-1}mol^{-1}] to determine the activation energy of conduction. The above-mentioned correlation is published using only standard cylinder samples (φ100 × 200 mm) and additional work is still needed for real-world resistivity measurements and also to consider the influence generated by moisture content as an additional factor. For standard curing conditions in water at 20°C, it was suggested that the effect of temperature can be eliminated [4]. Spragg et al. [43] stated that the resistivity measured using the same mature sample can differ by as much as 80% when the temperature of the sample fluctuates between 10°C and 45°C. Additionally, several other studies showed in detail a prominent effect on resistivity results due to temperature variation [46, 47, 51, 52]. In summary, when the temperature increases, electrons move faster causing higher electrical conductivity, thus lowering resistivity.

5.2.8. Effect of Moisture Content. Moisture content is one of the influencing factors that can inversely affect the concrete electrical resistivity measurements. Essentially, electrical conductivity increases with an increase in moisture content due to change in the ion mobility [41, 94]; however, there could be a difference in resistivity between immersed samples (either water or lime water) and those in high humidity. Larsen et al. [94] found that when moisture degree decreases from 88% to 77%, the resistivity increases by an average of two times and from 88% to 66% it increases by an average of 6 times. In the air-dry state, it was also reported that concrete had approximately 50% higher resistivity than that in saturated condition [4]. As the electrical current is carried by ion flowing through the pore solution in concrete, higher moisture content leads to easier electrical flow and thus the electrical resistivity reduced. However, at low resistivity values, moisture content has less effect on the accuracy of the resistivity measurements [46]. For quality control purposes, it is essential to ensure that the same moisture degree in different mixtures and also the resistivity measurements of the specimens with low moisture content are inappropriate [25]. Furthermore, it was experimentally found that poor surface saturation using pressurized water and static ponding can lead to misinterpretation of resistivity of over 30% compared to full laboratory saturation [86]. So, additional research is still required to simulate the on-site saturation by different systems and to understand how much time is required for water to penetrate through concrete surface in order to obtain constant moisture level through the bulk specimen.

6. Correlation between Concrete Resistivity and Its Durability Characteristics

Through following sections, relationship between electrical resistivity and the two main concrete durability characteristics including chloride ingress and corrosion of embedded reinforcements will be presented.

6.1. Correlation between Electrical Resistivity and Chloride Diffusivity. Diffusivity is the controlling parameter which determines the time it takes for chloride ions to diffuse into concrete and reach the critical chloride threshold for corrosion initiation. Typically, this can be measured through Rapid Chloride Migration (RCM) test, Rapid Chloride Permeability Test (RCPT), or Bulk Diffusion (BD) method [95]. However, such test methods are either time-consuming or expensive for evaluating the concrete permeability properties. Hence, electrical resistivity was found experimentally and theoretically to be an efficient method that can be applied indirectly to determine concrete chloride permeability.

In theory, the relationship between diffusivity of ion species i and its partial conductivity can be described by Nernst-Einstein equation [96]:

$$D_i = \frac{RT\sigma_i}{Z_i^2 F^2 C_i}, \quad (10)$$

where D_i is the diffusivity of ion i (m^2/s); σ_i is the partial conductivity of ion i (s/m); R is the gas constant (8,314 J/mol); T is absolute temperature (K); Z_i is the charge of ion i; F is Faraday's constant (96500 Coulombs/mole); and C_i is the concentration of ion i (mol/m^2). By applying Archie's law, the correlation between the bulk resistivity, pore solution resistivity, and porosity can be expressed by the following [37, 97]:

$$F = \frac{\rho}{\rho_0} = \frac{\sigma_0}{\sigma} = a \cdot \varphi^{-m}, \quad (11)$$

where F is formation factor; a and m are constants; and φ is porosity of concrete. m is dependent on tortuosity of concrete and has been found to be 1.5 to 3.2 [37, 97]. Combining the above equations gives the formation factor as a function of effective chloride diffusion coefficient of concrete (D_{eff}) and chloride ion diffusion coefficient in the pore solution (D_0):

$$F = \frac{D_0}{D_{\text{eff}}}. \quad (12)$$

Several researchers have also conducted various experiments to investigate the relationship between concrete resistivity and chloride diffusivity. A strong correlation between these two parameters has been reported for various concrete mixtures at different ages [4, 42, 47, 50–52]. Due to presence of chloride, experimental results showed that the electrical resistivity of mortar was reduced [17]. The concrete resistivity was inversely related to chloride ingress, where lower

Table 5: Classification of concrete permeability to surface resistivity values.

Chloride ion permeability	RCP test charged passed (Coulombs)	RCP test versus surface resistivity Surface Resistivity Test 4 × 8 cylinder (kΩ-cm) $a = 1.5$ $k = 1.8$ (measured)	6 × 12 cylinder (kΩ-cm) $a = 1.5$ $k = 1.8$ (measured)	Semi-infinite slab (real)	KDOT 4 × 8 cylinder (kΩ-cm) $a = 1.5$ $k = 1.8$ (measured)
High	>4000	<12	<9.5	<6.7	<7.0
Moderate	2000–4000	12–21	9.5–16.5	6.7–11.7	7.0–13.0
Low	1000–2000	21–37	16.5–29	11.7–20.6	13.0–24.3
Very low	100–1000	37–254	29–199	20.6–141.1	24.3–191
Negligible	<100	>254	>199	>141.1	>191

resistivity indicated the area where chloride penetration will be faster [7]; however, its effect on electrical resistivity readings is relatively small compared to carbonation process [7, 50]. Recent field experience indicated that the relationship between surface resistivity at 28 days and 56 days with RCP data was moderate [53]. Experimental investigation by Florida Department of Transportation (FDOT) on wet cured concrete specimens in a controlled environment or cured samples in lime water, using RCP test and 4-point Wenner method in conjunction, confirmed the inverse correlation of chloride resistivity [98, 99]. The summary of FDOT proposed relationship between resistivity and chloride can be found in Table 5. The table also shows an indication of the risk of chloride penetrability with regard to charge passed based on ASTM C1202 standard [100]. In RCP test, a 60 V potential is applied across the sample of 95 mm diameter and 50 mm thickness for 6 hours. Current readings are taken every 30 minutes, at minimum, and then the total charge passed in Coulombs is calculated by integrating the Current versus Time graph developed from the readings taken.

Similar experimental investigation conducted by the Louisiana Transportation Research Center (LTRC) also found a good correlation between resistivity and RCP test while the classified concrete permeability to surface resistivity values are equal to those proposed by FDOT as shown in Table 5 [31]. Following these experimental investigations, AASHTO published a provisional method TP 95-11 for indicating the concrete's ability to resist chloride ion penetration using surface resistivity measurement method [3]. However, Kansas Department of Transportation (KDOT) reported different value range for surface resistivity [54]. For instance, it is considered that chloride penetration is high when surface resistivity value is less than 7.0 kΩ-cm for 4 × 8 cylinder with 1.5-inch probe spacing which is a relatively small value compared to two other experimental results (Table 5) [54].

In addition to the experimental study on laboratory specimens, the correlation between electrical resistivity and apparent diffusivity coefficients (D_{app}) is reported by researchers for field specimens [53, 55]. Due to complexity of the exposure environments such as humidity difference, temperature, and the elevation from water level, results are scattered. However, a recent investigation on Florida bridges showed a better correlation between resistivity and apparent diffusivity once resistivity was measured under saturated condition [65]. The authors of this paper are currently involved in a study where the effect of full or partial saturation is being studied. To predict chloride threshold (Cl_{TH}) values, the electrical resistivity method was used periodically during 1000-day monitoring of reinforced concrete element exposure to marine environment [28]. After applying the geometrical factor, the relationship between chloride content, in terms of chlorides relative to the weight of cement, and resistivity (ρ) can be expressed as shown in the following [28]:

$$Cl_{TH}\,(\%) = 0.019\rho + 0.401. \tag{13}$$

A good correlation between RCM coefficients and electrical resistivity measured by two-electrode method was reported in European Union-Brite EuRam III experimental investigation [92]. Similarly, this correlation for resistivity measured by Wenner method is reported in Dhir et al.'s work [101]. Although resistivity measurements can be possibly a good alternative or replacement for the RCM test, more experiments need to be conducted to validate this correlation. The electrical resistivity test was suggested to be used as a quality control predictor of the chloride ingress resistance, but not as a predictor of diffusion behaviour of all kinds of concretes or as replacement of the long-term diffusion tests [56]. FDOT long-term monitoring of various concretes concluded that correlation between the RCM and the long-term tests was equal to or slightly better than those obtained by the RCP and surface resistivity due to less sensitivity of RCM test to presence of supplementary cementitious materials [102]. A linear trend exists between chloride diffusion coefficient (D_{RCM}) derived by RCM test and concrete electrical conductivity using two-electrode resistivity measurements [33, 46, 57, 58]. Additionally, to account for the retardation of chloride penetration due to the chloride reaction or binding with the cement phases, a factor needs to be applied to electrical resistivity measurements [59]. A reaction factor (r_{cl}) is defined as the ratio between the diffusion coefficients D_{app} and D_{eff} assuming lineal chloride binding

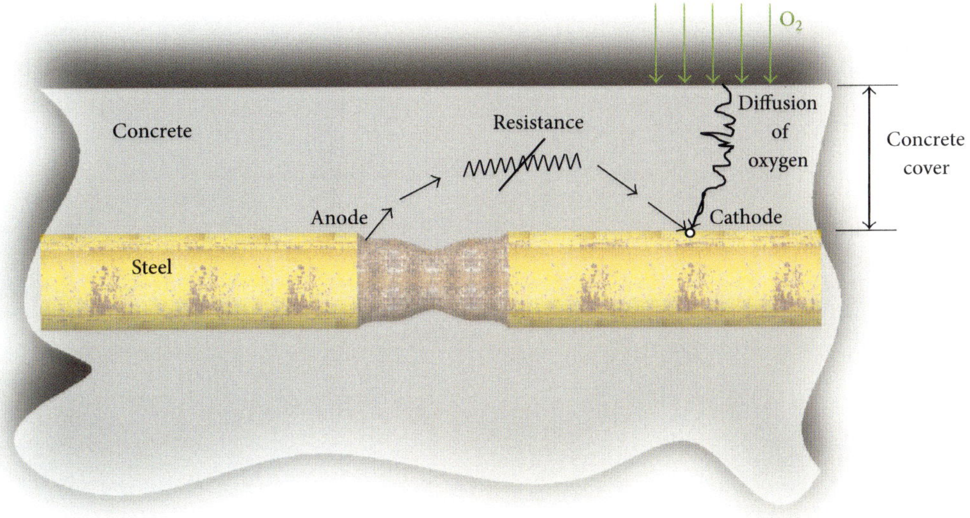

FIGURE 11: Schematic descriptions of factors which may affect corrosion rate of steel in concrete: (i) O_2 availability and (ii) electrical resistance of concrete (reproduced from [19]).

as shown in (14) where ρ_{app} and ρ_{eff} are apparent and effective electrical resistivity [59]:

$$r_{cl} = \frac{D_{eff}}{D_{app} \cdot \varphi} = \frac{\rho_{app}}{\rho_{eff}}. \quad (14)$$

6.2. Correlation between Electrical Resistivity and Reinforcement Corrosion. Once rebar is depassivated and corrosion is initiated by chloride ions or carbonation, corrosion rate will be the most influential parameter that determines how fast the reinforced concrete structure is deteriorating. It is dependent on many parameters including oxygen availability, ratio of anodic/cathodic area, relative humidity (RH), and concrete electrical resistivity (Figure 11) [19]. When enough oxygen is available to supply the anodic current especially under aerated state (such as the splashing area zone), cathodic control no longer exists. In this condition, the corrosion rate can be limited by controlling the flow of ionic current through concrete which is directly related to electrical resistivity of concrete [19]. Therefore, several researchers have attempted to correlate the corrosion probability and corrosion rate to concrete electrical resistivity [60, 103–108].

Among all the studies conducted so far, there is an agreement that corrosive environment in reinforced concrete and electrical resistivity of concrete have an inverse relationship. As the electrical resistivity of concrete decreases, the rate of steel reinforcement corrosion increases. A theory by Glass et al. [103] stated that the corrosion rate of reinforced concrete is under anodic control with the anodic reaction being limited by the mortar resistivity. Glass's anodic resistance theory is also supported by studies conducted by Morris et al. [60] and Bertolini and Polder [104]. In their investigations, it has been found that concrete resistivity affected not only corrosion rate, but also the corrosion potential [60, 104]. The only standardized test method for corrosion monitoring is the ASTM C876 half-cell potential mapping technique [105]. Hence, electrical resistivity measurements can be an effective standard indicator of steel reinforcement corrosion potential and rate.

Combining both half-cell potential and electrical resistivity measurements techniques makes it possible to examine corrosion probability and corrosion rate once it is initiated [106]. For high resistivity values, Carino [106] stated that even if the embedded rebar in concrete is actively corroding, detected by half-cell potential technique, the corrosion rate may be slow when steel rebar depassivates; thus the concrete electrical resistivity has better correlation with corrosion rate than half-cell potential. Experimental investigation on reinforced concrete specimens immersed in saline solution came to the conclusion that steel rebar is likely to reach a corrosion state while resistivity readings are lower than 10 kΩ·cm and reach a passive state when resistivity measurements are higher than 30 kΩ·cm [60]. The results also showed that the risk for chloride-induced corrosion and the chloride threshold value that will initiate the corrosion process can be estimated via a straightforward and nondestructive technique like concrete resistivity [28]. For various concrete mixture types, Polder and Peelen [50] indicated the corrosion initiation with respect to electrical resistivity under simulated deicing salt, but no relationship or effect was established. Similarly, Basheer et al. [57] and Smith et al. [46] used both electrical resistivity measurements and half-cell potential to evaluate the probability of corrosion in embedded steel rebar but, then again, no correlation was developed. It was also confirmed that corrosion potential can be detected by means of electrical resistivity but it was not specifically indicated how this can be done [107]. In addition, using the electrical resistivity or half-cell potential measurements was suggested to evaluate the deterioration caused by corrosion in the experimental study preformed on a concrete bridge deck [108]. In this study, it was stated that having no need of electrical connection to embedded rebar is the main advantage of electrical resistivity over half-cell potential technique. The effect of concrete cover as an influential parameter on

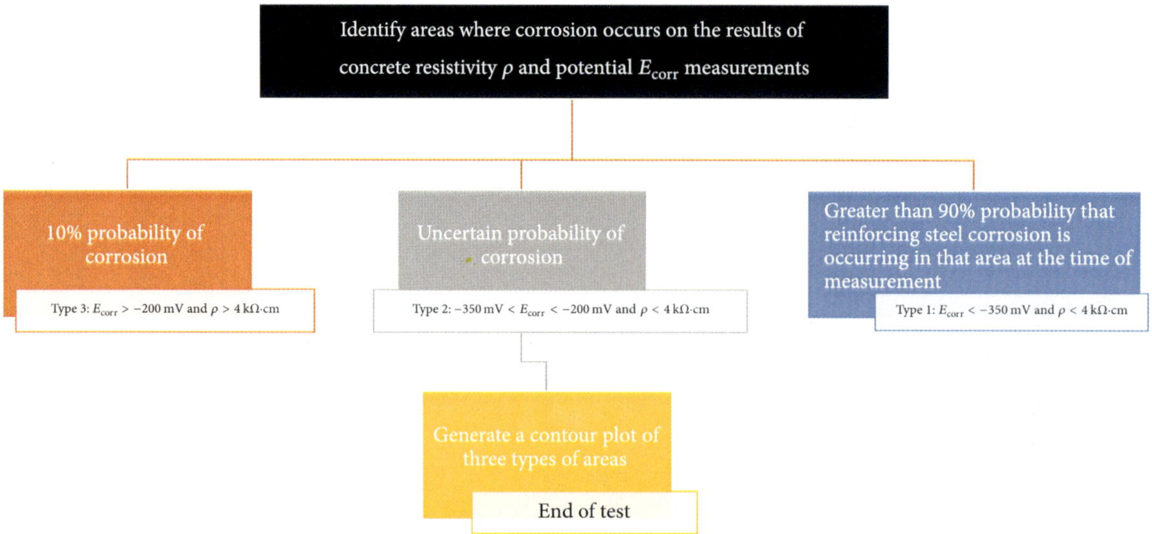

FIGURE 12: Assessment of corrosion probability in concrete slabs through half-cell potential and resistivity measurements [20].

TABLE 6: Concrete resistivity and risk of corrosion of steel reinforcement.

Corrosion risk	Resistivity values (kΩ·cm)		
	Polder [7]	Song and Saraswathy [66], Elkey and Sellevold [67]	Commercial Wenner probe instrument manuals [68, 69]
High	<10	<5	≤10
Moderate	10–50	5–10	10–50
Low	50–100	10–20	50–100
Negligible	>100	>20	≥100

corrosion detection using both electrical resistivity and half-cell potential measurements has been studied experimentally [75]. For small concrete cover thickness, it was observed that both measurements corresponded well together, but more discrepancy in measurements values was obtained when concrete cover became thicker [75]. From few published papers in this area that examined the possibility of corrosion by applying both electrical resistivity and half-cell potential technique, it was difficult to find a systematic methodology. As a rule of thumb, methodology proposed in Figure 12 by Sadowski [20] can be used when possible corrosion areas are identified by both half-cell potential and a four-point Wenner probe. Only identifying the probability of corrosion was suggested when conducting a corrosion assessment [20].

A number of researchers as well as commercial Wenner probe instrument manuals (Proceq and Giatec Scientific Inc. [68, 69]) provided a general guideline in terms of corrosion risk when interpreting electrical resistivity measurements, as seen in Table 6. The difference between provided resistivity values originates in the various experimental setup, concrete quality, concrete composition, and initial chloride concentration. For example, the corrosion state resistivity for passivation value reported in Polder's [7] review is higher than the one recorded in Morris et al.'s work [60] due to different experimental setup and test conditions.

A linear relationship between concrete electrical conductivity and corrosion rate has been found in several articles [104, 106, 109–111]. Experimental study on carbonated mortars also confirmed this linear trend and showed a different slope in case of chloride presence as the corrosion rate was affected by Cl^-/OH^- ratio [110]. Variation in concrete cover depth as well as concrete composition also showed an effect on the slope of linear trend between resistivity and corrosion rate [104]. Andrade and Alonso [112] proposed an empirical equation describing relationship between resistivity and corrosion rate:

$$I_{corr} \cong \frac{3 \times 10^3}{\rho}. \tag{15}$$

Two other similar models have been proposed by DuraCrete R17 [92] and LIFECON [113], as illustrated below:

$$I_{corr} = \frac{k_0}{\rho(t)} \times F_{cl} \times F_{Galv} \times F_{O_2}, \tag{16}$$

where I_{corr} is corrosion rate in $\mu A/cm^2$, k_0 is constant regression parameter in $\mu m \cdot \Omega m/a$, $\rho(t)$ is actual resistivity at time t in Ωm, F_{cl} accounts for the influence of the chloride content, F_{Galv} is influence of galvanic effect, and F_{O_2} is availability of oxygen.

In most cases, F_{cl} is dependent on the chloride concentration at the location that corrosion occurs and F_{Galv} and F_{O_2} equal 1. Due to microcell geometric arrangements of anodic and cathodic sites, concrete electrical resistivity plays an important role in controlling corrosion rates [111]. Similar to Glass et al.'s study [103], it was stated by Gulikers [111] that overall corrosion cell resistance is not inherently controlled by concrete electrical resistivity, but rather cathodic activation controls mostly. It should always be considered that the interpretation of electrical resistivity measurements is

FIGURE 13: Instrumented circular hollow columns being studied by authors to establish the relationship between electrical resistivity and durability characteristics.

a challenging task due to many variables affecting the results like moisture condition, salt content, and so forth.

In summary, it can be concluded that still large range and scatter exist for correlation between corrosion rate and concrete resistivity. Also, effect of moisture state and temperature as well as corrections to corrosion rate measurements should be considered during an investigation on finding correlation between resistivity and corrosion rate. Knowledge is still lacking in the literature to understand which mechanism dominates the corrosion process and how resistivity measurements are impacted. To practically determine the corrosion and resistivity relationship, more field data should be collected and analyzed. In order to address some of these issues, authors of this paper initiated a project with field specimens as shown in Figure 13. In this study, influence of various parameters (noted below) on resistivity is being studied (parameters: cover thickness, crystalline water-proofing admixtures effect, and cement type). In addition, authors are also studying the effect of self-sealing on resistivity of concrete. Previous work done in this area (self-sealing) is reported elsewhere [114, 115].

7. Correlation between Electrical Resistivity and Compressive Strength

One of the most important mechanical properties of concrete is compressive strength. It can be simply measured by compression testing machine, as load at the failure divided by area of specimen gives the compressive strength of concrete [56]. Thus, a number of researchers performed different experiments to understand the relation between compressive strength and electrical responses in cementitious materials. In the compressive strength, one of the key factors is the strength of interfacial transition zone (ITZ) that has no remarkable impact on concrete electrical resistivity. It was observed when different cementitious materials were used in concrete mixture that there was no sensible relationship between concrete resistivity and compressive strength ($R^2 = 0.413$) [56]. This is mainly linked to the fact that chemical compound of pore solution does not have a great impact on compressive strength while affecting the concrete resistivity significantly. On the other hand, for the case of similar cementitious materials, better correlations can be achieved between compressive strength and concrete resistivity due to relationship between permeability and compressive strength [56]. As the strength increases, the electrical resistivity increases correspondingly for the same concrete mixture design. The study on the electrical properties of concrete with White Portland Cement also found an approximately linear trend between these two properties as both directly depend on the porosity of the matrix at early age. As concrete matured, this relationship was affected by other properties such as the conductivity of the pore solution and the degree of concrete saturation [34]. Also, as the density and compactness of the concrete structure increase, its compressive strength and the electrical resistivity will both increase. It was reported that, at the same w/c ratio, the changes of the specimens' resistivities with different sizes (φ100 mm × 200 mm and φ150 mm × 300 mm) were close to those changes of the compressive strengths. For example, the differences between the resistivities or compressive strengths of the specimens were both around 93% at w/c of 0.6 and 95% at w/c of 0.4 [25]. No direct relationship between 56-day concrete electrical resistivity and 28-day compressive strength was found in Gudimettla and Crawford's work [53] due to variety of samples taken from different field projects. In terms of activation energy, the investigation done by Liu and Presuel-Moreno [44] showed that resistivity is noticeably higher than the reported activation energy for compressive strength. In conclusion, there is no practical correlation between electrical resistivity and compressive strength for various cementitious materials; however, for the same mixture design, electrical resistivity is linearly related to compressive strength. Generally, as concrete permeability decreases, both electrical resistivity and compressive strength increase.

8. Correlation between Surface and Bulk Electrical Resistivity

As it is discussed in previous sections, every electrical resistivity (ρ) measurement is composed of a resistance (R) and a geometry correction factor (k) that convert the resistance to resistivity which is a geometry independent intrinsic material property [21]. The correction factor (k) depends on the geometry and size of the sample and the electrodes location. Hence, for the same concrete material, variation in electrode geometries and sample size may result in different resistance values but eventually will yield the same resistivity using (17). Four common measurement electrode geometries that have been employed in several studies to conduct electrical test on cementitious cylinders are included: (1) uniaxial resistance (or bulk resistance), (2) surface resistance, (3) embedded sensors, and (4) disc method (one external electrode). In this section, researchers' review on correlation between the first two common (bulk and surface) resistivity measurement methods is summarized.

$$\rho = R \cdot k. \tag{17}$$

Theoretically, the ratio of surface and bulk resistance for standard size cylinder specimen (100 mm × 200 mm) and probe spacing of $a = 50$ mm can be computed in the following equation [18]:

$$\frac{R_1}{R_2} = \left(\frac{\rho_1}{\rho_2}\right)\left(\frac{1}{8}\right), \tag{18}$$

where R_1 is resistance measured by 4-point Wenner method (surface resistance), R_2 is resistance measured by uniaxial method (bulk resistance), ρ_1 is resistivity measured by 4-point Wenner method (surface resistivity), and ρ_2 is resistivity measured by uniaxial method (bulk resistivity).

The ratio of two different resistivity types (ρ_1 and ρ_2) is equal to 2.63 [18]. Therefore, the theoretical ratio of surface and bulk resistance can be computed as

$$\frac{R_1}{R_2} = \frac{2.63}{8} = 0.33. \tag{19}$$

Also, a number of researchers experimentally attempted to study this correlation between bulk and surface resistivity data. Studies performed by Ghosh and Tran [61, 62] showed the correlation between bulk and surface resistivity for different groups of binary and ternary mixtures. It was observed that for most groups of mixtures, the coefficient of determination (R^2) values for linear trend line was higher than 0.8 and sometimes close to 1 except some silica fume mixtures at early-ages where the coefficient became less than 0.80. The linear trend line varied from 0.29 to 0.47 at different concrete ages of 7, 14, 28, 56, 91, and 161 days. For probe spacing $a = 38.1$ mm, Spragg et al. [39] also reported a strong linear correlation with an $R^2 = 0.9986$, with surface resistivity measurements tending to be 1.86 times higher than bulk resistivity which supports previous work done by Morris et al. [18]. Several Department of Transportation (DOT) and other agencies participated in a round robin program (summarized by Indiana DOT). A recent study on specimens casted during actual concrete production in the field through different US states proved that the surface and bulk measurements correlate extremely well with each other at 28- and 56-day age [53]. It was observed that the surface resistivity results were typically 1.9 times higher than the bulk resistivity obtained data. Authors have found a similar correlation between bulk and surface resistivity ($R^2 = 0.979$).

Both surface and bulk resistivity measurement methods have been used to determine the presence of a heterogeneity problem [21]. To investigate the heterogeneous systems, the surface and uniaxial resistivity tests were performed on the cylindrical specimen with a 10-mm thick white cement paste on outer layer and the prismatic specimens with two separate layers of white and grey cement pastes setup in parallel and series configurations, as illustrated in Figure 14 [21]. For the uniaxial case, the total resistivity (ρ_c) can be calculated through the proposed equations (20a) and (20b) for parallel and series configurations. A correction factor (Γ) has been developed for heterogeneity which depends on the material of interest and an approach was proposed to compare the resistivities measured from a surface test and a bulk test. In this approach, if there is heterogeneity, the values from the two tests will be different. This difference can give an indication as to whether the surface layer is more or less resistive [21]. Similarly, correlation between surface and bulk resistivity found in other studies indicated a strong relationship between the resistivity values when the appropriate geometry factor was applied [8, 37]:

$$\frac{1}{\rho_c} = \sum_{i=1}^{n} \frac{A_i}{\rho_i} \quad \text{(Parallel configuration)}, \tag{20a}$$

$$\rho_c = \sum_{i=1}^{n} \rho_i \cdot L_i \quad \text{(Series configuration)}. \tag{20b}$$

In summary, the coefficient of determination value (R^2) for linear trend between bulk and surface resistivity in different literatures for various ages is shown in Table 7. Based on the reported data, it can be concluded that a strong relationship exists between resistivity measured by surface and bulk tests. However, in practical cases, caution is needed for multilayered cementitious composite systems as electrical current flows differently in these two techniques and for the types of layered electrical properties that can occur due to moisture gradients, chemical changes, and ionic gradients [21]. More complicated situations for resistivity measurement that arise during field measurements where seasonal and wet/drying might take place at the tidal section on partially

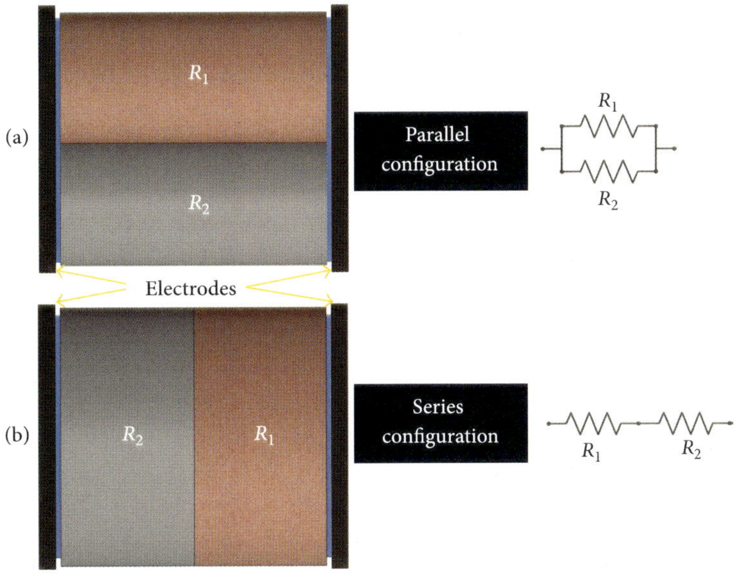

FIGURE 14: Schematic of the (a) parallel and (b) series models of heterogeneous systems (reproduced from [21]).

TABLE 7: Coefficient of determination (COD) value for linear trend between bulk and surface resistivity in the literatures.

Reference	Coefficient of determination value (R^2)
Sengul and Gjorv [37] (2008)	0.99
Sengul and Gjorv [8] (2009)	0.99
Spratt et al. [39] (2011)	0.9986
Ghosh and Tran [61] (2015)	0.82–0.95
Gudimettla and Crawford [53] (2016)	0.98
Authors' work (2016)	0.979

immersed structures have been studied by Presuel-Moreno et al. [65].

9. Summary and Conclusions

Through an extensive literature review, this paper identifies several factors which might have potential influence on the electrical resistivity of concrete. Effect of each parameter is briefly summarized below.

(i) In agreement with most studies, when there is an embedded rebar in the concrete, the electrical current field is distorted, and thus errors can result in the electrical resistivity measurements. To minimize this effect, it is suggested to place all electrodes perpendicular to the embedded rebar on the concrete surface and take at least five measurements, each a few millimetres in distance from one another. Also, a correction factor should be applied to resistivity measurements once rebar is present in concrete. However, effect of rebar presence on the resistivity measurements is well-understood; just a few studies could be found to identify the rebar presence effect and more field investigations are still needed in this area.

(ii) Presence of cracks in concrete was also identified as an influential parameter on electrical resistivity since it is initially presumed that concrete is homogenous, isotropic, and uncracked. Depth of crack, orientation of probes on crack, and type of crack (conductive or isolated) can individually affect the resistivity readings. As suggested for embedded rebar in the concrete, all electrodes should be placed in perpendicular direction to cracks. It was reported that higher resistivity readings were obtained from conductive cracks whereas lower resistivity values were gained for insulated cracks. However, no information is provided to show how much cracking induced by corrosion influences concrete resistivity. In addition when both insulated and conductive cracks are bridged together, their integrated influence on electrical resistivity is not well-understood.

(iii) The moisture state and temperature of concrete during resistivity measurements were also found to be of major influence on recorded data. As temperature increases, the ions mobility becomes faster; consequently, electrical conductivity of concrete also increases. To lessen the temperature effect on resistivity results, no practical correlation is still published for real-world conditions. Essentially, electrical resistivity reduces with an increase in moisture content as a result of changing in the ion movement. It is strongly recommended to take resistivity measurements when concrete is in Saturated Surface Dry (SSD) condition. Yet, more investigations are required to understand how much time is needed for water to infiltrate the concrete to obtain constant moisture level through the bulk sample.

(iv) The studies related to resistivity measurement test device confirmed that electrical signal shape and frequency, electrode contact with concrete surface, and probe spacing of surface resistivity measurement device can affect the resistivity results. To minimize the effect of signal shape and frequency, using low frequency range square-wave signal for surface resistivity and high frequency range for bulk resistivity is recommended. Proper contact should also be provided to not mislead resistivity readings. Using a saturated sponge between electrodes and concrete surface can reduce this effect. In 4-point Wenner probe method, as it is assumed that concrete is homogenous material, aggregates inside concrete also affect its homogeneity due to their higher resistivity. Therefore, providing enough wide spaces between electrodes is essential to diminish aggregates effect. As a rule of thumb, probe spacing 1.5 times higher than the maximum aggregate size should be considered.

(v) Aggregate content and type were identified to have an influence on concrete resistivity. Increase in aggregate content results in higher resistivity values due to their less porosity and lower electrical conductivity. Also, aggregates with rough surface texture were found to have higher resistivity as their tortuosity is higher. Therefore, the effect of aggregates content and type should be accounted for in resistivity measurements.

(vi) Carbonation process in aged concrete forms a multilayered system that results in various resistivity values through the concrete depth. So, its effect should be mitigated during resistivity measurements. Lower resistivity is also generated when w/b ratio is high due to higher percentage of porosity.

Correlation between concrete electrical resistivity and its certain durability characteristics such as chloride diffusivity, compressive strength, and corrosion potential/rate was discussed in this paper and is summarized below.

(i) Concrete resistivity is inversely related to chloride ingress, where lower resistivity indicated the area where chloride diffusion will be faster. A retardation of chloride can be taken into account through the introduced reaction factor (r_{cl}). More research is still required to better simulate sea site conditions.

(ii) Furthermore, a strong correlation can be found between increasing electrical resistivity of concrete and the corrosion rate. The relationship can be seen when corrosion has initiated (active conditions). It will not be valid in the case of saturated concrete, where although the resistivity is low, the corrosion rate will be small because of lack of oxygen. Field data was considered in just a few investigations and, thus, it is of high interest to gather more field experience.

(iii) In addition, concrete compressive strength and its electrical resistivity have a direct relationship with each other as both directly depend on the porosity of the matrix at early age. However, at a later age, the conductivity of the pore solution and the degree of concrete saturation both influence this relationship. No practical relationship was identified in the literatures between compressive strength and electrical resistivity.

(iv) Four common measurement electrode geometries that have been employed in several studies to conduct electrical test on cementitious cylinders were introduced in this article. Among them, a strong direct linear correlation between two common methods of resistivity measurements (concrete surface and bulk resistivity) was presented. However, more attention should be paid to multilayered cementitious composite systems in the field as electrical current flows differently in these two techniques and for the types of layered electrical properties that can happen because of moisture gradients, chemical changes, and ionic gradients.

Abbreviations

CON: Concrete
CEM: Cement paste
MOR: Mortar
MK: Metakaolin
SF: Silica fume
RHA: Rice Husk Ash
OPC: Ordinary Portland Cement
SLG: Slag
FA: Fly ash
NP: Natural pozzolan
UPV: Ultrasonic Pulse Velocity
WPC: White Portland Cement
PFA: Pulverized-Fuel Ash
MS: Microsilica
LPR: Linear Polarization Resistance technique
TNVM: Total Number of Various Mixtures
w/b: Water/binder ratio
NR: Not reported
CR: Counter electrode
RE: Reference electrode
SC: Slag cement
CA: Crystalline admixtures
PLC: Portland Limestone Cement.

Conflicts of Interest

The authors declare that there are no conflicts of interest regarding the publication of this paper.

References

[1] P. K. Mehta and P. J. M. Monteiro, *Concrete: microstructure, properties, and materials*, McGraw-Hill, New York, NY, USA, 3rd edition, 2006.

[2] ASTM, "Standard test method for bulk electrical conductivity of hardened concrete," ASTM C1760-12, ASTM International, 2012.

[3] AASHTO, "Method of test for surface resistivity indication of concrete's ability to resist chloride ion penetration," AASHTO TP 95, Am. Assoc. State Highw. Transp. Off., 2011.

[4] O. Sengul, "Use of electrical resistivity as an indicator for durability," *Construction and Building Materials*, vol. 73, pp. 434–441, 2014.

[5] F. Rajabipour, J. Weiss, and D. M. Abraham, "Insitu electrical conductivity measurements to assess moisture and ionic transport in concrete (A discussion of critical features that influence the measurements)," in *Proceedings of the International RILEM Symposium on Concrete Science and Engineering: A Tribute to Arnon Bentur*, 2004.

[6] H. Layssi, P. Ghods, A. R. Alizadeh, and M. Salehi, "Electrical resistivity of concrete," *Concrete International*, pp. 41–46, 2015.

[7] R. B. Polder, "Test methods for on site measurement of resistivity of concrete—a RILEM TC-154 technical recommendation," *Construction and Building Materials*, vol. 15, no. 2-3, pp. 125–131, 2001.

[8] O. Sengul and O. E. Gjorv, "Effect of embedded steel on electrical resistivity measurements on concrete structures," *ACI Materials Journal*, vol. 106, no. 1, 2009.

[9] M. Salehi, P. Ghods, and O. Burkan Isgor, "Numerical investigation of the role of embedded reinforcement mesh on electrical resistivity measurements of concrete using the Wenner probe technique," *Materials and Structures*, vol. 49, no. 1-2, pp. 301–316, 2016.

[10] J. F. Lataste, C. Sirieix, D. Breysse, and M. Frappa, "Electrical resistivity measurement applied to cracking assessment on reinforced concrete structures in civil engineering," *NDT & E International*, vol. 36, no. 6, pp. 383–394, 2003.

[11] N. Wiwattanachang and P. H. Giao, "Monitoring crack development in fiber concrete beam by using electrical resistivity imaging," *Journal of Applied Geophysics*, vol. 75, no. 2, pp. 294–304, 2011.

[12] K. Hornbostel, C. K. Larsen, and M. R. Geiker, "Relationship between concrete resistivity and corrosion rate—a literature review," *Cement and Concrete Composites*, vol. 39, pp. 60–72, 2013.

[13] D. A. Whitting and M. A. Nagi, *Electrical Resistivity of Concrete*, Portland Cement Association, Skokie, Ill, USA, 2003.

[14] RILEM TC 154-EMC, "Recommendations of RILEM TC 154-EMC: electrochemical techniques for measuring metallic corrosion half-cell potential measurements—potential mapping on reinforced concrete structures," *Materials and Structures*, vol. 36, no. 261, pp. 461–471, 2003.

[15] F. Wenner, "A method of measuring earth resistivity," *Bulletin of the Bureau of Standards*, vol. 12, no. 4, pp. 469–478, 1916.

[16] R. Ranade, J. Zhang, J. P. Lynch, and V. C. Li, "Influence of micro-cracking on the composite resistivity of Engineered Cementitious Composites," *Cement and Concrete Research*, vol. 58, pp. 1–12, 2014.

[17] K. R. Gowers and S. G. Millard, "Measurement of concrete resistivity for assessment of corrosion severity of steel using wenner technique," *ACI Materials Journal*, vol. 96, no. 5, 1999.

[18] W. Morris, E. I. Moreno, and A. A. Sagüés, "Practical evaluation of resistivity of concrete in test cylinders using a Wenner array probe," *Cement and Concrete Research*, vol. 26, no. 12, pp. 1779–1787, 1996.

[19] A. Bentur, S. Diamond, and B. N. Steven, *Steel Corrosion in Concrete: Fundamentals and Civil Engineering Practice*, EFN Spon, London, UK, 1st edition, 1997.

[20] L. Sadowski, "Methodology for assessing the probability of corrosion in concrete structures on the basis of half-cell potential and concrete resistivity measurements," *The Scientific World Journal*, vol. 2013, Article ID 714501, 8 pages, 2013.

[21] R. Spragg, C. Villani, J. Weiss et al., "Surface and uniaxial electrical measurements on layered cementitious composites having cylindrical and prismatic geometries," in *Proceedings of the International Conference on the Durability of Concrete Structures*, West Lafayette, Ind, USA, 2014.

[22] D. P. Bentz, K. A. Snyder, and A. Ahmed, "Anticipating the setting time of high-volume fly ash concretes using electrical measurements: feasibility studies using pastes," *Journal of Materials in Civil Engineering*, vol. 27, no. 3, Article ID 4014129, 2015.

[23] R. Weydert and C. Gehlen, "Electrolytic resistivity of cover concrete: relevance, measurement and interpretation," in *Proceedings of CIB W078 Workshop on Information Technology in Construction*, Rotterdam, Netherlands, 1999.

[24] F. Presuel-Moreno, Y. Liu, and M. Paredes, *Understanding the Effect of Rebar Presence and/or Multilayered Concrete Resistivity on the Apparent Surface Resistivity Measured via the Four Point Wenner Method*, NACE International, 2009.

[25] C.-T. Chen, J.-J. Chang, and W.-C. Yeih, "The effects of specimen parameters on the resistivity of concrete," *Construction and Building Materials*, vol. 71, pp. 35–43, 2014.

[26] A. J. Garzon, J. Sanchez, C. Andrade, N. Rebolledo, E. Menéndez, and J. Fullea, "Modification of four point method to measure the concrete electrical resistivity in presence of reinforcing bars," *Cement and Concrete Composites*, vol. 53, pp. 249–257, 2014.

[27] Y.-C. Lim, T. Noguchi, and C.-G. Cho, "A quantitative analysis of the geometric effects of reinforcement in concrete resistivity measurement above reinforcement," *Construction and Building Materials*, vol. 83, pp. 189–193, 2015.

[28] W. Morris, A. Vico, and M. Vázquez, "Chloride induced corrosion of reinforcing steel evaluated by concrete resistivity measurements," *Electrochimica Acta*, vol. 49, no. 25, pp. 4447–4453, 2004.

[29] U. M. Angst and B. Elsener, "On the applicability of the wenner method for resistivity measurements of concrete," *ACI Materials Journal*, vol. 111, no. 6, pp. 661–672, 2014.

[30] M. Goueygou, O. Abraham, and J.-F. Lataste, "A comparative study of two non-destructive testing methods to assess near-surface mechanical damage in concrete structures," *NDT and E International*, vol. 41, no. 6, pp. 448–456, 2008.

[31] T. Rupnow and P. Icenogle, "Evaluation of surface resistivity measurements as an alternative to the rapid chloride permeability test for quality assurance and acceptance," Tech. Rep. 2290, Performing Organization Name and Address Louisiana Transportation Research Center, Baton Rouge, La, USA, 2012.

[32] F. Presuel-Moreno and Y. Liu, "Temperature effect on electrical resistivity measurements on mature saturated concrete," in *Proceedings of the NACE—International Corrosion Conference Series 7*, Salt Lake City, Utah, USA, 2012.

[33] R. Van Noort, M. Hunger, and P. Spiesz, "Long-term chloride migration coefficient in slag cement-based concrete and resistivity as an alternative test method," *Construction and Building Materials*, vol. 115, pp. 746–759, 2016.

[34] A. Lübeck, A. L. G. Gastaldini, D. S. Barin, and H. C. Siqueira, "Compressive strength and electrical properties of concrete with white Portland cement and blast-furnace slag," *Cement and Concrete Composites*, vol. 34, no. 3, pp. 392–399, 2012.

[35] J. Su, C. Yang, W. Wu, and R. Huang, "Effect of moisture content on concrete resistivity measurement," *Journal of the Chinese Institute of Engineers*, vol. 25, no. 1, pp. 117–122, 2002.

[36] M. Saleem, M. Shameem, S. E. Hussain, and M. Maslehuddin, "Effect of moisture, chloride and sulphate contamination on the electrical resistivity of Portland cement concrete," *Construction and Building Materials*, vol. 10, no. 3, pp. 209–214, 1996.

[37] O. Sengul and O. E. Gjorv, "Electrical resistivity measurements for quality control during concrete construction," *ACI Materials Journal*, vol. 105, no. 6, 2008.

[38] M. D. Newlands, M. R. Jones, S. Kandasami, and T. A. Harrison, "Sensitivity of electrode contact solutions and contact pressure in assessing electrical resistivity of concrete," *Materials and Structures*, vol. 41, no. 4, pp. 621–632, 2008.

[39] R. Spratt, J. Castro, T. Nantung, E. Paredes, and W. J. Weiss, "Variability Analysis of the Bulk Resistivity Measured Using Concrete Cylinders," Tech. Rep., Purdue University Press, West Lafayette, Ind, USA, 2011.

[40] J. W. McCarter, G. Starrs, S. Kandasami, M. R. Jones, and M. Chrisp, "Electrode configurations for resistivity measurements on concrete," *ACI Materials Journal*, vol. 106, no. 3, 2009.

[41] Y. Liu and F. J. Presuel-Moreno, "Normalization of temperature effect on concrete resistivity by method using Arrhenius law," *ACI Materials Journal*, vol. 111, no. 4, pp. 433–442, 2014.

[42] J. W. Bryant Jr., R. E. Weyers, and J. M. Garza, "In-place resistivity of bridge deck concrete mixtures," *ACI Materials Journal*, vol. 106, no. 2, 2009.

[43] R. Spragg, C. Villani, K. Snyder, D. Bentz, J. Bullard, and J. Weiss, "Factors that influence electrical resistivity measurements in cementitious systems," *Journal of the Transportation Research Board*, vol. 2342, pp. 90–98, 2013.

[44] Y. Liu and F. Presuel-Moreno, "Effect of elevated temperature curing on compressive strength and electrical resistivity of concrete with fly ash and GGBS," *ACI Materials Journal*, vol. 111, no. 5, 2014.

[45] W. J. McCarter, T. M. Chrisp, G. Starrs, P. A. M. Basheer, and J. Blewett, "Field monitoring of electrical conductivity of cover-zone concrete," *Cement and Concrete Composites*, vol. 27, no. 7-8, pp. 809–817, 2005.

[46] K. M. Smith, A. J. Schokker, and P. J. Tikalsky, "Performance of supplementary cementitious materials in concrete resistivity and corrosion monitoring evaluations," *ACI Materials Journal*, vol. 101, no. 5, 2004.

[47] H. Minagawa and M. Hisada, "Consideration about chloride ion diffusion coefficient estimated by electric resistivity of concrete exposed in tidal zone," in *Proceedings of the 3rd International Conference on Sustainable Construction Materials and Technologies*, Kyoto, Japan, 2013.

[48] F. Presuel-Moreno, Y.-Y. Wu, and Y. Liu, "Effect of curing regime on concrete resistivity and aging factor over time," *Construction and Building Materials*, vol. 48, pp. 874–882, 2013.

[49] E. Güneyisi, T. Özturan, and M. Gesoğlu, "A study on reinforcement corrosion and related properties of plain and blended cement concretes under different curing conditions," *Cement and Concrete Composites*, vol. 27, no. 4, pp. 449–461, 2005.

[50] R. B. Polder and W. H. A. Peelen, "Characterisation of chloride transport and reinforcement corrosion in concrete under cyclic wetting and drying by electrical resistivity," *Cement and Concrete Composites*, vol. 24, no. 5, pp. 427–435, 2002.

[51] T. Y. D. Lim, S. Teng, S. D. Bahador, and O. E. Gjørv, "Durability of very-high-strength concrete with supplementary cementitious materials for marine environments," *ACI Materials Journal*, vol. 113, no. 1, pp. 95–103, 2016.

[52] R. J. Kessler, R. G. Powers, E. Vivas, M. A. Paredes, and Y. P. Virmani, "Surface resistivity as an indicator of concrete chloride penetration resistance," in *Proceedings of the Concrete Bridge Conference*, p. 18, 2008.

[53] J. Gudimettla and G. Crawford, "Resistivity tests for concrete—recent field experience," *ACI Materials Journal*, vol. 113, no. 4, 2016.

[54] A. Jenkins, "Surface resistivity as an alternative for rapid chloride permeability test of hardened concrete," Tech. Rep. FHWA-KS-14-15, Kansas Department of Transportation Bureau of Research, Topeka, Kan, USA, 2015.

[55] Y. Liu, F. J. Presuel-Moreno, and M. A. Paredes, "Determination of chloride diffusion coefficients in concrete by electrical resistivity method," *ACI Materials Journal*, vol. 112, no. 5, pp. 631–640, 2015.

[56] A. A. Ramezanianpour, A. Pilvar, M. Mahdikhani, and F. Moodi, "Practical evaluation of relationship between concrete resistivity, water penetration, rapid chloride penetration and compressive strength," *Construction and Building Materials*, vol. 25, no. 5, pp. 2472–2479, 2011.

[57] P. A. M. Basheer, P. R. V. Gilleece, A. E. Long, and W. J. Mc Carter, "Monitoring electrical resistance of concretes containing alternative cementitious materials to assess their resistance to chloride penetration," *Cement and Concrete Composites*, vol. 24, no. 5, pp. 437–449, 2002.

[58] D. T. Y. Lim, B. Sabet Divsholi, D. Xu, and S. Teng, "Evaluation of high performance concrete using electrical resistivity technique," in *Proceedings of the Our World in Concrete and Structures*, Singapore, 2011.

[59] C. Andrade, R. d'Andrea, and N. Rebolledo, "Chloride ion penetration in concrete: the reaction factor in the electrical resistivity model," *Cement and Concrete Composites*, vol. 47, pp. 41–46, 2014.

[60] W. Morris, A. Vico, M. Vazquez, and S. R. de Sanchez, "Corrosion of reinforcing steel evaluated by means of concrete resistivity measurements," *Corrosion Science*, vol. 44, no. 1, pp. 81–99, 2002.

[61] P. Ghosh and Q. Tran, "Correlation between bulk and surface resistivity of concrete," *International Journal of Concrete Structures and Materials*, vol. 9, no. 1, pp. 119–132, 2015.

[62] P. Ghosh and Q. Tran, "Influence of parameters on surface resistivity of concrete," *Cement and Concrete Composites*, vol. 62, pp. 134–145, 2015.

[63] A. L. G. Gastaldini, G. C. Isaia, T. F. Hoppe, F. Missau, and A. P. Saciloto, "Influence of the use of rice husk ash on the electrical resistivity of concrete: a technical and economic feasibility study," *Construction and Building Materials*, vol. 23, no. 11, pp. 3411–3419, 2009.

[64] C. Andrade, M. Castellote, and R. D'Andrea, "Measurement of ageing effect on chloride diffusion coefficients in cementitious matrices," *Journal of Nuclear Materials*, vol. 412, no. 1, pp. 209–216, 2011.

[65] F. Presuel-Moreno, A. Suares, and Y. Liu, *Characterization of New and Old Concrete Structures Using Surface Resistivity Measurements*, Florida Department of Transportation, Dania Beach, Fla, USA, 2010.

[66] H.-W. Song and V. Saraswathy, "Corrosion monitoring of reinforced concrete structures-a review," *International Journal of Electrochemical Science*, vol. 2, pp. 1–28, 2007.

[67] W. Elkey and E. J. Sellevold, *Electrical Resistivity of Concrete*, supplement 80, Norwegian Road Research Laboratory, Oslo, Norway, 1995.

[68] "Operating Instructions Resipod Family Manual." Proceq.

[69] SURF™, *Surface Electrical Resistivity of Concrete Manual*, Giatec Scientific Inc., https://www.giatecscientific.com/concrete-resistivity/surf/.

[70] J. Weiss, K. Snyder, J. Bullard, and D. Bentz, "Using a saturation function to interpret the electrical properties of partially saturated concrete," *Journal of Materials in Civil Engineering*, vol. 25, no. 8, pp. 1097–1106, 2013.

[71] S. G. Millard, "Reinforced concrete resistivity measurement techniques," *Institution of Civil Engineers*, vol. 91, pp. 71–88, 1991.

[72] F. Presuel-Moreno, Y. Liu, and Y.-Y. Wu, "Numerical modeling of the effects of rebar presence and/or multilayered concrete resistivity on the apparent resistivity measured via the Wenner method," *Construction and Building Materials*, vol. 48, pp. 16–25, 2013.

[73] M. Salehi, *Numerical investigation of the role of embedded reinforcement mesh on electrical resistivity measurements of concrete using the Wenner probe technique [M.S. thesis]*, Carleton University, Ottawa, Canada, 2013.

[74] J. Sanchez, C. Andrade, J. Torres, N. Rebolledo, and J. Fullea, "Determination of reinforced concrete durability with on-site resistivity measurements," *Materials and Structures*, vol. 50, no. 1, 2017.

[75] M. Morales, "Experimental investigation of the effects of embedded rebar, cracks, chloride ingress and corrosion on electrical resistivity measurements of reinforced concrete," *Master, Oregon State University*, 2015.

[76] A. A. Shah and Y. Ribakov, "Non-destructive measurements of crack assessment and defect detection in concrete structures," *Materials and Design*, vol. 29, no. 1, pp. 61–69, 2008.

[77] M. Chouteau and S. Beaulieu, "An investigation on application of the electrical resistivity tomography method to concrete structures," in *Proceedings of the Geophysics 2002. The 2nd Annual Conference on the Application of Geophysical and NDT Methodologies to Transportation Facilities and Infrastructure*, Washington DC, USA, 2002.

[78] E. Taillet, J. F. Lataste, P. Rivard, and A. Denis, "Non-destructive evaluation of cracks in massive concrete using normal dc resistivity logging," *NDT and E International*, vol. 63, pp. 11–20, 2014.

[79] M. Salehi, P. Ghods, and O. B. Isgor, "Numerical study on the effect of cracking on surface resistivity of plain and reinforced concrete elements," *Journal of Materials in Civil Engineering*, vol. 27, no. 12, Article ID 04015053, 2015.

[80] A. J. Ewins, "Resistivity Measurements in Concrete," *British Journal of NDT*, vol. 32, no. 3, pp. 120–126, 1990.

[81] R. Henkensiefken, J. Castro, D. Bentz, T. Nantung, and J. Weiss, "Water absorption in internally cured mortar made with water-filled lightweight aggregate," *Cement and Concrete Research*, vol. 39, no. 10, pp. 883–892, 2009.

[82] M. D. A. Thomas, "The use of conductive gel," 2008.

[83] J. D. Shane, *Electrical conductivity and transport properties of cement-based material measured by impedance spectroscopy [ph.D. thesis]*, Nothwestern University, Evanston, Ill, USA, 2000.

[84] R. Spragg, Y. Bu, K. Snyder, D. Bentz, and J. Weiss, "Electrical testing of cement-based materials: role of testing techniques, sample conditioning," Tech. Rep. FHWA/IN/JTRP-2013/28, Purdue University, 2013.

[85] A. Shahroodi, *Development of test methods for assessment of concrete durability for use in performance-based specifications [M.S. thesis]*, University of Toronto, Toronto, Canada, 2010.

[86] J. M. S. Marquez, *Influence of saturation and geometry on surface electrical resistivity measurements [M.S. thesis]*, Concorida University, Montreal, Canada, 2015.

[87] M. Castellote, C. Andrade, and M. C. Alonso, "Standardization, to a reference of 25°C, of electrical resistivity for mortars and concretes in saturated or isolated conditions," *ACI Materials Journal*, vol. 99, no. 2, 2002.

[88] Y. A. Villagrán Zaccardi, J. F. García, P. Huélamo, and Á. A. Di Maio, "Influence of temperature and humidity on Portland cement mortar resistivity monitored with inner sensors," *Materials and Corrosion*, vol. 60, no. 4, pp. 294–299, 2009.

[89] S. Poyet, "Experimental investigation of the effect of temperature on the first desorption isotherm of concrete," *Cement and Concrete Research*, vol. 39, no. 11, pp. 1052–1059, 2009.

[90] S. G. Millard, J. A. Harrison, and K. R. Gowers, "Practical measurement of concrete resistivity," *British Journal Non-Destructive Testing*, vol. 33, pp. 59–63, 1991.

[91] T. M. Chrisp, G. Starrs, W. J. McCarter, E. Rouchotas, and J. Blewett, "Temperature-conductivity relationships for concrete: an activation energy approach," *Journal of Materials Science Letters*, vol. 20, no. 12, pp. 1085–1087, 2001.

[92] DuraCrete R17, "Probabilistic performance based durability design of concrete structures, includes general guidelines for durability design and redesign," Tech. Rep. BE95-1347/R17, The European Union – Brite EuRam III, CUR, Gouda.

[93] M. Pour-Ghaz, O. B. Isgor, and P. Ghods, "The effect of temperature on the corrosion of steel in concrete. Part 1: Simulated polarization resistance tests and model development," *Corrosion Science*, vol. 51, no. 2, pp. 415–425, 2009.

[94] C. K. Larsen, E. J. Sellevold, J.-M. Østvik, and Ø. Vennesland, "Electrical resistivity of concrete—Part II: influence of moisture content and temperature," in *Proceedings of the 2nd International RILEM Symposium on Advances in Concrete through Science and Engineering*, 2006.

[95] Y. Liu, *Accelerated curing of concrete with high volume pozzolans -resistivity, diffusivity and compressive strength [ph.D. thesis]*, Florida Atlantic University, Boca Raton, Fla, USA, 2012.

[96] X. Lu, "Application of the Nernst-Einstein equation to concrete," *Cement and Concrete Research*, vol. 27, no. 2, pp. 293–302, 1997.

[97] K. R. Backe, O. B. Lile, and S. K. Lyomov, "Characterizing curing cement slurries by electrical conductivity," *SPE Drilling and Completion*, vol. 16, no. 4, pp. 201–207, 2001.

[98] FDOT, "Florida Method of Test For Concrete Resistivity as an Electrical Indicator of its Permeability," FM 5-578, 2004.

[99] R. J. Kessler, R. G. Powers, and M. A. Paredes, "Resistivity measurements of water saturated concrete as an indicator of permeability," in *Proceedings of the Corrosion*, Houston, Tex, USA.

[100] ASTM, "Standard test method for electrical indication of concrete's ability to resist chloride ion penetration," ASTM International C1202-12, 2012.

[101] R. K. Dhir, M. J. McCarthy, and M. D. Newlands, Eds., *Challenges of Concrete Construction: Concrete for Extreme Conditions: Proceedings of the International Conference held at the University of Dundee, on 9–11 September 2002*, vol. 6, Thomas Telford Publishing, Scotland, UK.

[102] E. Vivas, A. Boyd, and H. R. Hamilton III, "Permeability of concrete-comparison of conductivity and diffusion methods,"

Tech. Rep. 4910 45 04 992-12 (0026899), Florida Department of Transportation, Tallahassee, Fla, USA, 2007.

[103] G. K. Glass, C. L. Page, and N. R. Short, "Factors affecting the corrosion rate of steel in carbonated mortars," *Corrosion Science*, vol. 32, no. 12, pp. 1283–1294, 1991.

[104] L. Bertolini and R. Polder, Concrete resistivity and reinforcement corrosion rate as a function of temperature and humidity of the environment TNO 97-NaN-R0574, Netherlands Organisation for Applied Scientific Research, Delft, The Netherlands, 1997.

[105] ASTM, "Test Method for Corrosion Potentials of Uncoated Reinforcing Steel in Concrete," ASTM International C876, 2015.

[106] N. J. Carino, "Nondestructive techniques to investigate corrosion status in concrete structures," *Journal of Performance of Constructed Facilities*, vol. 13, no. 3, pp. 96–106, 1999.

[107] N. Gucunski, F. Romero, S. Kruschwitz, R. Feldmann, A. Abu-Hawash, and M. Dunn, "Multiple complementary nondestructive evaluation technologies for condition assessment of concrete bridge decks," *Transportation Research Record: Journal of the Transportation Research Board*, no. 2201, pp. 34–44, 2010.

[108] N. Gucunski, A. Imani, F. Romero et al., *Nondestructive testing to identify concrete bridge deck deterioration*, Transportation Research Board, Washington DC, USA, 2012.

[109] C. Alonso, C. Andrade, and J. A. González, "Relation between resistivity and corrosion rate of reinforcements in carbonated mortar made with several cement types," *Cement and Concrete Research*, vol. 18, no. 5, pp. 687–698, 1988.

[110] C. Andrade and C. Alonso, "Corrosion rate monitoring in the laboratory and on-site," *Construction and Building Materials*, vol. 10, no. 5, pp. 315–328, 1996.

[111] J. Gulikers, "Theoretical considerations on the supposed linear relationship between concrete resistivity and corrosion rate of steel reinforcement," *Materials and Corrosion*, vol. 56, no. 6, pp. 393–403, 2005.

[112] C. Andrade and C. Alonso, "Test methods for on-site corrosion rate measurement of steel reinforcement in concrete by means of the polarization resistance method," *RILEM*, vol. 37, pp. 623–643, 2004.

[113] E. Vesikari and M.-K. Soderqvist, "Life cycle management of concrete infrastructures for improved sustainability," in *Proceedings of the 9th International Bridge Management Conference*, pp. 15–28, 2003.

[114] R. Gupta and A. Biparva, "Innovative test technique to evaluate 'self-sealing' of concrete," *Journal of Testing and Evaluation*, vol. 43, no. 5, pp. 1091–1098, 2015.

[115] R. Gupta and A. Biparva, "Do crystalline water proofing admixtures affect restrained plastic shrinkage behavior of concrete?" *Revista ALCONPAT*, vol. 7, no. 1, p. 15, 2017.

13

Stiffness-Displacement Correlation from the RC Shear Wall Tests of the SAFE Program: Derivation of a Capacity Line Model

Francisco J. Molina,[1] Pierre Pegon,[1] and Pierre Labbé[2]

[1] *European Commission, Joint Research Centre (JRC), Institute for the Protection and Security of the Citizen (IPSC), 21027 Ispra, Italy*
[2] *Université Paris-Est, RENON (IRC-ESTP, IFSTTAR), IRC-ESTP, 28 avenue du Président Wilson, 94234 Cachan, France*

Correspondence should be addressed to Francisco J. Molina; francisco.molina@jrc.ec.europa.eu

Academic Editor: Santiago Garcia-Granda

The response of 13 reinforced concrete shear walls submitted to successive seismic tests has been postprocessed to produce time histories of secant stiffness and displacement oscillation amplitude. For every wall an envelope curve of displacement amplitude versus stiffness is identified which is fairly modelled by a straight line in double logarithmic scale. This relatively simple model, when used as a capacity line in combination with the demand response spectrum, is able to predict in an approximate manner the maximum response to the applied earthquakes. Moreover, the graphic representation of the demand spectrum and a unique model capacity line for a group of equal walls with different assumed design frequencies on them gives a visual interpretation of the different safety margins observed in the experiments for the respective walls. The same method allows as well constructing vulnerability curves for any design frequency or spectrum. Finally, the comparison of the different identified line models for the different walls allows us to assess the qualitative effect on the behaviour of parameters such as the reinforcement density or the added normal load.

1. Introduction

The seismic tests performed on reinforced concrete (RC) shear walls within the SAFE program were aimed at observing the effect of several parameters in the response. Particularly, as discussed by Labbé et al. [1], the study of the effect of the design frequency on the safety margin was one of the objectives. The current work takes advantage of those experimental results in order to derive a structural degradation model that potentially may be used for predicting in an approximate manner the maximum response of those walls to any given seismic excitation.

The description of the degradation in the structural behaviour as a function of the damage, and particularly by means of a reduction in the stiffness (or eigenfrequency) that is a function of the maximum performed displacement, is present in many studies and we will refer here just to a few of them that have especial common points with the current work.

In the study of Benedetti and Limongelli [2], shaking-table tests results on masonry buildings were used to plot "effective modal stiffness" values as a function of the displacement regarding single cycles in the response to successive earthquakes on the same specimen. They observed how for those structures the stiffness K was dependent only on the maximum previously reached displacement X at every cycle. By keeping in this representation only the envelope points of growing displacement, they constructed a model $K(X)$ that reproduced the experimental envelope curve. The model was based on the combination of brittle-elastic elements and elastoplastic elements and was defined by four parameters that were identified from the experimental results.

In the work of Brun et al. [3], results from the seismic tests on the RC shear walls of the SAFE program were reproduced by a finite element model based on the fixed smeared crack concept. The dynamic results of such finite element model to a series of excitations were used for representing the fundamental frequency associated with every level of maximum displacement reached. There, the fundamental frequency f was derived from the secant stiffness of single cycles. After doing similar observations to the ones of Benedetti and Limongelli [2], this envelope curve $f(X)$ from the numerical simulations was substituted by a formula based on a combination of continuous functions and defined by eight parameters that were identified. Then, a simplified dynamic model was used to produce the time response of a wall by just step-by-step integration of the linear 1-DoF equation of motion, but with variable natural frequency at every step, decided from the recorded maximum displacement according to the previously obtained curve $f(X)$. The damping ratio for this simplified dynamic model was chosen fixed at 7% and considered as an average of the experimental values.

On the other hand, in the more recent work of Brun et al. [4], the finite element model based on fixed smeared crack that had been used to produce dynamic response was now used to produce static pushover curves from which to extract the secant stiffness and derive the fundamental frequency at every level of displacement. Moreover, the derived $f(X)$ curve for every wall was successfully compared with the equivalent envelope curve derived from the experimental results of the SAFE program. This time the simplified dynamic model was implemented by using directly the $f(X)$ function, either observed from the numerical pushover simulations (first case) or from the PsD tests for every wall (second case), without additional simplification by a formula and with the use of damping values derived from the experimental ones. The computed responses of the simplified model were similar between the two cases and also with the experimental ones for most of the studied walls.

In the current work, as in the mentioned studies, the variation of the stiffness with respect to the displacement level is observed from the experimental results. The special methods used for deriving an instant secant stiffness and displacement amplitude, as well as the rule for defining the envelope curve $K(X)$ or $f(X)$, are introduced and applied for all the tested walls of the SAFE program. Then, as a step forward with respect to other works, such curve is used as a capacity line that can be directly checked with the demand spectrum [5] in order to predict the maximum response to a given earthquake. Moreover, for every single tested wall, or for groups of them in some cases, the envelope curve is adjusted to a very simple straight line in double logarithmic scale, which is also used as the capacity line instead of the experimental curves. Interestingly, this simple model (characterised by only two parameters) is able to reproduce the order of magnitude of the maximum response to the tested inputs in most of the cases. It is also effective for producing vulnerability lines that give justification to the observations made by Labbé et al. [1] regarding the influence of the design frequency in the safety margin, for example. Moreover, this study gives one more graphical illustration of the influence of two main factors on the safety margin, which are, according to Labbé [6], the type of structural stiffness degradation and the shape of the demand spectrum at the performed frequencies.

2. SAFE Program Data Processing

2.1. Testing Campaign. As described in other publications [1, 3, 4, 7], the wall specimens of the SAFE program were 13 with names T01$\cdots$T13 and they were all seismically tested in pure shear at the ELSA laboratory by means of the pseudodynamic (PsD) method, which is a hybrid testing technique by which the inertial forces are modelled numerically [8].

For all the walls, the length was

$$L = 3\,\mathrm{m} \tag{1}$$

and the height was

$$H = 1.2\,\mathrm{m}. \tag{2}$$

The main parameters with different value among the walls are displayed in Tables 1 and 2. The thickness of the walls was

$$t_{\mathrm{T}i} = 0.16\,\mathrm{m} \tag{3}$$

for the first four specimens T01$\cdots$T04 and

$$t_{\mathrm{T}i} = 0.2\,\mathrm{m} \tag{4}$$

for the remaining ones T05$\cdots$T13. Consequently, the section of every wall was

$$S_{\mathrm{T}i} = Lt_{\mathrm{T}i}. \tag{5}$$

Even though the concrete used was of two types with different actual capacities, the design of the walls assumed a conventional capacity value of

$$f_{c28}^{d} = 30\,\mathrm{MPa} \tag{6}$$

corresponding to a design shear modulus of

$$G_d = 1.424 \cdot 10^{10}\,\mathrm{Pa} \tag{7}$$

for a Poisson's ratio 0.2 (according to Labbé et al. [1]), and the design stiffness

$$K_{d\mathrm{T}i} = \frac{G_d S_{\mathrm{T}i}}{H}. \tag{8}$$

In the SAFE program, the design frequency $f_{d\mathrm{T}i}$ was made a key parameter that was imposed with several values among the different walls in order to observe the effect of it in the response. As shown in Table 1, its value was

$$f_{d\mathrm{T}i} = 4,\ 8 \text{ or } 12\,\mathrm{Hz}, \tag{9}$$

whereas the design damping ratio was adopted constant

$$\xi_d = 7\% \tag{10}$$

for all the walls.

Table 1 also contains the reinforcement density in horizontal $\rho_{h\mathrm{T}i}$ and vertical $\rho_{V\mathrm{T}i}$ directions. These are understood as nominal ratio of the steel section divided by the concrete section transversal to the bars. The conventional design

TABLE 1: Parameters of tested walls I.

Specimen Ti	$t_{\mathrm{T}i}$ m	$S_{\mathrm{T}i}$ m^2	$K_{d\mathrm{T}i}$ GN/m	$f_{d\mathrm{T}i}$ Hz	$M_{\mathrm{T}i}$ ton	$\rho_{h\mathrm{T}i}$# %	$\rho_{v\mathrm{T}i}$# %
Wall T01	0.16	0.48	$	$	$	0.8 (0.770)	0.8 (0.770)
Wall T02	0.16	0.48	$	$	$	0.8 (0.770)	0.8 (0.770)
Wall T03	0.16	0.48	5.7	4	9017.5	0.8 (0.770)	0.8 (0.770)
Wall T04	0.16	0.48	5.7	12	1001.9	0.8 (0.785)	0.8 (0.785)
Wall T05	0.2	0.6	7.12	8	2818	0.8 (0.767)	0.8 (0.767)
Wall T06	0.2	0.6	7.12	12	1252.4	0.6 (0.628)	0.4 (0.402)
Wall T07	0.2	0.6	7.12	4	11272	0.6 (0.628)	0.4 (0.402)
Wall T08	0.2	0.6	7.12	12	1252.4	0.4 (0.402)	0.4 (0.402)
Wall T09	0.2	0.6	7.12	4	11272	0.4 (0.402)	0.4 (0.402)
Wall T10	0.2	0.6	7.12	4	11272	0.6 (0.628)	0.6 (0.628)
Wall T11	0.2	0.6	7.12	4	11272	0.4 (0.402)	0.4 (0.402)
Wall T12	0.2	0.6	7.12	4	11272	0.11 (0.113)	0.11 (0.113)
Wall T13	0.2	0.6	7.12	12	1252.4	0.4 (0.402) + CFRP	0.4 (0.402) + CFRP

$The tests on walls T01 and T02 did not follow the same design scheme as the rest of the walls and their results will be used in this study only in a limited manner.
#Actual values of reinforcement density implemented in the experiments are shown in brackets.

TABLE 2: Parameters of tested walls II.

Specimen Ti	$\sigma_{nd\mathrm{T}i}$ MPa	$\sigma_{n\mathrm{T}i}$ (actual) MPa	$\tau_{d\mathrm{T}i}$ MPa	$\gamma_{d\mathrm{T}i}$ mrad	$u_{d\mathrm{T}i}$ mm	$^s\mathrm{psa}_0(f_{d\mathrm{T}i}, \xi_d)$ m/s^2	$\beta_{\mathrm{T}i}$
Wall T01	0	0.37	4.00	0.281	0.337	$	$
Wall T02	0	0.37	4.00	0.281	0.337	$	$
Wall T03	0	0.37	4.00	0.281	0.337	2.55	0.083
Wall T04	0	0.37	4.00	0.281	0.337	1.49	1.286
Wall T05	0	0.32	4.00	0.281	0.337	2.01	0.424
Wall T06	1	1.01	3.00	0.211	0.253	1.49	0.965
Wall T07	1	1.01	3.00	0.211	0.253	2.55	0.063
Wall T08	0	0.32	2.00	0.140	0.169	1.49	0.643
Wall T09	0	0.32	2.00	0.140	0.169	2.55	0.042
Wall T10	0	0.32	3.00	0.211	0.253	2.55	0.063
Wall T11	0	0.32	2.00	0.140	0.169	2.55	0.042
Wall T12	1	1.01	1.55	0.109	0.131	2.55	0.032
Wall T13	0	0.32	2.00	0.140	0.169	1.49	0.643

$The tests on walls T01 and T02 did not follow the same design scheme as the rest of the walls and their results will be used in this study only in a limited manner.

criterion applied to the SAFE walls [1] was based on the design shear stress defined as

$$\tau_{d\mathrm{T}i} = \rho_{v\mathrm{T}i} f_{ed} + \sigma_{nd\mathrm{T}i}, \tag{11}$$

where the nominal capacity of the steel was

$$f_{ed} = 500\,\mathrm{MPa} \tag{12}$$

and the nominal normal stress was

$$\sigma_{nd\mathrm{T}i} = 0\,\mathrm{MPa}\ (\text{"low"})\ \text{ or }\ 1\,\mathrm{MPa}\ (\text{"high"}) \tag{13}$$

depending on the wall as shown in Table 2. Note that the actual value of normal stress implemented in the tests was between 0.32 and 0.37 ("low") and 1.01 MPa ("high") but the design was based on the nominal values 0 or 1. Similarly, for the reinforcement densities, the value entered in (11) was the nominal one, independently of its difference with the actual value also shown in Table 1 in brackets for completeness.

The associated design drift or shear deformation was

$$\gamma_{d\mathrm{T}i} = \frac{\tau_{d\mathrm{T}i}}{G_d} \tag{14}$$

and the design displacement at the top of the wall was

$$u_{d\mathrm{T}i} = H\gamma_{d\mathrm{T}i}. \tag{15}$$

The reference input motion $g_0(t)$ was an artificial accelerogram normalised to have Peak Ground Acceleration (PGA) of 1 m/s^2 (Figure 1). The sampling period of this specified accelerogram was

$$\Delta t = 0.01\,\mathrm{s}. \tag{16}$$

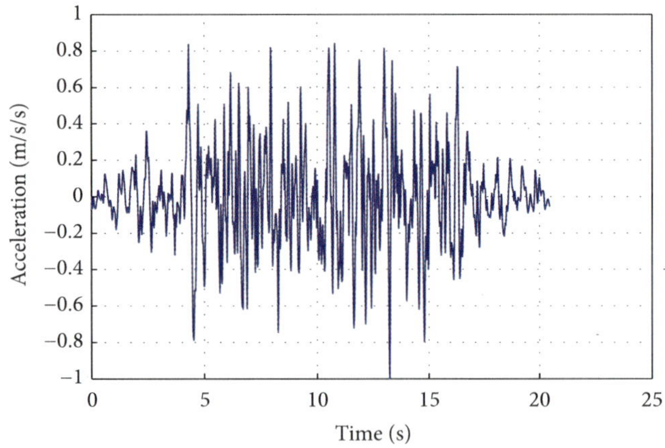

FIGURE 1: Reference input motion artificial normalised accelerogram $g_0(t)$.

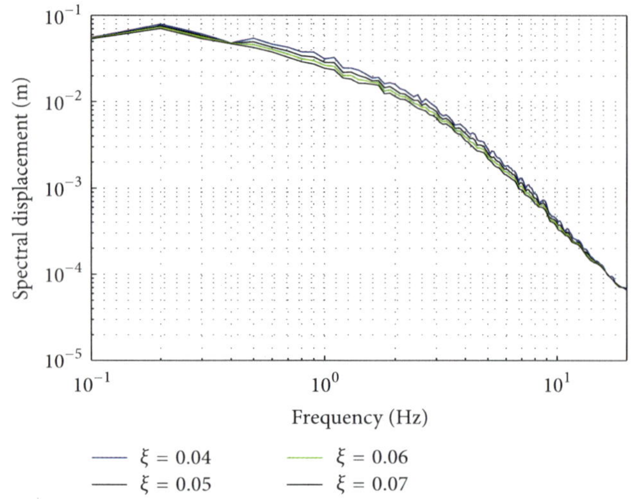

FIGURE 2: Displacement response spectrum ${}^{s}u_0(f,\xi)$ of reference input.

Figure 2 shows the displacement response spectrum (RS) ${}^{s}u_0(f,\xi)$ of the reference input and Figure 3 shows the associated pseudoacceleration spectrum

$$^{s}\mathrm{psa}_0(f,\xi) = {}^{s}u_0(f,\xi)(2\pi f)^2. \quad (17)$$

for damping ratio values of

$$\xi = 4, 5, 6, 7\%. \quad (18)$$

The applied design criterion in the SAFE program was to match the design shear stress (11), and consequently associated drift (14) and displacement (15), at the design spectrum, for the design frequency and damping by introducing a scaling factor $\beta_{\mathrm{T}i}$; that is,

$$u_{d\mathrm{T}i} = \beta_{\mathrm{T}i}\,{}^{s}u_0(f_{d\mathrm{T}i}, \xi_d). \quad (19)$$

Thus, the scaling factor for every wall is defined as

$$\beta_{\mathrm{T}i} = \frac{u_{d\mathrm{T}i}}{{}^{s}u_0(f_{d\mathrm{T}i}, \xi_d)} = \frac{\tau_{d\mathrm{T}i}H/G_d}{{}^{s}u_0(f_{d\mathrm{T}i}, \xi_d)} \quad (20)$$

with values shown in Table 2. Correspondingly, the design input motion of wall Ti is defined as

$$g_{d\mathrm{T}i}(t) = \beta_{\mathrm{T}i}g_0(t) \quad (21)$$

which nominally should induce the design stress in the wall.

Within the SAFE testing campaign, every wall was submitted to several successive earthquake runs with increasing intensity:

$$g_{\mathrm{T}ij}(t) = \alpha_{\mathrm{T}ij}g_{d\mathrm{T}i}(t), \quad (22)$$

where $\alpha_{\mathrm{T}ij}$ was the scaling factor of wall Ti for experiment run j as displayed in Table 3.

The chosen design frequency for every wall was achieved by adopting a theoretical mass

$$M_{\mathrm{T}i} = \frac{K_{d\mathrm{T}i}}{(2\pi f_{d\mathrm{T}i})^2} \quad (23)$$

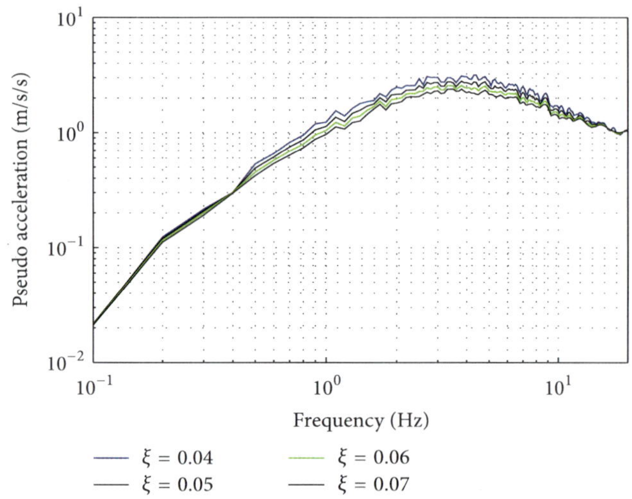

FIGURE 3: Pseudo acceleration response spectrum ${}^{s}\mathrm{psa}_0(f,\xi)$ of reference input.

to oscillate with the displacement of the wall in the equation of motion integrated within the PsD testing method (Table 1).

Note that in Table 3, apart from the test name in the format "Ti − j," a test reference code "fNN" is also shown for data traceability in the laboratory database. The last six columns in this table contain identified parameters that will be discussed in the following sections.

2.2. Signals Used for the Current Study. The current study is based on the identification of instantaneous values of secant stiffness, displacement amplitude, and other variables from the quasistatic pseudo-dynamic response at every experiment in order to characterise the behaviour of every wall specimen.

Within the applied PsD testing method, the step-by-step time-integrated displacement is sent to the controller of the actuators as a reference displacement that has to be reproduced at the feedback displacement transducer. This is

Table 3: Parameters of the performed tests.

Wall	β_{Ti}	Test	Test reference	α_{Tij}	PGA m/s^2	u^*_{Tij} mm	$\overline{f}^*_{Tij}$ Hz	$\overline{\xi}^*_{Tij}$ %	$\overline{G}^*_{Tij}$ GPa	$\overline{\overline{\xi}}_{Ti}(\overline{G}^*_{Tij})$ %	ξ^{op}_{Ti} %
T01	$	T01-1	f03	$	$	$	$	$	$	$	$
		T01-2	f05	$	$	$	$	$	$	$	$
T02	$	T02-1	f13	$	$	$	$	$	$	$	$
T03	0.083	T03-1	f22	1	0.0850	1.43	1.71	4.9	2.6144	4.7	5.3
		T03-2	f23	2	0.1700	4.67	1.27	4.7	1.4442	5.3	2.9
		T03-3	f24	3	0.2550	6.18	1.13	5.4	1.1438	5.5	6.0
		T03-4	f25	5	0.4250	11.74	0.86	15.2	0.6549	12.6	7.1
T04	1.286	T04-1	f32	1	1.3061	8.06	3.74	4.6	1.3866	4.5	2.8
		T04-2	f33	1.3	1.6979	14.02	2.33	19.4	0.5383	24.2	9.6
		T04-3	f34	1.5	1.6470	12.13	1.71	14.9	0.2898	42.8	42.0
T05	0.424	T05-1	f50	1	0.4293	3.85	2.72	4.4	1.6480	5.0	5.5
		T05-2	f51	1.3	0.5581	6.50	2.51	5.1	1.4001	4.7	4.2
		T05-3	f52	1.5	0.6438	7.18	2.37	4.5	1.2530	4.8	4.5
		T05-4	f53	2	0.8585	10.32	1.96	8.2	0.8561	8.4	7.0
		T05-5	f54	2	0.7216	18.63	0.88	14.9	0.1735	10.9	7.3
T06	0.965	T06-1	f44	1	0.9797	3.10	4.27	4.5	1.8021	4.6	6.8
		T06-2	f45	1.3	1.2736	6.22	3.52	5.6	1.2270	6.5	6.9
		T06-3	f46	1.5	1.4695	8.79	3.33	7.3	1.0972	7.0	5.9
		T06-4	f47	1.8	1.7634	14.43	2.37	13.4	0.5568	13.3	9.1
T07	0.063	T07-1	f58	1	0.0635	0.71	2.36	5.6	4.9715	5.7	4.6
		T07-2	f59	2	0.1270	1.86	1.71	5.3	2.6030	5.3	7.8
		T07-3	f60	5	0.3176	7.70	1.16	7.5	1.1956	7.3	4.6
		T07-4	f61	10	0.6351	17.45	0.43	14.6	0.1629	17.0	8.0
T08	0.643	T08-1	f65	1	0.6530	2.56	4.02	5.6	1.5975	5.6	6.0
		T08-2	f66	1.4	0.9142	5.42	3.34	3.8	1.1052	4.8	5.9
		T08-3	f67	1.8	1.1754	11.85	1.80	13.1	0.3197	9.4	15.2
T09	0.042	T09-1	f72	1	0.0423	0.58	2.34	5.4	4.8871	5.6	2.4
		T09-2	f74	3	0.1268	2.25	1.51	5.0	2.0225	5.2	5.9
		T09-3	f75	6	0.2536	5.74	1.15	4.2	1.1751	4.2	6.1
		T09-4	f76	10	0.4227	15.77	0.57	26.5	0.2864	28.7	8.0
T10	0.063	T10-1	f84	1	0.0635	0.78	2.05	5.7	3.7418	6.0	6.1
		T10-2	f85	3	0.1905	5.43	1.27	5.0	1.4330	4.5	2.6
		T10-3	f86	6	0.3811	11.57	0.99	9.5	0.8742	8.3	4.3
		T10-4	f87	10	0.5339	14.72	0.78	16.1	0.5414	15.7	7.9
T11	0.042	T11-1	f79	1	0.0423	0.46	2.36	3.8	4.9365	4.0	5.1
		T11-2	f80	6	0.2536	6.47	1.08	5.9	1.0302	6.0	5.6
		T11-3	f81	10	0.4227	13.38	0.75	9.0	0.4994	7.3	5.9
T12	0.032	T12-1	f37	1	0.0306	0.25	3.23	3.1	9.3013	3.5	2.9
		T12-2	f38	3	0.0917	0.85	2.19	5.0	4.2524	5.0	9.2
		T12-3	f39	5	0.1528	2.60	1.45	4.9	1.8587	5.2	6.8
		T12-4	f40	10	0.3056	7.80	1.03	7.0	0.9413	6.2	6.1
		T12-5	f41	15	0.4585	18.92	0.51	28.4	0.2328	29.2	6.9
T13	0.643	T13-1	f90	1	0.6530	0.40	8.83	4.2	7.7074	5.0	6.3
		T13-2	f91	1.8	1.1754	7.16	3.37	4.6	1.1242	6.0	5.4
		T13-3	f92	2.6	1.6977	15.58	1.35	12.1	0.1807	13.7	31.0

[$]The tests on walls T01 and T02 did not follow the same design scheme as the rest of the walls and their results will be used in this study only in a limited manner.

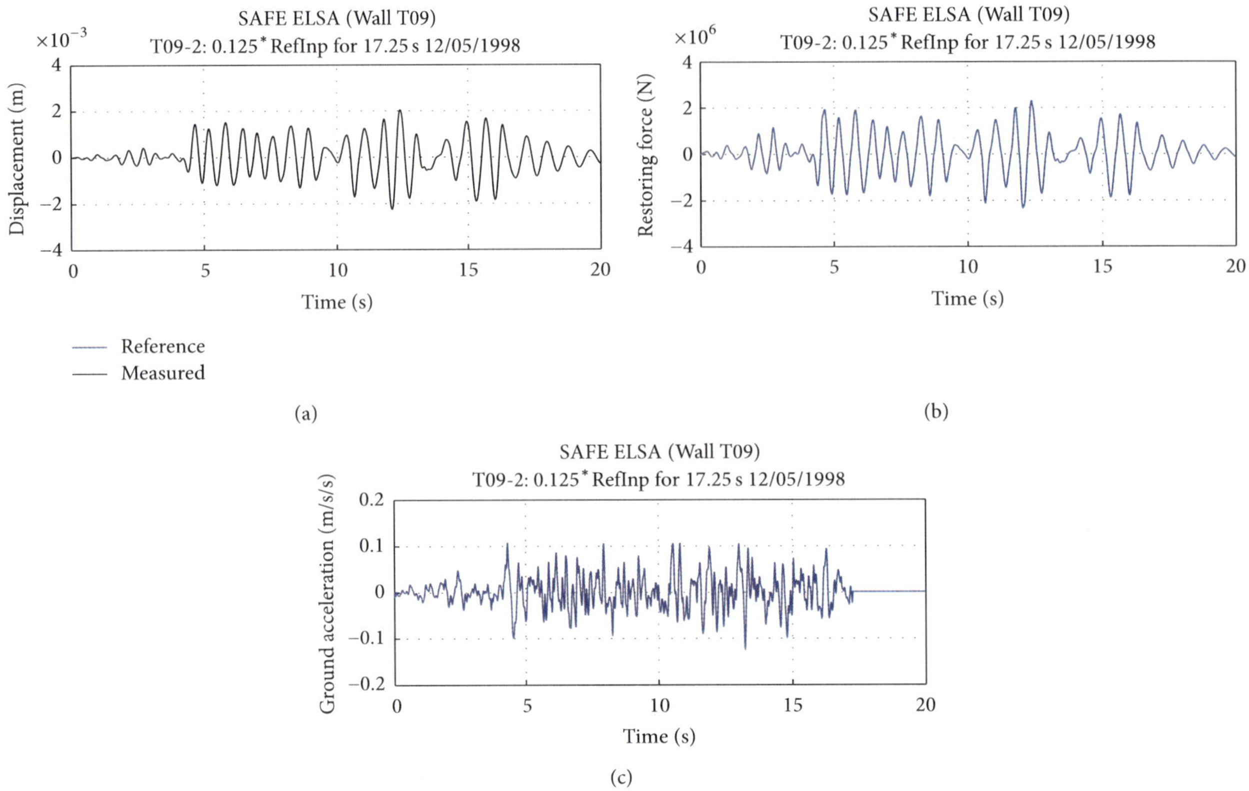

FIGURE 4: Example of earthquake response in the 2nd test on wall T09.

normally achieved with a small discrepancy which is called control error. In cases of inadequate testing conditions and, particularly for fast movement of the actuators, the amount of error may be large enough to affect significantly the apparent behaviour of the experimental model and its response to the specified input. This is due to the fact that the PsD equation of motion is based on measured forces that do not correspond to the integrated reference displacement [9, 10]. This problem was present in the SAFE program for the tests on the first wall T01 and in a discrete manner on the second wall T02. However, even for those tests with an erroneous response to the earthquake, the synchronous measures of force and displacement always give reliable information about the real behaviour of the specimen as from any cyclic test, without considering the discrepant reference displacement at the controller. In addition, the tests performed on walls T01 and T02 did not follow the same method as the rest of the walls for the definition of the theoretical mass and the input acceleration. Thus, in this study, they will be considered only as cyclic tests and their results will be used only for the identification of the deformation-stiffness correlation.

As an example for a test response in one of the walls (wall T09, 2nd earthquake), in Figure 4, the time histories of reference and measured displacements (a), restoring force (b), and ground acceleration (c) are plotted. For most of the tests the input accelerogram was set to zero for a final period of time before the end of the test in order to observe the free response of the system. Note that for this experiment the accelerogram was set to zero from 17.25 s on. For the same wall T09, the measured force-displacement cycle for the four earthquakes applied to it is shown in Figure 5.

2.3. Spatial Model for Frequency and Damping Identification. In contrast with other studies, such as the ones mentioned in the Introduction, the identification of stiffness (or eigenfrequency) and damping is not based here on a cycle-by-cycle analysis. It is instead based on a moving observation window of constant duration. Nevertheless, the authors think that this choice would not affect much the proposed models and conclusions.

The Spatial Model method [9, 11–13] identifies the linear-equivalent stiffness and damping coefficients that best fit the displacement and measured (quasistatic) restoring force. The identification is performed based, firstly, on the reference displacement and its derived velocity and, secondly, on the measured displacement and its derived velocity. In both cases the derivatives of the displacement are computed by central differences based on the recorded time increment that coincides with the sampling period of the accelerogram (16). In order to avoid the influence of the control errors in this study, the identification based on the measured displacement is used by default for further computation of parameters.

Since the linear identified model assumes a time-invariant system (without degradation), the identification is repeated at every recording time increment of 0.01 s and is based on a window of data of limited duration that has constant length for all the earthquakes on the same specimen

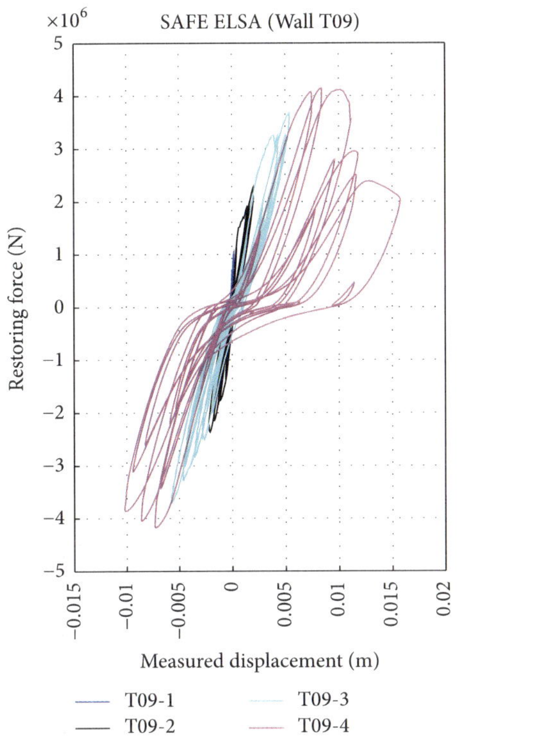

FIGURE 5: Force-displacement cycles for all the tests performed on wall T09.

Ti. That window is chosen to cover one "expected maximum" period of oscillation

$$T_{\mathrm{T}i}^{\max} = 2\pi\sqrt{\frac{M_{\mathrm{T}i}}{(G^{\min}S_{\mathrm{T}i}/H)}} \tag{24}$$

or at least 40 time samplings (0.4 s), where it was assumed

$$G^{\min} = 5 \cdot 10^{8}\ \mathrm{Pa}. \tag{25}$$

Note that in general the adopted time window has to be narrow enough so that the system does not change strongly inside of it (due to degradation or change of oscillation amplitude), but, at the same time, it has to contain enough data to allow the compensation of different existing data noises. Owing to the width of the symmetric window around time t, this identification cannot be done centred at the beginning or the end of the record.

From the computed histories of identified stiffness $\overline{K}_{\mathrm{T}i}(t)$ and damping $\overline{C}_{\mathrm{T}i}(t)$, the eigenfrequency

$$\overline{f}_{\mathrm{T}i}(t) = \frac{1}{2\pi}\sqrt{\frac{\overline{K}_{\mathrm{T}i}(t)}{M_{\mathrm{T}i}}} \tag{26}$$

and damping ratio

$$\overline{\xi}_{\mathrm{T}i}(t) = \frac{\overline{C}_{\mathrm{T}i}(t)}{\sqrt{4\overline{K}_{\mathrm{T}i}(t)\,M_{\mathrm{T}i}}} \tag{27}$$

are derived based on the theoretical mass (23). The computed values, either from the reference displacement or from the measured displacement, are shown as an example for one of the tests in Figure 6. The comparison between both identified values allows us to know the magnitude of the effect of the control error on the identified parameters and presumably on the response. Such effect is negligible in the graphs within this figure.

The tendency of the frequency is to decrease with the time as a consequence of the larger amplitudes and the accumulated damage. On the other hand, the evolution of the damping is much more erratic and difficult to interpret.

2.4. Displacement Oscillation Amplitude. In this study, the instant value of identified stiffness or damping will be correlated with the amplitude of oscillation of the displacement rather than with the instant value of the displacement itself. To do so, the displacement amplitude is computed here by using the Hilbert Transform as computed through the Fast Fourier Transform (FFT) [14]. This technique has the advantage of being easy to automate. However, in the same way as for the identification of the stiffness and damping, a cycle-by-cycle analysis approach could have been adopted instead.

For the identification of the displacement oscillation amplitude $\overline{u}_{\mathrm{T}i}(t)$, in order to have results that do not depend on the control errors, the measured displacement is used, unless stated otherwise. See the example of identified displacement amplitude in Figure 7(a).

Because the automatic identification of the displacement amplitude is based on the FFT, in cases in which the final amplitude was large in comparison to the initial one, it may give aberrant large amplitude values at the beginning of the record. This was avoided by adding a fictitious period of zeros (5 s) at the end of the signals before applying the FFT. Then, back in the time domain, the computed amplitude was cut out the final 5 s. This technique was used in a uniform manner for all the tests and was successful for avoiding the mentioned aberration, except for experiment T02-1. In fact this experiment was special because it showed very few oscillations before an extreme deformation was introduced collapsing the wall. In order to have a consistent amplitude curve, only for this experiment, a special exponential weighting window, starting just before the collapse of the specimen, was applied for the FFT.

Purely as a check, it is interesting to compare, on the one hand, the force amplitude derived from the identified stiffness multiplied by the displacement amplitude and, on the other hand, the real measured force history during the test. Such derived force amplitude (called projected amplitude in Figure 7(b)) appears to be a good approximation on the effective force amplitude, demonstrating the consistency of the identified stiffness as a secant stiffness.

2.5. Envelope of Drift Amplitude with respect to Shear Modulus. Before exploring the correlation between stiffness and displacement amplitude, another set of variables will be introduced for this study. In order to be able to compare walls of different length, thickness, and height or associated mass, for every earthquake, an identified equivalent shear modulus

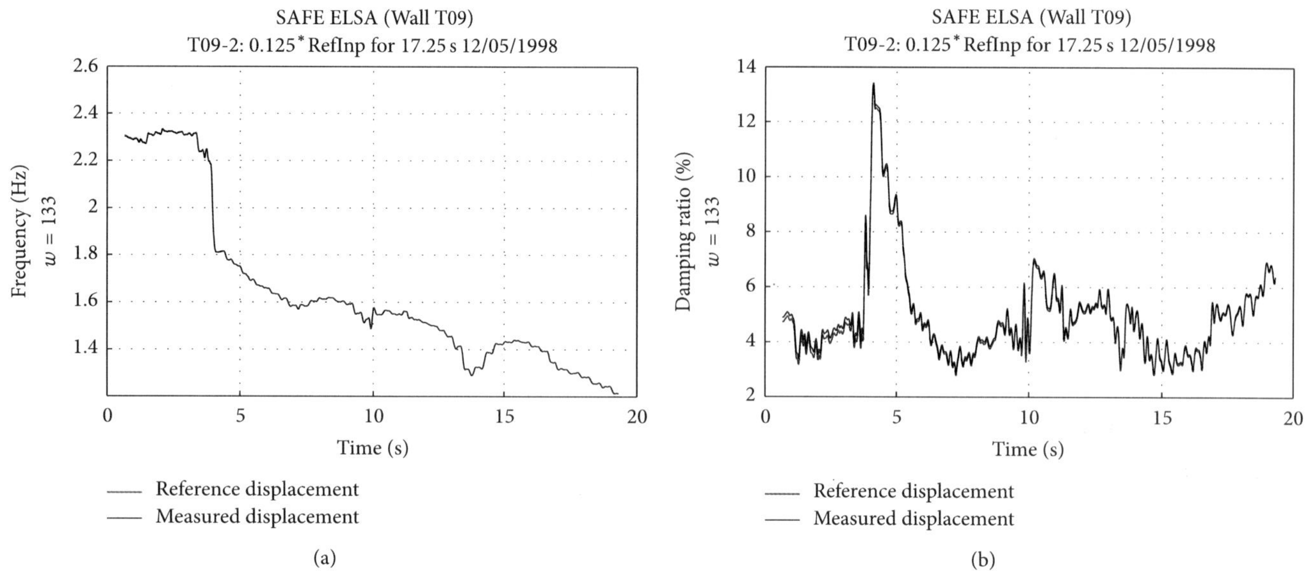

FIGURE 6: Example of identified frequency and damping ratio from 2nd earthquake response on wall T09. The number of sampling points used for the identification window was $w = 133$ for this wall.

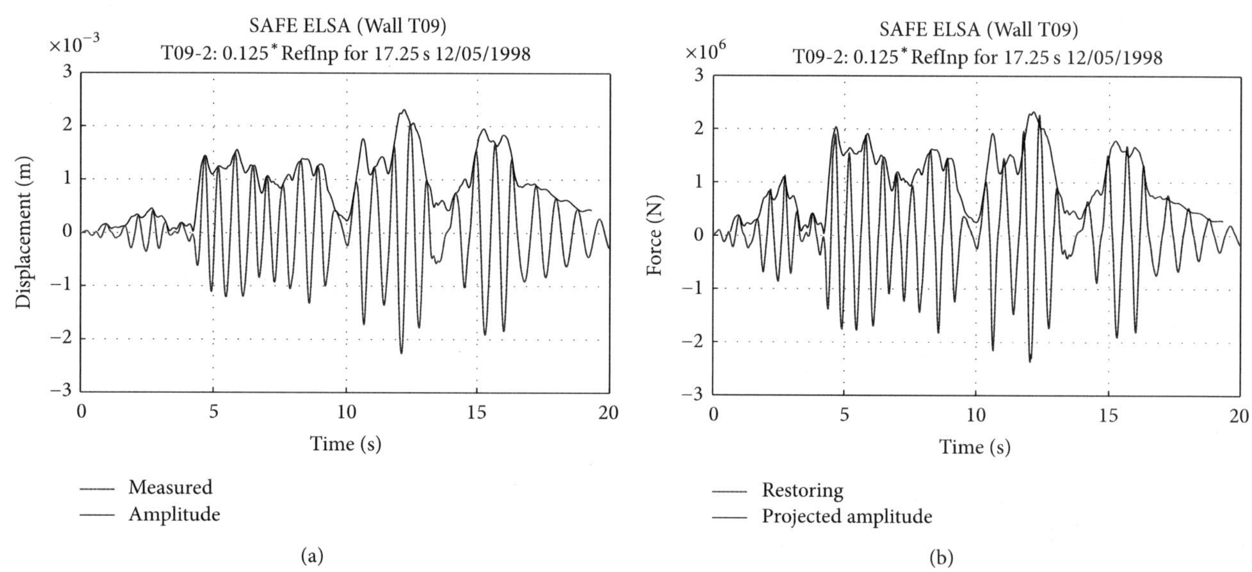

FIGURE 7: Example of identified displacement amplitude and projected force amplitude from 2nd earthquake response on wall T09.

history is derived from the stiffness (or squared frequency by mass) history, as identified by the Spatial Model, by dividing it by the section of the wall and multiplying by its height:

$$\overline{G}_{Ti}(t) = \frac{H}{S_{Ti}}\overline{K}_{Ti}(t) = \frac{M_{Ti}H}{S_{Ti}}\left(2\pi\overline{f}_{Ti}(t)\right)^2. \quad (28)$$

Similarly, the identified amplitude of drift strain is simply computed as the amplitude of displacement divided by the height of the wall:

$$\overline{\gamma}_{Ti}(t) = \frac{\overline{u}_{Ti}(t)}{H}. \quad (29)$$

For all the earthquakes on a single wall, the synchronous values of $\overline{G}_{Ti}(t)$ and $\overline{\gamma}_{Ti}(t)$ can be represented in the (G, γ) axes. For example, this is done in double logarithmic scale in Figure 8(a) with a different colour for the response trajectory $(\overline{G}_{Ti}(t), \overline{\gamma}_{Ti}(t))$ for every earthquake applied on wall T09 (tests T09-1$\cdots$4 in Table 3). For the progressively growing intensity input, there is a tendency to irreversibly decrease the identified equivalent shear modulus whenever the drift amplitude in the response reaches new higher values. The meaning of this is interpreted by assuming that new damage is introduced only when a new largest displacement amplitude is recorded. Then, a monotonic envelope line can be constructed by processing point by point chronologically from the successive trajectories but retaining only some selected points. More precisely, the first envelope point $t_1^{\times}$ for the wall is selected as the first recorded point instant t_1 from its first

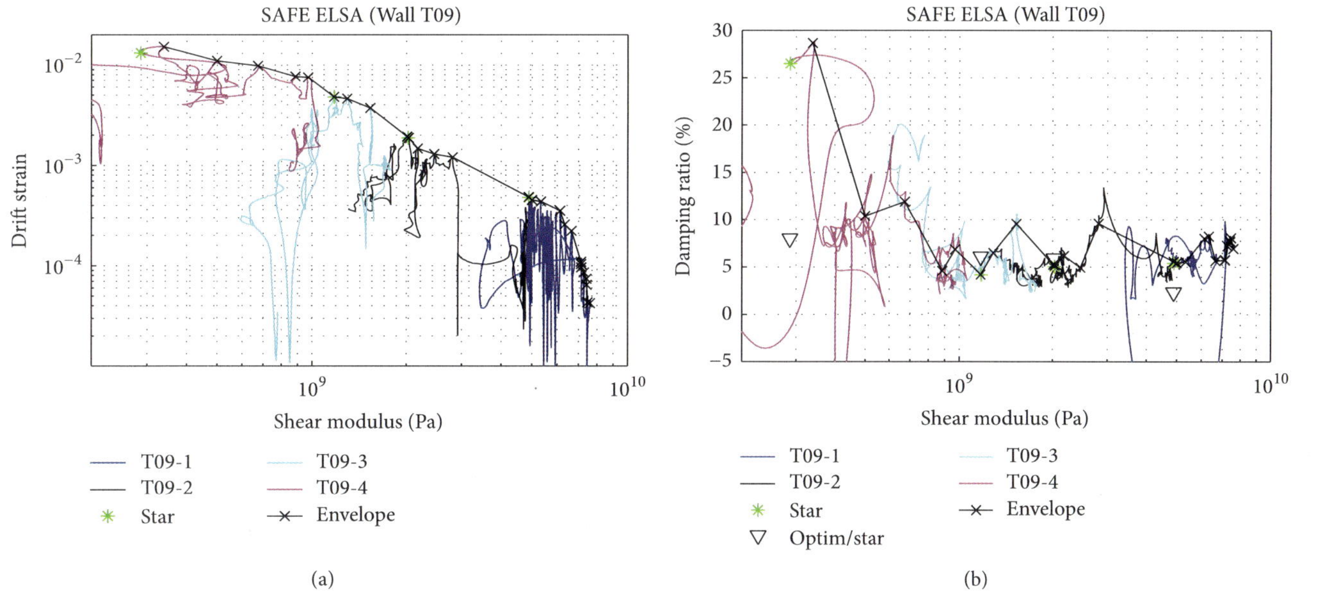

FIGURE 8: Identified values of drift amplitude and damping ratio as a function of the identified shear modulus from all the successive earthquake responses on wall T09.

test. Afterwards, a successive time instant is taken as a new selected point $t_{n+1}^{\times}$ for the envelope only when, with respect to the last previous selected point $t_n^{\times}$, it represents a positive drift amplitude increment $\overline{\gamma}_{\mathrm{T}i}(t_{n+1}^{\times}) > \overline{\gamma}_{\mathrm{T}i}(t_n^{\times})$ associated with a decrement of the modulus $\overline{G}_{\mathrm{T}i}(t_{n+1}^{\times}) < \overline{G}_{\mathrm{T}i}(t_n^{\times})$. The obtained monotonic envelope is denoted by $\overline{\overline{\gamma}}_{\mathrm{T}i}(\overline{\overline{G}}_{\mathrm{T}i})$ and in Figure 8 it is represented with a black line joining the selected points $(\overline{G}_{\mathrm{T}i}(t_n^{\times}), \overline{\gamma}_{\mathrm{T}i}(t_n^{\times}))$ marked by crosses $\times$.

Note that there exist also cases of reversible decrease of the secant modulus associated with a decrease of the cycles amplitude. This is visible in Figure 8 and, by looking at Figure 5, it can be interpreted as a pinching behaviour associated with a gap at the central part of the cycles that is created by the previous existing cracks that close and reopen. Such decrease of the secant modulus is temporary: it does not seem to introduce new damage and does not contribute new points to the defined envelope.

In a similar manner, Figure 8(b) shows in the (G, ξ) axes the synchronous identified values of $\overline{G}_{\mathrm{T}i}(t)$ and $\overline{\xi}_{\mathrm{T}i}(t)$ for the same earthquakes in semilogarithmic scale. Then, considering only the time instants $t_n^{\times}$ of the points that have been selected for the $\overline{\overline{\gamma}}_{\mathrm{T}i}(\overline{\overline{G}}_{\mathrm{T}i})$ envelope in Figure 8(a), the value of the viscous-equivalent damping ratio $\overline{\xi}_{\mathrm{T}i}(t_n^{\times})$ allows creating a curve $\overline{\overline{\xi}}_{\mathrm{T}i}(\overline{\overline{G}}_{\mathrm{T}i})$ of damping as a function of the modulus $\overline{G}_{\mathrm{T}i}(t_n^{\times})$ based on the contemporary points of the $\overline{\overline{\gamma}}_{\mathrm{T}i}(\overline{\overline{G}}_{\mathrm{T}i})$ envelope.

The maximum response absolute drift measured for every earthquake run j on wall Ti

$$\gamma_{\mathrm{T}ij}^{*} = \left|\gamma_{\mathrm{T}i}\left(t_{\mathrm{T}ij}^{*}\right)\right| \tag{30}$$

happens at a time instant called $t_{\mathrm{T}ij}^{*}$ and can be represented in the same axes by associating with it a shear modulus

$$\overline{G}_{\mathrm{T}ij}^{*} = \overline{G}_{\mathrm{T}i}\left(t_{\mathrm{T}ij}^{*}\right) \tag{31}$$

and a damping ratio

$$\overline{\xi}_{\mathrm{T}ij}^{*} = \overline{\xi}_{i}\left(t_{\mathrm{T}ij}^{*}\right) \tag{32}$$

both taken to happen at the same time instant $t_{\mathrm{T}ij}^{*}$ as the maximum drift. These defined "star" values are represented in the same axes in Figure 8 with a green "$*$" mark for every earthquake run on wall T09. In general, these star points are close to the envelopes $\overline{\overline{\gamma}}_{\mathrm{T}i}(\overline{\overline{G}}_{\mathrm{T}i})$ and $\overline{\overline{\xi}}_{\mathrm{T}i}(\overline{\overline{G}}_{\mathrm{T}i})$ but do not fall exactly on them, because, firstly, they derive from the instantaneous absolute drift $|\gamma_{\mathrm{T}i}(t)|$ and not from the drift amplitude $\overline{\gamma}_{\mathrm{T}i}(t)$ (see, e.g., the difference for the displacement in Figure 7) and, secondly, the values of shear modulus and damping correspond to different time instants from the ones selected for the envelope curves.

3. Shear Wall Behaviour Modelling

3.1. Interpretation of the Monotonic Envelope as a Capacity Curve. The value of the equivalent shear modulus at every moment can be assumed a function of the damage state in a wall, which is hypothetically considered a function of exclusively the maximum experienced drift amplitude in the specimen. With these assumptions, the previously derived monotonic envelope $\overline{\overline{\gamma}}_{\mathrm{T}i}(\overline{\overline{G}}_{\mathrm{T}i})$ can be interpreted in general in Figure 9 as a capacity curve. For a given current value of equivalent shear modulus G_1 in a specimen, assuming

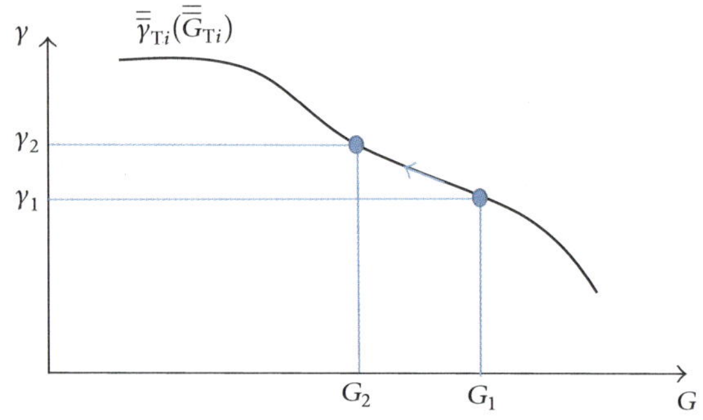

FIGURE 9: Interpretation of the monotonic envelope as a capacity curve.

constant stiffness behaviour for simplification, such capacity curve allows knowing the maximum value of drift deformation $\gamma_1 = \overline{\overline{\gamma}}_{Ti}(G_1)$ that can be reached by that wall before introducing new damage in it. When that level of deformation γ_1 is exceeded up to a larger value γ_2, new damage is introduced allowing us to pass to higher values of strain corresponding to lower values of modulus in the capacity curve following from $\gamma_1 = \overline{\overline{\gamma}}_{Ti}(G_1)$ up to $\gamma_2 = \overline{\overline{\gamma}}_{Ti}(G_2)$. After the introduction of this new damage, the current value of the modulus is G_2 and it will not change until a deformation larger than γ_2 is reached.

It must be noticed that the pinching effect is ignored by this modelling. The pinching effect was observed in the studied experiments and, as described in the previous section, is characterised by a temporary (mostly reversible) variation of the modulus associated with a reduction of the displacement oscillation amplitude. Because of such effects that make the secant stiffness change without introduction of new damage, the current concept of a capacity curve cannot be reliably used for predictions regarding deformation amplitudes smaller than the maximum one previously reached.

A further interpretation of such capacity curve model would be to use it for predicting the maximum response deformation to a given input accelerogram by means of the linear RS for a chosen value of damping ratio. According to this procedure, the estimated maximum response to that earthquake would be at the intersection point of the capacity curve $\overline{\overline{\gamma}}_{Ti}(\overline{\overline{G}}_{Ti})$ with the RS represented in the axes (G, γ). The concept of capacity curve is, in fact, the base for some existing design methods, even though the representation used here is different from the typical acceleration-displacement diagram [5]. Taking as a source the spectral displacement ${}^{s}u_{Tij}(f, \xi)$, in order to transform it to the current axes, these formulas are applied for every value of the frequency f and displacement ${}^{s}u_{Tij}$:

$$G = \frac{H}{S_{Ti}} K_{Ti} = \frac{H}{S_{Ti}} M_{Ti} (2\pi f)^2, \quad (33)$$

$$^{s}\gamma_{Tij} = \frac{^{s}u_{Tij}}{H}. \quad (34)$$

Note that, using (23) and (8), expression (33) can also be written as

$$G = G_d \left(\frac{f}{f_{dTi}} \right)^2. \quad (35)$$

The way to select the damping ratio for the computation of the RS is not established and, for research purposes in this section, we have used several values that result in several curves of the demand spectrum for a given earthquake to be checked against the capacity curve. In Figure 10 the RS for every earthquake is represented for three different values of damping ratio as follows:

(1) A design fixed value defined as $\xi_d = 7\%$ according to the design parameter (10).

(2) The interpolated value in the defined envelope damping line at $G = \overline{G}^*_{Tij}$: that is, $\overline{\xi}^*_{Tij} = \overline{\overline{\xi}}_{Ti}(\overline{\overline{G}}_{Ti})|_{\overline{G}^*_{Tij}}$ (called ξ_{env} in the graph legend), where $\overline{G}^*_{Tij}$ is the modulus identified value at the time instant of the maximum experimental response (31). Point $(\overline{G}^*_{Tij}, \overline{\xi}^*_{Tij})$ is represented as a green "$*$" in Figure 8(b) so that its value can be compared with the other values of damping.

(3) The optimal value ξ_{op} that makes the value of the spectrum at $G = \overline{G}^*_{Tij}$ coincide with the experimental maximum drift (30): that is, ${}^{s}\gamma_{Tij}(\overline{G}^*_{Tij}, \xi_{op}) = \gamma^*_{Tij}$. This means that the spectrum is forced to pass exactly by the "star" point (green "$*$" in Figure 10) of the experimental maximum response for that test. Point $(\overline{G}^*_{Tij}, \xi_{op})$ is represented as a red triangle in Figure 8(b) so that its value can be compared with the other values of damping as well.

Finally, the square symbol in Figure 10 corresponds to the design point (design stress and design stiffness) from which the beta parameter (20) (fixing the intensity of the design earthquake) has been chosen so that the corresponding spectrum ($\alpha = 1$) passes through it at G_d. This point nominally corresponds to the response to the design earthquake, but typically it is not on the capacity curve because the effective initial stiffness of the wall is lower (in average by 0.7 according to Labbé et al. [1]) than the design one and because the design deformation (14) is typically reached for a certain level of damage with a degraded shear modulus. Note also that in some cases the design point might not be exactly on the response spectrum of any implemented test accelerogram, due to the difference regarding the applied duration of the signal (see the remarks done in Section 2.2).

3.2. Modelling the Monotonic Envelope by a Polygonal Curve. A generic experimental monotonic envelope of a wall linking drift deformation and shear modulus $\overline{\overline{\gamma}}_{Ti}(\overline{\overline{G}}_{Ti})$, as introduced in the previous sections, is represented in double logarithmic scale in Figure 11 by using a black line. We propose here to model such envelope by means of a piecewise line curve. Starting from the lowest values of deformation, the model may include up to three straight-line segments as follows:

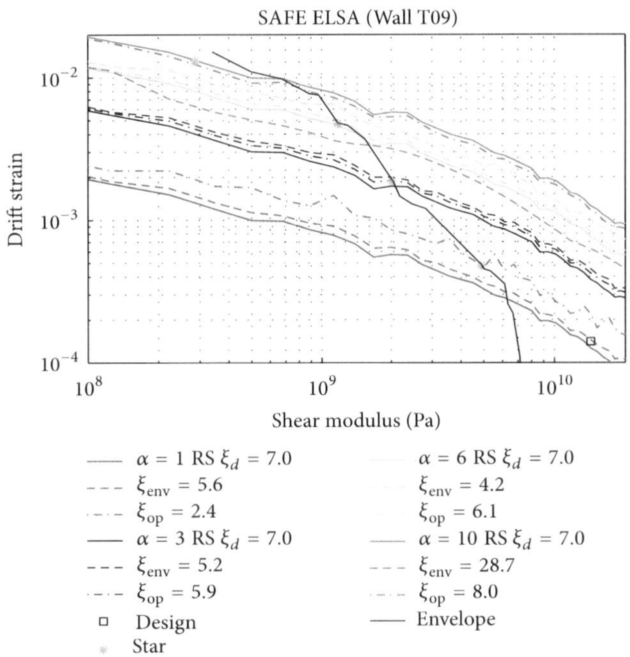

FIGURE 10: Matching of the monotonic envelope with the earthquake response spectra.

(i) Low Degradation Line, where the stiffness diminishes moderately when the deformation increases.

(ii) Regular Degradation Line (RDL), where the stiffness diminishes in a regular manner for a wide range of deformation amplitude (roughly for drift between $2 \cdot 10^{-4}$ and 10^{-2} rad).

(iii) Massive Degradation Line, where the stiffness accelerates its fall near the collapse of the wall.

The maximum response for most of the executed experiments in the SAFE program falls within the RDL segment, which is considered to extend between points L and M in Figure 11. The axis of the RDL is defined by the formula

$$\frac{\gamma}{\gamma_R} = \left(\frac{G}{G_R}\right)^m \tag{36}$$

which is a straight line in the double logarithmic scale that is defined by its slope m and by passing by the reference point R: (G_R, γ_R).

In the following sections this model will be identified for single walls or groups of walls and then it will be used as a simpler substitute of the envelope capacity line for predicting the maximum response to given earthquakes.

3.3. Polygonal Curve Model for Every Wall Specimen and Prediction of Maximum Response to Every Earthquake. In this section, from the monotonic envelope based on the experimental results on every wall, a polygonal curve model will be identified. For all the walls, the identification of the RDL segment, as represented in Figure 11 and (36), will be based on the envelope points for which two conditions apply:

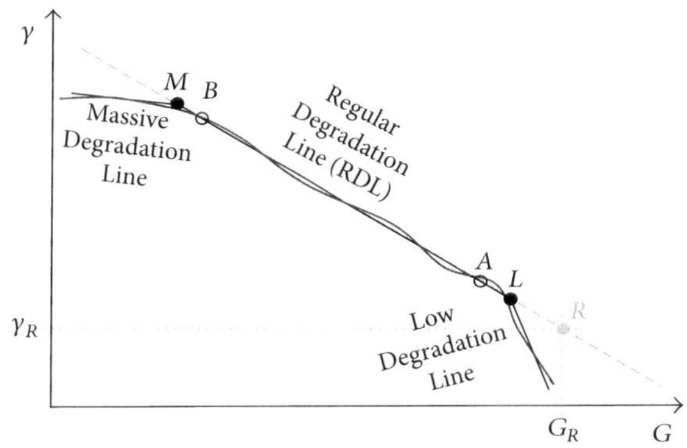

FIGURE 11: Piecewise line model of the monotonic envelope in double logarithmic scale.

$$\gamma > 2.5 \cdot 10^{-4}, \qquad G > 7.0 \cdot 10^{8}\ \text{Pa} \tag{37}$$

in order to make sure that the initial low degradation and final massive degradation phases are excluded. Since the abscissa of the reference point R for the RDL axis is in principle arbitrary, here it has been chosen fixed as

$$G_R = G_d = 1.424 \cdot 10^{10}\ \text{Pa} \tag{38}$$

so that for simplification it coincides with the design shear modulus (7). The coefficients in (36) have been identified after writing such formula in logarithmic form

$$\log(\gamma) = m \log(G) - m \log(G_R) + \log(\gamma_R) \tag{39}$$

and applying linear regression, using log(G) as independent variable and log(γ) as ordinate, taking the input values from the points of the experimental monotonic envelope $\overline{\overline{\gamma}}_{Ti}(\overline{\overline{G}}_{Ti})$. The identified parameters m and γ_R for every wall are collected in Table 4.

In order to simplify the model as much as possible in the current application, we will assume that the abscissa of point L (Figure 11), determining the start of the regular degradation in the model, equals the design shear modulus

$$G_L = G_d = 1.424 \cdot 10^{10}\ \text{Pa} \tag{40}$$

and, thus, points L and R are coincident in this case. Note that a more accurate position of the abscissa of point L could in practice be dependent on the casting method and manipulation of the specimen that always introduces an amount of initial damage. At the same time, the response to significant earthquakes typically falls rather in the regular degradation behaviour after the point L. Consequently, for practical reasons, the model does not pretend at this stage to be reliable regarding the position of the transition point L between low and regular degradation and a fixed simple criterion has been chosen instead. The Low Degradation Line is chosen to be vertical with constant shear modulus.

TABLE 4: For every wall specimen, identified parameters of RDL with maximum and standard errors of the model predictions by using the response spectrum of the applied earthquakes with damping ratios of $\xi = 4, 5, 6$, or 7%.

Wall	m	γ_R $(*1E-6)$	Max err% (ξ = 4%)	Max err% (ξ = 5%)	Max err% (ξ = 6%)	Max err% (ξ = 7%)	Std err% (ξ = 4%)	Std err% (ξ = 5%)	Std err% (ξ = 6%)	Std err% (ξ = 7%)
T01	−1.7962	61.97	$	$	$	$	$	$	$	$
T02	−1.7593	72.67	$	$	$	$	$	$	$	$
T03	−1.7804	66.09	18	**30**	47	64	14	**16**	26	40
T04	−1.8360	78.39	7	**32**	64	117	6	**29**	58	95
T05	−1.7216	85.01	34	**13**	29	50	24	**8**	19	38
T06	−1.5513	124.79	89	**41**	9	70	71	**29**	6	34
T07	−1.6183	115.28	32	**15**	24	44	22	**8**	12	25
T08	−1.6039	69.84	84	**31**	34	46	53	**19**	21	36
T09	−1.6691	77.37	30	**31**	56	82	20	**20**	31	46
T10	−1.7660	70.31	29	**29**	47	65	20	**17**	25	35
T11	−1.7113	63.14	21	**5**	14	31	17	**5**	11	26
T12	−1.4469	123.32	81	**84 (57#)**	104 (41#)	124 (57#)	48 (41#)	**45 (32#)**	47 (30#)	53 (34#)
T13	−1.4699	142.31	79 (48#)	**141 (36#)**	147 (46#)	152 (84#)	51 (32#)	**66 (19#)**	71 (28#)	92 (57#)
		Average err:	46 (43#)	**41 (29#)**	52 (37#)	77 (65#)	31 (29#)	**24 (18#)**	30 (24#)	47 (42#)

$The tests on walls T01 and T02 did not follow the same design scheme as the rest of the walls and their results are used in this table only for the identification of the model and not for determination of the prediction error.
#By using $G_L = 1.0 \cdot 10^{10}$ Pa, instead of $G_L = 1.424 \cdot 10^{10}$ Pa, in the model (see Section 3.8), the obtained error was different in the cases shown in brackets in this table.

Regarding point M in (Figure 11) and the Massive Degradation Line, also a simple position is adopted by setting its abscissa at

$$G_M = 0 \tag{41}$$

since there is no high reproducibility in the observed behaviour in that phase and, anyway, it is hardly expected to be able to provide accurate predictions near the collapse. This means that the Regular Degradation Line (36) is assumed as the only model after point L is overpassed (this simplifies the implementation of the model). Therefore, the current model cannot be reliable in the phases of low and massive degradation.

For example, in Figure 12 for Wall T09, the experimental monotonic envelope is represented by the thick yellow line and the obtained model is represented by the thick black line. The response spectra corresponding to the seismic experiments performed on that specimen are represented by different thin lines using damping ratio of 5% in (a) and 7% in (b). The figure shows the predicted maximum model response to every earthquake as the intersection of the spectrum with the model line (black rings). The same figure also shows the "star" points of maximum experimental response to the earthquake as introduced by (30).

In order to assess the reliability of the model predictions, its error has been quantified. The error of the model is defined here as the difference between the predicted maximum drift and the experimental one for the same earthquake. Such error has been computed in a special manner in Table 4 for every wall (walls T01 and T02 are not included in the computation of the error for the reasons explained in Section 2.2). The "maximum error" for every wall is obtained by computing the maximum, for the applied earthquakes, of the absolute difference between the logarithms of predicted and experimental drift. Alternatively, the "standard error" is obtained by computing the square root of the average, for the applied earthquakes, of the square of the difference between the logarithms of predicted and experimental drift. The derived maximum (or alternatively standard deviation) logarithm difference is then converted to error percentage by applying the antilogarithm, subtracting the unit and multiplying by 100. The obtained error values in percentage are shown in the table for every wall and for four different cases of damping ratio used for the spectrum in the prediction ($\xi = 4, 5, 6$ or 7%). In the last row of the table, among all the walls, the average value of every column is computed. The minimum value of such average is 41% for the maximum error (with $\xi = 5\%$) or 24% for the standard error (also with $\xi = 5\%$) as shown in bold. This means that $\xi = 5\%$ is in general the most recommendable common value of damping ratio to introduce in the response spectrum in order to minimize, in the average, the prediction error regarding these experimental data and the applied model.

It should be mentioned that for the cases in which the experimental response of the wall was around 1% of drift, the behaviour entered in the massive degradation area of the envelope (Figure 11) and the model predictions for successive earthquakes were potentially unreliable. Due to this reason, in Table 4, the results from tests T04-3 and T05-5 (see Table 3) have not been included for the computation of the prediction

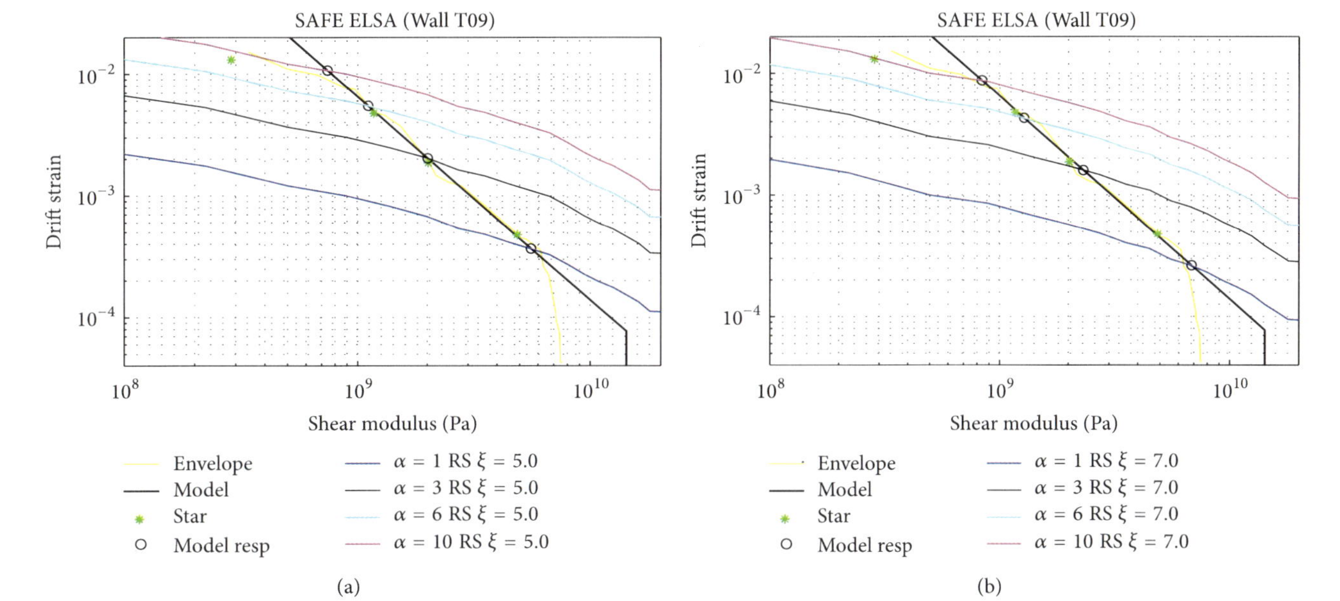

FIGURE 12: Piecewise line model of the monotonic envelope for Wall T09 and response spectra of the experimentally applied earthquakes for damping ratio, $\xi = 5\%$ (a) and $\xi = 7\%$ (b).

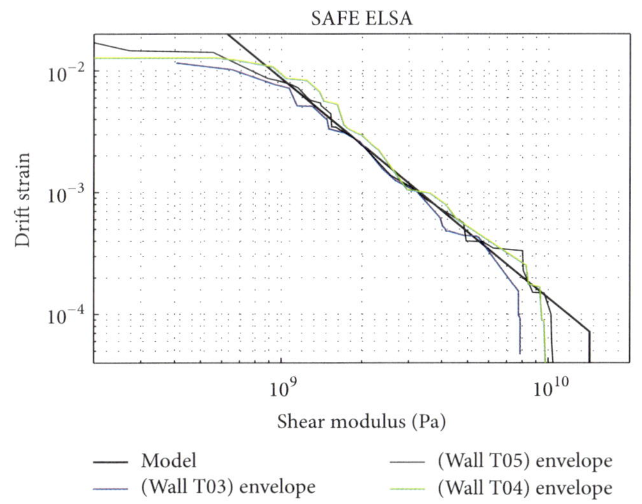

FIGURE 13: Identification of a unique model in the (G, γ) axes by combining the data of walls T03, T04, and T05 (group "$\rho = 0.8r$").

error (even though they have been used for the identification of the model).

3.4. Example of Model Identification from a Group of Walls and Derivation of Respective Vulnerability Curves. In this section, the same definition of polygonal curve model of the previous section will be applied with (37) to (41), but using for the identification the experimental data from several walls at the same time, instead of from a single wall.

According to Tables 1 and 2, walls T03, T04, and T05 were constructed with (nominally) equal vertical and horizontal reinforcement parameters and were all tested with low vertical load. However, the first two of them had a thickness of 16 cm and the third one had a thickness of 20 cm. Another difference in their tests was the specified design frequency f_{dTi} which, respectively, was 4, 12, and 8 Hz as displayed in Table 1. The experimental envelope in the (G, γ) axes for the three walls is represented in Figure 13. In the same figure, the piecewise line model identified from the data of the three envelopes together is represented by a thicker black line with identified parameters:

$$m = -1.8031, \qquad \gamma_R = 7.181 \cdot 10^{-5}. \tag{42}$$

Indeed, the degradation characteristics of the three walls are very similar in this representation, independently of the different thickness and specified design frequency for the tests. This suggests applying a unique model line for all of them. Moreover, their experimentally observed different behaviour in terms of safety margin can still be well explained by a unique model by considering their different implemented design frequencies. To do so, in Figure 14 the (f, u) axes can be used instead of the ones used for the identification of the model in Figure 13. The conversion of the points of the model is done taking into account equations (34) and (35). Thus, the RDL definition by (36) transforms into

$$\frac{u}{u_R} = \left(\frac{f}{f_R}\right)^{2m} \tag{43}$$

or equivalently

$$\log(u) = 2m\log(f) - 2m\log(f_R) + \log(u_R) \tag{44}$$

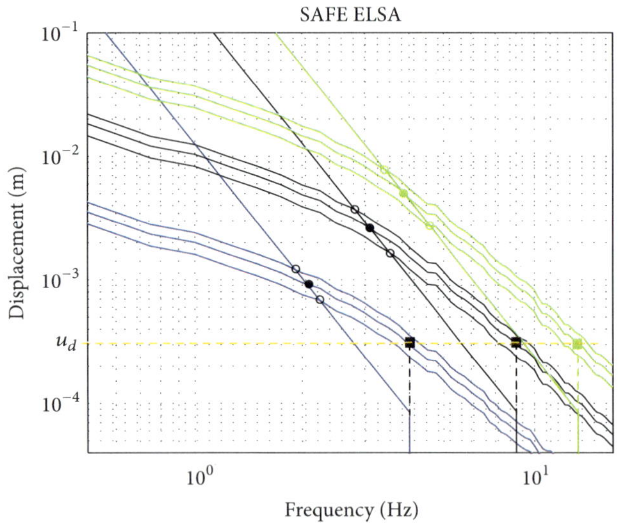

FIGURE 14: Representation of one unique model of the group of walls T03, T05, and T04 in the (f, u) axes for the different respective design frequencies of 4 (blue), 8 (red), and 12 Hz (green). The response spectrum of the accelerogram is represented (with the respective colour) at the corresponding design intensity of every wall (PGA = β_{Ti}) as well as for an increment and decrement of 20%, with respect to that intensity.

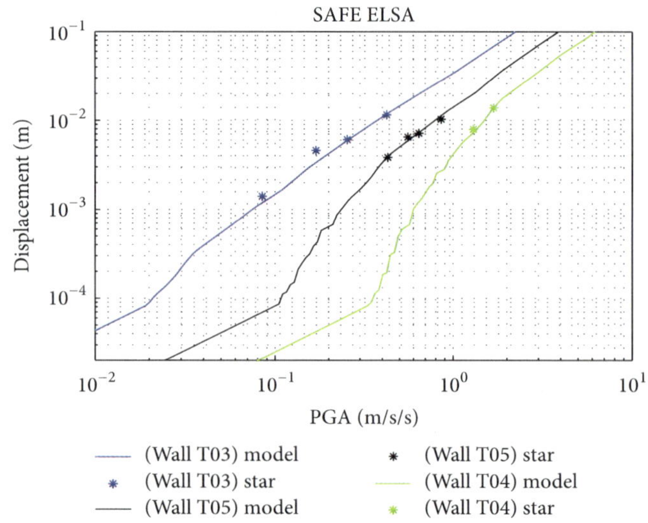

FIGURE 15: Predicted vulnerability line for walls T03, T04, and T05, according to one unique piecewise line model, in terms of displacement for every level of PGA based on the response spectrum for 5% damping ratio. Discrete ("star") points represent the recorded maximum experimental response at the applied earthquakes.

with

$$f_L = f_R = f_{d\mathrm{T}i}\left(\frac{G_R}{G_d}\right)^{1/2} = f_{d\mathrm{T}i},$$
$$u_L = u_R = H\gamma_R = 1.2 \times 7.181 \cdot 10^{-5} = 8.617 \cdot 10^{-5}\,\mathrm{m} \quad (45)$$

which is based on (38) and (42).

Thus, in Figure 14 one unique model based on parameters (42) appears as three response lines with corner point $f_L = f_{d\mathrm{T}i}$ at the design frequency for each one of the three walls, that is, 4 Hz (T03, blue line), 8 Hz (T05, red line), and 12 Hz (T04, green line). With the same respective colour for each wall, the response spectrum of the specified accelerogram at the respective design PGA is represented (for the design damping ratio of 7%). More precisely, three lines are used for the input spectrum for every wall, the central one corresponding to the design level (PGA = β_{Ti}), as indicated in Table 2, and the higher and lower ones corresponding to an increment and decrement of 20%. This is done in order to emphasize graphically what the effect of a fixed percent variation of PGA can be on the predicted response. The design displacement for each wall was $u_d = 3.37 \cdot 10^{-4}$ m according to Table 2, which corresponds to the value of the design spectrum at the design frequency for a damping ratio of 7%. It is marked in the figure by a square symbol with the respective colours. However, the predicted displacement by using the piecewise line model is given by the intersection of the design spectrum with the RDL line of every wall and this is indicated by the solid circle in the figure. The predicted response for the incremented and decremented spectra is indicated by the ring symbols. In this manner, the representation done in Figure 14 allows us to interpret how several walls with the same mechanical characteristics may exhibit a clearly different behaviour depending on their design frequencies. In particular, two effects are explained here.

Firstly, since the design displacement parameter u_d falls at a higher level than the model for the design frequency, the response to the design earthquake is predicted for a certain level of damage corresponding to a degraded frequency and thus at a higher displacement. This effect of increased response is inversely related to the difference in the global slopes of the RDL line and of the spectrum in the interval limited by the design frequency and the predicted degraded frequency. Thus, in the SAFE program, because the shape of the spectrum has steeper slope at higher frequencies, the higher the design frequency is, the higher the response displacement to the design earthquake becomes.

Secondly, for the represented fixed increment (+20% and −20%) of the intensity of the spectra, the amount of variation in the predicted response is different for every wall depending on its design frequency, being stronger for the case of higher design frequency in this study. This effect is again inversely related to the difference in the slopes between the RDL line and the spectrum, but now regarding only the local area around the predicted degraded frequency.

For different values of PGA, the predicted maximum response displacement is always obtained at the intersection of the spectrum with the piecewise line model. Thus, by applying this model, it is possible to derive the vulnerability curve of every wall for the specified accelerogram. The representation of the predicted displacement as a function of the imposed PGA is done in Figure 15 in log-log scale for every one of the three walls T03, T04, and T05. It must be observed that, for displacement values under ~10^{-4} m, the prediction falls in the low degradation constant-frequency

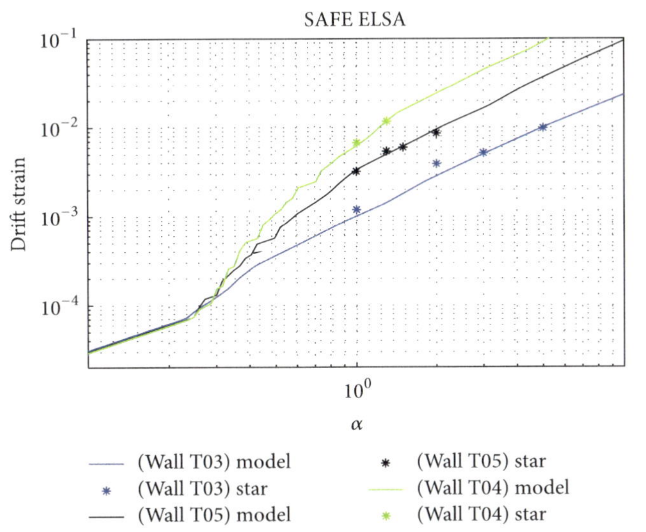

FIGURE 16: Predicted vulnerability line for walls T03, T04, and T05, according to one unique piecewise line model, in terms of drift for every level of design margin α based on the response spectrum for 5% damping ratio. Discrete ("star") points represent the recorded maximum experimental response at the applied earthquakes.

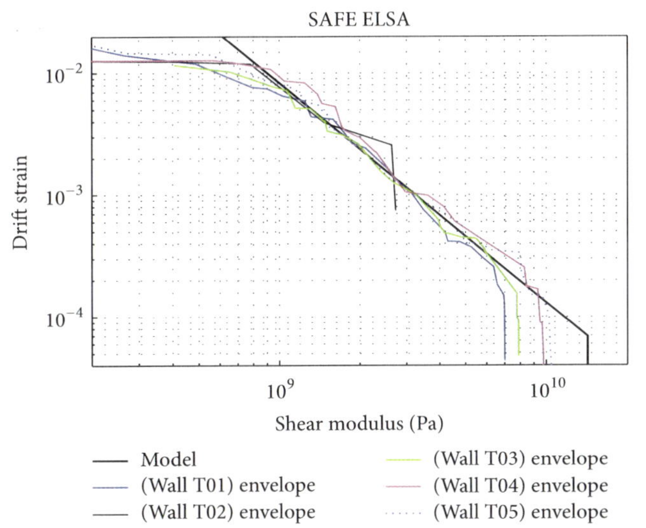

FIGURE 17: Identification of one unique model in the (G, γ) axes by combining the data of walls T01 to T05 (group "$\rho = 0.8$").

branch of the model that corresponds to a range with low reliability due to the large variability of results between the specimens. These vulnerability curves are derived in this occasion from the spectra for a damping ratio of 5% in order to minimize the prediction error. The "$*$" symbols represented with the respective colour of every wall correspond to the experimental maximum response ("star" point as introduced by (30)) for every applied earthquake on the concerned wall.

Finally in Figure 16, the same vulnerability relationships (based on the response spectrum for 5% damping ratio) and experimental data are represented but using the drift deformation and the design margin α (22) as variables in the axes. This representation shows again the two important effects of the change of the design frequency in this study that have been mentioned also for Figure 14: for higher design frequency, first, the larger response to the design earthquake (for $\alpha = 1$) and, second, the successive larger relative increase of the response for every increase of α.

It should be mentioned that, in Figures 15 and 16, the experimental "star" points for tests T04-3 and T05-5 are not represented since they took the walls to the total collapse and beyond the intended reliable range of applicability of the model.

Just for the purpose of completeness, even though T01 and T02 were not tested following the same criteria about the definition of the mass and the input, Figure 17 shows the experimental envelopes of these walls together with the ones already shown of T03, T04, and T05. All the walls in this figure had the same nominal reinforcement densities and low additional normal load. In this case, the piecewise line model identified from all the data of the five envelopes together is represented by the thick black line with identified parameters:

$$m = -1.8032,$$
$$\gamma_R = 6.912 \cdot 10^{-5}. \tag{46}$$

3.5. Model Identification for Different Groups of Walls. In the previous section, two models have been identified from two groups of walls. The current section is dedicated to perform similar identifications for a whole list of groups as shown in Table 5. In fact, the first two groups listed in this table are the ones described in the previous section with their respective identified parameters (42) and (46). As for Table 4, the current table also includes (for $\xi = 4$, 5, 6, and 7%) the values of the defined maximum and standard errors, which in this case are computed having into account all the experimental earthquakes applied to the respective group of walls (excluding T01 and T02). For the same reasons explained in the previous section, the results of tests T04-3 and T05-5 have not been used for the computation of the errors shown in Table 5.

As for Table 4, the columns referring to $\xi = 5\%$ have been shown in bold as well in Table 5 since it seems that this choice of damping, as a common parameter value, offers the best compromise for avoiding large errors.

The graphical representation of the models listed in Table 5 is done in Figures 13 and 17 (already discussed) and from Figures 18 to 23. These figures include the experimental envelopes of the affected walls and the single model line for the group identified by using the data of those envelopes. Finally, Figure 24 includes together just the identified models for the most representative groups from Table 5.

3.6. Effect of Reinforcement and Normal Load on Characteristic Drift Values. In this section, the positions of the identified RDL line model (Figure 11) for all the walls will be compared in order to analyse the qualitative effects of the existing variability in reinforcement and normal load among the walls. To do so, instead of defining the RDL in terms of m and γ_R (with values displayed in Table 4), two

TABLE 5: For every wall group, identified parameters of RDL with maximum and standard errors of the model predictions by using the response spectrum of the applied earthquakes with damping ratios of $\xi = 4, 5, 6$, or 7%.

Group	Walls	m	γ_R $(*1E-6)$	Max err% ($\xi = 4\%$)	Max err% ($\xi = 5\%$)	Max err% ($\xi = 6\%$)	Max err% ($\xi = 7\%$)	Std err% ($\xi = 4\%$)	Std err% ($\xi = 5\%$)	Std err% ($\xi = 6\%$)	Std err% ($\xi = 7\%$)
$\rho = 0.8$	T01$\cdots$T05$^\$$	−1.8032	69.12	34	**35**	52	70	21	**13**	23	40
$\rho = 0.8r$	T03$\cdots$T05	−1.8031	71.81	32	**37**	54	73	19	**13**	26	44
$\rho_h = 0.6$ $\rho_v = 0.4$	T06 T07	−1.5865	119.65	81	**25**	24	76	44	**14**	9	35
$\rho = 0.4$	T08 T09 T11	−1.6471	70.67	37	**23**	47	106	23	**14**	24	47
$\rho = 0.6$	T10	−1.7660	70.31	29	**29**	47	65	20	**17**	25	35
$\rho = 0.11$	T12	−1.4469	123.32	81	**84** **(57$^\#$)**	104 (41$^\#$)	124 (57$^\#$)	48 (41$^\#$)	**45** **(32$^\#$)**	47 (30$^\#$)	53 (34$^\#$)
CFRP	T13	−1.4699	142.31	79 (48$^\#$)	**141** **(36$^\#$)**	147 (46$^\#$)	152 (84$^\#$)	51 (32$^\#$)	**66** **(19$^\#$)**	71 (28$^\#$)	92 (57$^\#$)
ALL	T01$\cdots$T13$^\$$	−1.6360	90.82	368	**226**	158	324	49	**37**	39	52 (50$^\#$)

$^\$$The tests on walls T01 and T02 did not follow the same design scheme as the rest of the walls and their results are used in this table only for the identification of the model and not for determination of the prediction error.
$^\#$By using $G_L = 1.0 \cdot 10^{10}$ Pa, instead of $G_L = 1.424 \cdot 10^{10}$ Pa, in the model (see Section 3.8), the obtained error was different in the cases shown in brackets in this table.

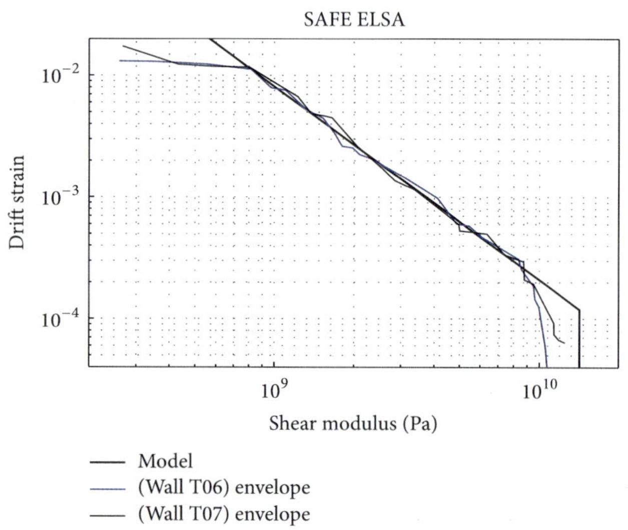

FIGURE 18: Identification of a unique model by including the data of walls T06 and T07 (group "$\rho_h = 0.6$, $\rho_v = 0.4$").

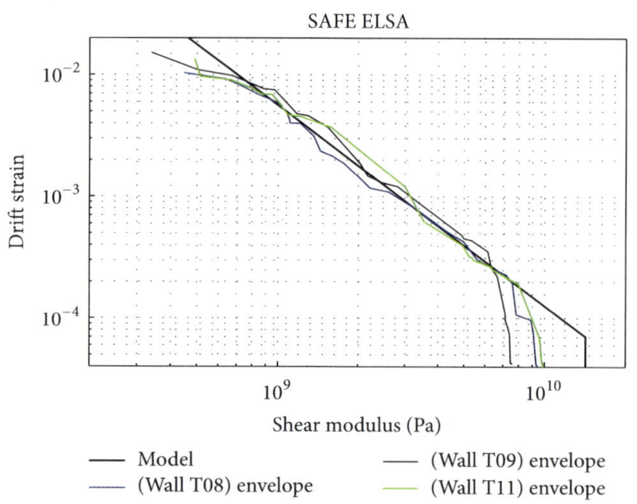

FIGURE 19: Identification of a unique model by including the data of walls T08, T09, and T11 (group "$\rho = 0.4$").

alternative parameters with an easier physical interpretation will be introduced. The new parameters of the model will be the ordinates of the characteristic points A and B that are shown in Figure 11, that is, respectively, as given by (36),

$$\gamma_A = \gamma_R \left(\frac{G_A}{G_R} \right)^m \tag{47}$$

which is the drift capacity at a reference "initial" modulus G_A and

$$\gamma_B = \gamma_R \left(\frac{G_B}{G_R} \right)^m \tag{48}$$

which is the drift capacity at a reference "near-collapse" modulus G_B. By looking at Figure 23, we have adopted the following fixed values of the reference modulus:

$$\begin{aligned} G_A &= 6.0 \cdot 10^9 \text{ Pa}, \\ G_B &= 1.0 \cdot 10^9 \text{ Pa} \end{aligned} \tag{49}$$

in order to have them within the experimental RDL for all the walls but close to the ends of it.

Note that, according to the model, if a wall has one of these characteristic drift values higher, this means that for the corresponding level of damage (initial or near-collapse

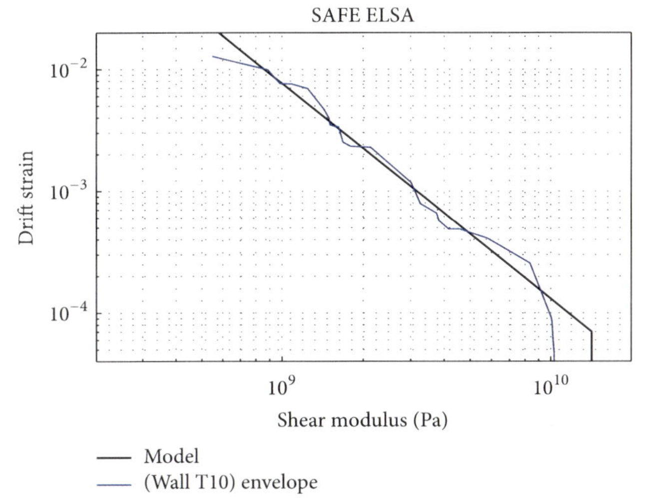

FIGURE 20: Identification of a unique model by including the data of wall T10 (group "$\rho = 0.6$").

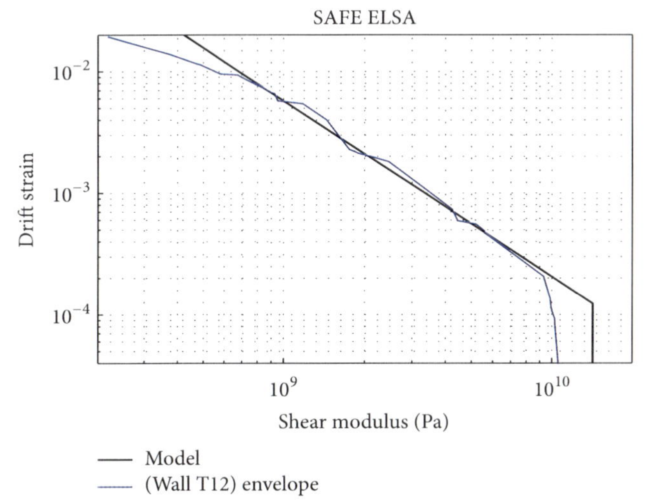

FIGURE 21: Identification of a unique model by including the data of wall T12 (group "$\rho = 0.11$").

modulus) there is a higher capacity of deformation and, consequently according to the model, a higher PGA earthquake can be resisted without new damage by that wall.

For every single wall, by using (47), (48), and (49) and the model parameters from Table 4, the corresponding values of the characteristic capacity drifts have been computed and displayed in Table 6. Table 6 also contains for completeness the average between horizontal and vertical actually implemented value of reinforcement density as well as the vertical normal stress taken out of Tables 1 and 2.

In Figure 25 the model characteristic drifts γ_A and γ_B that define each one of those models listed in Table 6 are used as ordinates (in logarithmic scale), whereas the reinforcement density of every wall is used as abscissa (in linear scale). The walls tested with lower normal load (between 0.32 and 0.37 MPa) are represented by blue circles and the ones tested with higher normal load (1.01 MPa) are represented by red circles. The wall with added CFRP reinforcement

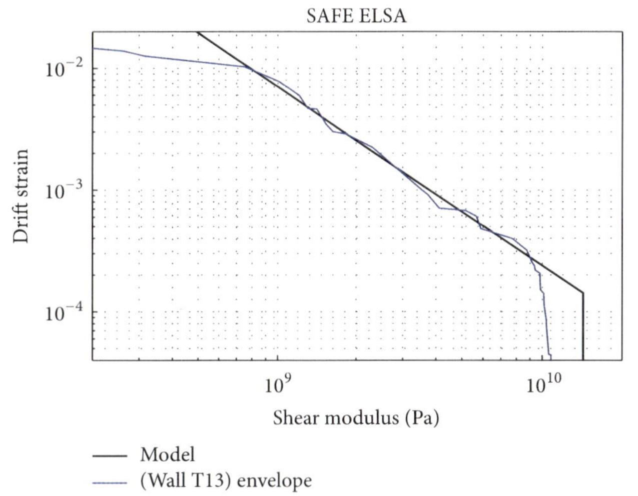

FIGURE 22: Identification of a unique model by including the data of wall T13 (group "CFRP").

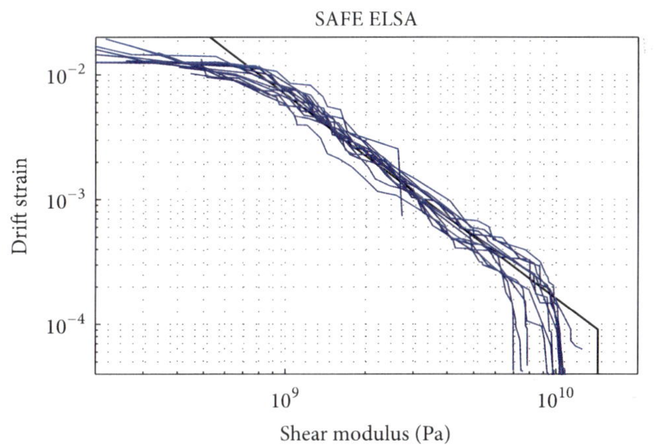

FIGURE 23: Identification of a unique model (thick black line) by including the data of walls T01 to T13 (group "ALL").

(T13) is represented by a green circle. The purpose of these graphs in Figure 25 is to illustrate the qualitative effects of reinforcement and normal load on the characteristic drift values as follows:

(i) Point A, Figure 25(a): since the first part of the envelope is almost vertical, corresponding to elastic response of the noncracked wall, γ_A should be closely related to the first cracking (degradation) phenomenon. As expected, this phenomenon is delayed (higher γ_A) by either a higher vertical stress (red circles) or an artificially increased tensile capacity (CFRP effect represented by the green circle). Otherwise, it should be kept in mind that, in reinforced concrete, the reinforcement should not play an actual role before the concrete is cracked. Therefore, as expected, γ_A is not sensitive to the reinforcement ratio as confirmed by the graph. The observed dispersion could be due to the existing variability in the tension capacity among the specimens.

TABLE 6: For every wall specimen, identified parameters of RDL (copied from Table 4) and derived characteristic capacity drifts γ_A and γ_B.

Wall	ρ (reinf. dens. %)	σ_n (MPa)	m	γ_R $(*1E-6)$	γ_A $(*1E-6)$	γ_B $(*1E-6)$
T01	0.770	0.37	−1.7962	62	293	7312
T02	0.770	0.37	−1.7593	73	332	7775
T03	0.770	0.37	−1.7804	66	308	7479
T04	0.785	0.37	−1.8360	78	383	10283
T05	0.767	0.32	−1.7216	85	376	8230
T06	0.515	1.01	−1.5513	125	477	7685
T07	0.515	1.01	−1.6183	115	467	8482
T08	0.402	0.32	−1.6039	70	279	4945
T09	0.402	0.32	−1.6691	77	327	6515
T10	0.628	0.32	−1.7660	70	324	7658
T11	0.402	0.32	−1.7113	63	277	5947
T12	0.113	1.01	−1.4469	123	431	5755
T13	0.402 + CFRP	0.32	−1.4699	142	507	7059

(ii) Point B, Figure 25(b): on the other hand, the ultimate shear capacity of a wall should be highly dependent on the reinforcement ratio. Effectively, (b) reveals that the associated drift value γ_B is an increasing function of the reinforcement. A light positive effect can be also attributed to the normal load and very marginally to the external reinforcement with CFRP. Regarding the effect of the vertical load, we want to point out here that, in the spirit of the design rules presented by Labbé et al. [1], a lower vertical reinforcement ratio in walls T06 and T07 could be compensated by a larger vertical stress, making them comparable to T10 in terms of shear capacity. Then, as a confirmation, it is interesting to observe in Table 6 and in Figure 25(b) that these three specimens actually exhibit comparable γ_B values.

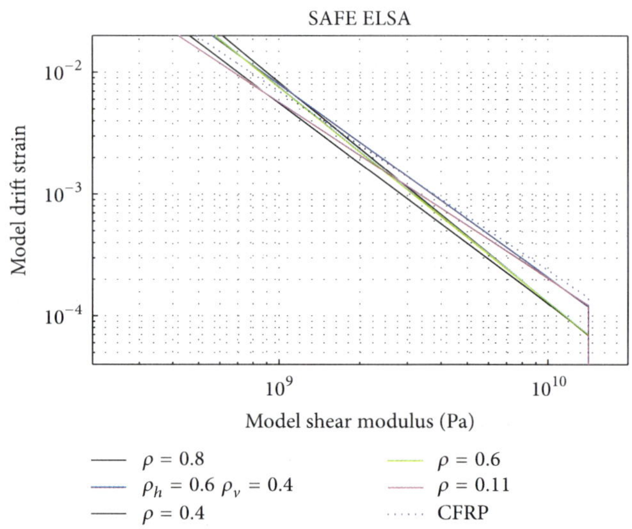

FIGURE 24: Comparison among the identified models for different groups.

3.7. *Effect of the Reinforcement and the Normal Load on Crack Width.*

In order to assess in this section the effects on the crack width, the identified characteristic values of drift for the tested walls (previous section) will be converted into correlated crack width. This will be done by using the formula

$$w = (0.1 + 133\gamma)\, C(\rho), \tag{50}$$

where w is the magnitude of the crack width in mm, γ is the drift in rad, and

$$C(\rho) = 0.2 + \frac{0.9}{\rho + 0.5} \tag{51}$$

is the term that considers the effect of the average reinforcement density ρ in %. These empirical relations (50) and (51) were identified by Labbé et al. [1] from the crack width measurements recorded during the tests of SAFE. Thus, by applying these expressions (50) and (51) to each one of the tested walls (excluding Wall T13 which is not validated for these formulas), the associated crack width capacities

$$w_A = (0.1 + 133\gamma_A)\, C(\rho) \tag{52}$$

at the reference initial modulus G_A and

$$w_B = (0.1 + 133\gamma_B)\, C(\rho) \tag{53}$$

at the reference near-collapse modulus G_B can be computed, where γ_A and γ_B are taken from Table 6. The computed values of these characteristic crack widths are displayed in Table 7 together with the actual values of average reinforcement density and normal load for the walls.

Now, as in the previous section, in Figure 26 the computed characteristic values are represented in logarithmic scale as a function of the reinforcement density and marked with different colours the different cases. The obtained graphs in Figure 26, together with the previous ones in Figure 25, can be interpreted as follows:

(i) Point A, Figure 26(a): once the cracking phenomenon has appeared, it is well known that the crack spacing

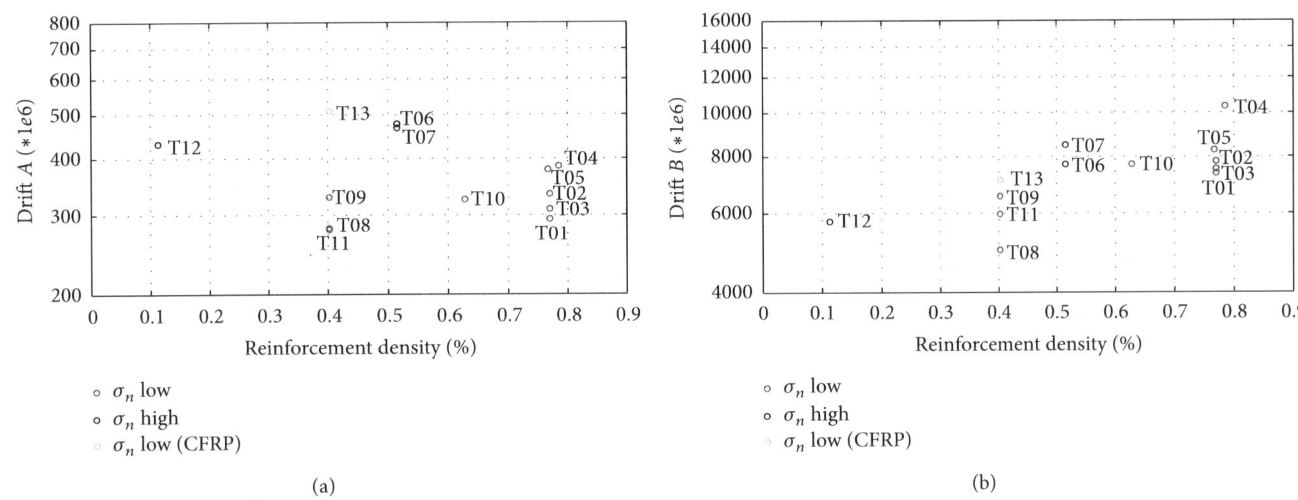

FIGURE 25: Comparison of the identified characteristic drifts for all the walls as a function of reinforcement and normal load. Drift A (γ_A, (a)) and Drift B (γ_B, (b)) are, respectively, the characteristic values at a reference initial stiffness and at a reference near-collapse stiffness.

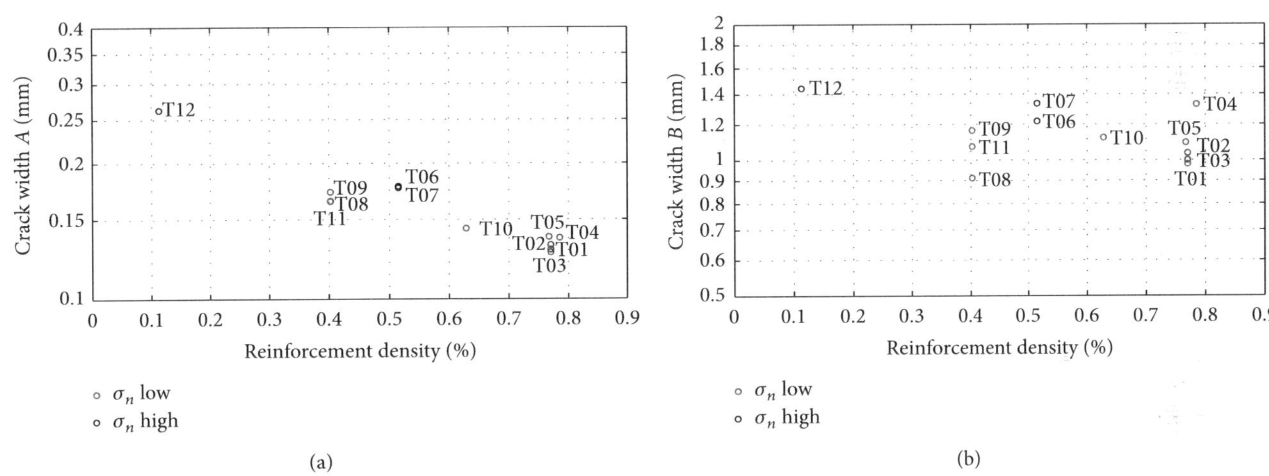

FIGURE 26: Comparison of the identified characteristic crack widths for all the walls as a function of reinforcement and normal load. Crack width A (w_A, (a)) and Crack width B (w_B, (b)) are, respectively, the characteristic values at a reference initial stiffness and at a reference near-collapse stiffness.

and crack width are sensitive to the reinforcement (the product of these two values being controlled by the drift). As expected, the larger the reinforcement is, the smaller the crack width develops (either for red and blue dots). Additional analyses would be necessary to discuss crack widths observed in the SAFE walls in comparison with formulas of the building codes and with the state of the art [15].

(ii) Point B, Figure 26(b): it is hard to recognise an effect of the reinforcement or the vertical stress on w_B, meaning basically that, for all the walls, Point B corresponds to a certain common level of crack width, which can be associated just with the degraded shear modulus G_B selected for this point.

3.8. Additional Remarks. The first remark concerns a more general application of the model described here. Independently of the practical reasons explained in Section 3.3, the choice, according to (40), of the corner of the polygonal line model at such a high stiffness as the one given by the design modulus would not be appropriate for the prediction of the response to nonsevere earthquakes on a virgin wall. This can be seen in all the figures comparing the actual envelopes with the corresponding model. For such purpose, alternatively, a value such as

$$G_L = 1.0 \cdot 10^{10}\ \text{Pa} \tag{54}$$

can be recommended having into account the observations reported by Labbé et al. [1] regarding the average initial modulus found to be around 0.67 times the design modulus. Even though the reproducibility of the experimental envelope at this corner point was not high and the model predictions cannot be very reliable in that range of deformation, the latter choice of this corner modulus (54) does change the earthquake response predictions for the experiments regarding specimens T12 and T13. This effect is strong and it is

TABLE 7: For every wall specimen, derived characteristic crack widths w_A and w_B.

Wall	ρ (reinf. dens. %)	σ_n	$C(\rho)$	w_A (mm)	w_B (mm)
T01	0.770	0.37	0.91	0.13	0.97
T02	0.770	0.37	0.91	0.13	1.03
T03	0.770	0.37	0.91	0.13	0.99
T04	0.785	0.37	0.90	0.14	1.32
T05	0.767	0.32	0.91	0.14	1.09
T06	0.515	1.01	1.09	0.18	1.22
T07	0.515	1.01	1.09	0.18	1.33
T08	0.402	0.32	1.20	0.16	0.91
T09	0.402	0.32	1.20	0.17	1.16
T10	0.628	0.32	1.00	0.14	1.12
T11	0.402	0.32	1.20	0.16	1.07
T12	0.113	1.01	1.67	0.26	1.44

reflected in the alternative values of the prediction error at the corresponding rows in Tables 4 and 5. Indeed the prediction error diminishes for these two walls by using the reduced initial modulus (54) in the model. The reason for which relatively lower intensity earthquakes were applied to these walls, and not for the other ones, could apparently be that their adopted design was too conservative. Particularly, for the case of T12 with a relatively low steel density, Labbé et al. [1] already mention that, by disregarding the tensile strength of the concrete in the formula of the design strength (11), the wall capacity might be underestimated. On the other hand, for the case of T13, the presence of the CFRP, which was not considered in its design strength parameter, could also justify this phenomenon.

The second remark concerns the onset of damage. In principle, the first crack appears when the tensile capacity of the concrete is exceeded. The Eurocode 2 rule to derive the tensile capacity from the compression capacity is

$$f_{t28} = 0.3 \left(f_{c28}\right)^{2/3} \text{ (in MPa)}. \tag{55}$$

The individual f_{c28} of walls T03 to T12 as presented by Labbé et al. [1], showing a noticeable variability, when converted to tension capacity through this formula (55), lead to an average value

$$\overline{f_{t28}} = 3.6\,\text{MPa}. \tag{56}$$

Now, with the average initial shear modulus value (54) and assuming a nil vertical stress, it leads to a drift strain of $3.6 \cdot 10^{-4}$. In the case of pure elastic behaviour in shear without any cracking, disregarding the reinforcement, this value should deliver an average approximation of the lower corner drift γ_L. However, it is normally reached by the envelopes shown in Figure 23 only after some degradation is already present. In average the vertical part of the envelope terminates rather at around $2 \cdot 10^{-4}$. This fact could be regarded as an indication that some degradation may develop in the walls before the first crack reaches it limit stress and presumably becomes visible.

4. Conclusions

The SAFE program consisted on seismic PsD tests on 13 RC shear wall specimens submitted to the same accelerogram for several increasing intensities. The parameters that changed among the walls were the thickness, the reinforcement, the normal load, and the design frequency, which was controlled by the theoretical mass introduced in the equation of motion for running the PsD tests. In fact, the design frequency was observed to have a strong influence on the damage at the respective design intensity of the wall and on the safety margin for higher intensities [1]. Some of the aspects that interact to produce this effect are made clearer by the current study.

The proposed model is based on identification methods applied to the observed experimental response. The most important identified parameters at every time instant are the secant stiffness, coming from the Spatial Model, and the displacement oscillation amplitude, coming from the Hilbert Transform. The trajectory of these two variables, or equivalently of the shear modulus and the drift amplitude, shows a clear envelope during the successive earthquakes on the same wall. This envelope can be associated with the idea that the stiffness remains roughly constant until new damage is introduced by reaching a displacement amplitude that is larger than any amplitude experienced before. Based on this idea that was already present in some models available in the literature, the envelope line computed from the identified variables for a tested wall can be interpreted as a capacity line that is able to predict response frequency values associated with displacement oscillation amplitudes or vice versa.

Moreover, the intersection of such capacity line with the demand response spectrum of an applied earthquake predicts the response maximum displacement and associated frequency. Further on, in the double logarithmic representation for the chosen variables of stiffness and deformation, we propose to substitute the experimentally derived envelope by a piecewise line curve, with a main straight branch called Regular Degradation Line (RDL). Another two branches with different slope could be required for the very small or very large deformations, but the reproducibility of the experimental data was not so high in these areas.

It is shown that the RDL identified for every wall is able to predict the maximum response to the applied earthquakes with standard error of 25%, average among the different walls, when using the response spectrum based on 5% damping ratio. If the RDL of the specific wall is replaced by a common RDL for all the walls (disregarding the different reinforcement and normal load of each one), the standard error is 40%. Other values of damping ratio from 4% to 7% have been found to give worse global results. Alternatively, identified values of instantaneous damping identified by the Spatial Model have been considered in this study but for the moment they do not lead to better general interpretations.

For walls of the SAFE program with the same reinforcement and normal load, very similar experimental envelope was derived and a unique RDL can be used to model the whole group, even if the different walls inside the group had been tested with different assigned design frequency. Then it is realized that the different positioning of the design frequency with respect to the spectrum, changes, not only the computed design intensity (that was based on a constant response frequency method), but also very drastically the design margins predicted according to the current model. The effect is directly related to the closeness of the local slope of the spectrum to the slope of the RDL. For the adopted spectrum within the SAFE program, the slope is always decreasing with the frequency and becoming closer to the slope of the capacity line. This is translated in shorter safety margins for the cases with larger assigned design frequency as it had been observed from the experiments and is now explained by the model.

By analysing the changes on the identified RDL model with respect to the different parameters of the walls, some effects can be recognised. Particularly, for the tested walls in the SAFE program, the characteristic drift at the initial shear modulus state is mostly dominated by the added normal load and the presence of the external CFRP, whereas the characteristic drift at the near-collapse state is mostly determined by the different amounts of steel reinforcement.

By using the empirical correlation between drift and crack width that was proposed in a previous study, the effects can also be assessed on the crack width. It is observed that at the initial shear modulus state, the crack width is smaller for the walls with higher reinforcement, whereas at the near-collapse state the crack width does not show a clear correlation with the amount of reinforcement.

Competing Interests

The authors declare that they have no competing interests.

Acknowledgments

The experimental program SAFE was financially supported by Electricité de France and COGEMA and executed at the ELSA laboratory of the JRC of the European Commission. Many colleagues from the collaborating institutions contributed to the quality of the results that has made possible several studies including the current one. From the ELSA laboratory we are particularly grateful to the contributions of Guido Verzeletti, Georges Magonette, and Vito Renda.

References

[1] P. Labbé, P. Pegon, J. Molina, C. Gallois, and D. Chauvel, "The SAFE experimental research on the frequency dependence of shear wall seismic design margins," *Journal of Earthquake Engineering*, vol. 20, no. 1, pp. 101–128, 2016.

[2] D. Benedetti and M. P. Limongelli, "A model to estimate the virgin and ultimate effective stiffnesses from the response of a damaged structure to a single earthquake," *Earthquake Engineering and Structural Dynamics*, vol. 25, no. 10, pp. 1095–1108, 1996.

[3] M. Brun, J. M. Reynouard, and L. Jezequel, "A simple shear wall model taking into account stiffness degradation," *Engineering Structures*, vol. 25, no. 1, pp. 1–9, 2003.

[4] M. Brun, P. Labbe, D. Bertrand, and A. Courtois, "Pseudo-dynamic tests on low-rise shear walls and simplified model based on the structural frequency drift," *Engineering Structures*, vol. 33, no. 3, pp. 796–812, 2011.

[5] P. Fajfar, "Capacity spectrum method based on inelastic demand spectra," *Earthquake Engineering & Structural Dynamics*, vol. 28, no. 9, pp. 979–993, 1999.

[6] P. B. Labbé, "Categorization of seismically-induced stresses for civil and mechanical engineering," *Nuclear Engineering and Design*, vol. 255, pp. 240–247, 2013.

[7] P. Pegon, G. Magonette, F. J. Molina et al., "Programme SAFE: rapport du test T4," Note Technique du CCR no. I.98.32, 1998.

[8] P. Pegon, F. J. Molina, and G. Magonette, "Continuous pseudo-dynamic testing at ELSA," in *Hybrid Simulation; Theory, Implementation and Applications*, V. E. Saouma and M. V. Sivaselvan, Eds., pp. 79–88, Taylor & Francis/Balkema, 2008.

[9] F. J. Molina, G. Magonette, P. Pegon, and B. Zapico, "Monitoring damping in pseudo-dynamic tests," *Journal of Earthquake Engineering*, vol. 15, no. 6, pp. 877–900, 2011.

[10] F. J. Molina, F. Marazzi, B. Viaccoz, and A. Bosi, "Stability and accuracy in a hybrid test example," *Earthquake Engineering and Structural Dynamics*, vol. 42, no. 3, pp. 469–475, 2013.

[11] F. J. Molina, P. Pegon, and G. Verzeletti, "Time-domain identification from seismic pseudodynamic test results on civil engineering specimens," in *Proceedings of the 2nd International Conference on Identification in Engineering Systems*, The Cromwell Press, 1999.

[12] F. J. Molina, G. Verzeletti, P. Pegon et al., *Ensayos Pseudodinámicos a Cortante en Muros de Hormigón Armado. Evolución de la Frecuencia y el Amortiguamiento durante el Terremoto*, Primer Congreso Nacional de Ingeniería Sísmica, AEIS, Murcia, Spain, 1999.

[13] F. J. Molina, Spatial and Filter Models, MATLAB Functions, Matworks File Exchange, The MathWorks, Natick, Mass, USA, 2011, http://www.mathworks.com/matlabcentral/fileexchange/32634.

[14] H. Herlufsen, "Dual channel FFT analysis (Part II)," Brüel & Kjaer Technical Reviews 2-1984, 1984.

[15] E. Gallitre, *Fissuration des voiles courts en béton armé soumis à des solicitations sismiques, evaluation des ouvertures de fissures [Ph.D. thesis]*, Institut National des Sciences Appliquées, Lyon, France, 2008.

14

Numerical Analysis to Enhance Delamination Strength around Bolt Holes of Unidirectional Pultruded Large Smart Composite Platform

Sheedev Antony, Monssef Drissi-Habti, and Venkadesh Raman

PRES LUNAM IFSTTAR, CS4 Route de Bouaye, 44344 Bouguenais, France

Correspondence should be addressed to Monssef Drissi-Habti; monssef.drissi-habti@ifsttar.fr

Academic Editor: Julian Wang

As a part of the DECID2 French National Project (2008–2012), construction of a large platform entirely made of smart composites was carried out and two demonstrators were installed. During a previous study, an ABAQUS model of smart composite platform was set up to perform numerical simulations that predict the mechanical behaviour of bolt-fastened platform under static three-point bending load, and a stress concentration is observed around the bolt holes. Unidirectional composites are subjected to delamination at very low stress, and this cannot be tolerated because most of the applications of pultruded structures are in civil engineering, which involves human safety. Therefore, it is essential to study the correlation of delamination onset and find a technology to enhance delamination strength. In this study, a numerical analysis was carried out to enhance the delamination strength around the bolt holes. Bidirectional fiberglass cloths were wrapped around unidirectional pultruded beams to reinforce the structure and to prevent delamination. Due to the high cost of these cloths, a study was also carried out to select an appropriate cloth material by taking into account two aspects: density and number of layers of cloths. The selected cloth was compared with conventional continuous fibreglass mat, and it was found that bidirectional fibreglass cloth material performs well in preventing delamination, even though it is expensive.

1. Introduction

Fiber-reinforced polymer composites are materials that are composed of polymer-based resin and fibres as reinforcement. Composite material properties are mainly determined based on fibres and resin properties, fibre volume fraction, and geometry and orientation of fibres. Utilization of polymer matrix composites is increasing day by day in many industries due to their better mechanical properties, corrosion resistance, and strength-to-weight ratio, but their durability remains an issue. The introduction of smart composite material structures with structural health monitoring (SHM) techniques is one solution to these problems. In SHM, sensors will detect strain at different locations of composite structures and monitor the health of the structure. The detailed studies of smart composite materials are discussed in the previous work [1].

The main goal of DECID2 project (2008–2012) was the construction of mechanically oriented platforms made of pultruded composite materials (Figure 1). Two demonstrators of dimensions $20\,\mathrm{m} \times 3.5\,\mathrm{m}$ were installed in IFSTTAR Nantes and Technocampus EMC2. The parts were attached using single and double lap-bolted joints. Different studies were carried out to reduce delamination and to study the behaviour of bolted joints in composite structures [2, 3]. It was found that high-speed drilling and low feed rates may reduce the intensity of delamination [2], but DECID2 project has been started prior to the publishing of these studies, and conventional drilling methods were used during construction. Strain sensors and ultrasonic sensors were embedded in pultruded composite structures. This technique is called structural heath monitoring where strain will be continuously monitored and warned when it reaches the strain limit. Several possible sensors for monitoring large smart composite

Figure 1: DECID2 platform.

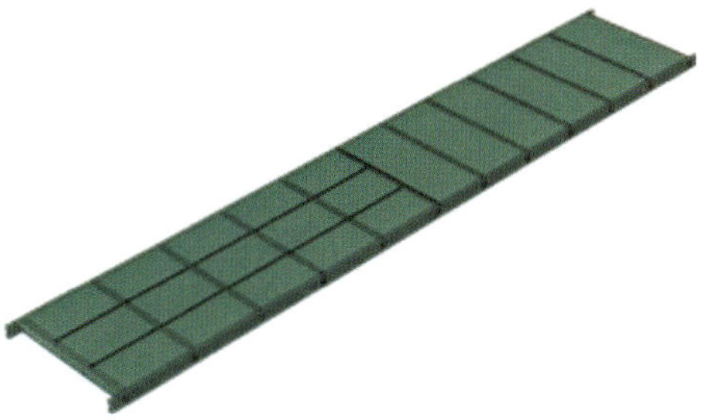

Figure 2: Smart composite platform CAD model [7].

structure have been studied previously [4, 5], and further works were extended to the study of UV radiation defect on mechanical behaviour of the DECID2 smart composite platform [6]. For further studies, a CAD model of large smart composite platform was set up, and a numerical simulation to predict the mechanical behaviour of the structure under static load has been performed (Figure 2). As a result of this work, it was observed that there was a stress concentration around bolt holes (Figure 3) [7].

Pultrusion is a continuous manufacturing process of composite materials with constant cross sections. In this technology, parts are made by the matrix extrusion process and by pulling the unidirectional fibres. This process will allow the fibres to align well before polymerization of the matrix. It is a highly automated, cost-effective process and produces finished parts with high volume fraction. Pultrusion is used to make products like beams, pipes, and tubes, which are used in civil structures like bridges, light poles, and towers (Figure 4) [8].

But in many cases, unidirectional composites are subjected to delamination at very low stress, and these issues cannot be tolerated in some demanding applications. In [11], it is mentioned that debonding (Figure 5) occurs prior to other local defects, and this will lead to delamination (Figure 6). As mentioned earlier, most of the applications of pultruded materials are in civil engineering industries, which involve human safety. Therefore, it is essential to study the correlation of delamination onset and find a technology to enhance delamination strength as high as possible. In this technology, unidirectional composite will be wrapped with bidirectional fiberglass cloth with appropriate density and number of layers, which can enhance the delamination strength and decrease the risk of catastrophic failure of the structure. In order to validate the use of bidirectional fiberglass cloth over the cheap continuous fiberglass mat, a comparative study will be also carried out.

The aim of this work is to suggest a technique to enhancing delamination strength around bolt holes of unidirectional pultruded composite. All the experimental validation tests of the platform have been carried out, except those relating to the local analysis of stainless steel-bolted joints. Indeed, we were forced to complete the project in the allotted time, and there remain some scientific and technological points to complete numerically. Experimental validation is unfortunately no longer possible because the composite material platform is already manufactured and cannot be dismantled for testing. In this paper, we have performed a numerical analysis of the assembly areas, and we have demonstrated numerically how these areas could be reinforced by fiberglass fabrics. On the one hand, the innovation lies more in the proposal of the concept of local extra thickness to be applied to the zones of assemblies, and on the other hand, the calculations made it possible to optimize the type and the density of fibreglass fabrics Our assumptions are strongly below the critical threshold of matrix cracking. We have therefore taken into account this mechanism which is even the source of this work. Firstly, a composite micromechanical model will be set up in ABAQUS. The bidirectional fibreglass cloth wrapping method (Figure 7) will be used to reduce delamination around bolt holes, and this method will be validated. Due to the high cost of bidirectional fibreglass cloths, a study will also be carried out to select an appropriate cloth material by taking into account two aspects: density and number of layers of cloths. The selected bidirectional fibreglass cloths will be compared with conventional continuous fiberglass mat to evaluate the difference in damage reduction performance.

2. Technology Proposed and Numerical Modelling

2.1. Wrapping Technology. In the wrapping technique, unidirectional pultruded beams will be wrapped using bidirectional fiberglass cloths or continuous fiberglass mats. In this study, thin bidirectional fiberglass cloths with E-glass vinylester material was used due to their better mechanical properties (Tables 1 and 2). The orientation of bidirectional fiberglass cloths will be set to 0° and 90°, and thicknesses will be scaled down for the following simulations.

Bolted joint fasteners and drilling method are the most commonly used techniques in composite structure to join parts. While drilling a unidirectional pultruded composite structure (Figure 8), the fibres will start to separate near the bolt holes and this will lead to a huge defect in the whole structure. The separation will start from the interface between fibre and matrix, which is called debonding. This debonding in microscale will lead to delamination, and it is very essential to avoid or decrease the intensity of this defect. This fibre separation can be reduced to an extent by wrapping these unidirectional pultruded fibres using bidirectional fiberglass cloths. This will reinforce composite structure and protect them from delamination by holding unidirectional fibres together. Bidirectional fiberglass cloth will also allow the structure to be reinforced in both directions while it was before reinforced in unidirectional only.

2.2. Micromechanical Model. To study delamination, a composite micromechanical model of pultruded unidirectional

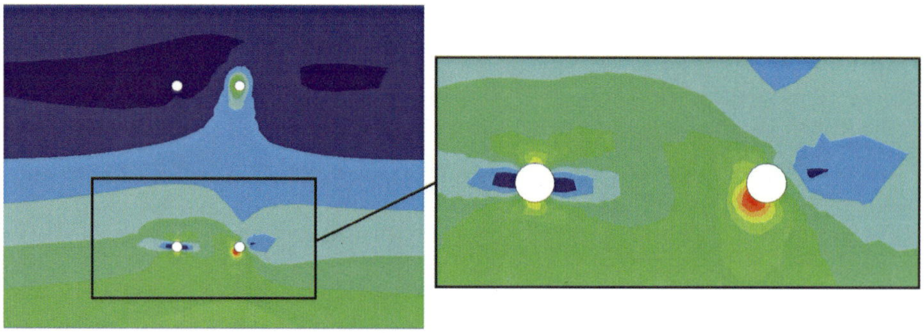

FIGURE 3: Stress concentration around the bolt holes (red contour indicating high stress) [7].

(a) (b)

FIGURE 4: Pultruded retaining wall in Azerbaijan (a) and pultruded railway bridge barriers in England (b) [9, 10].

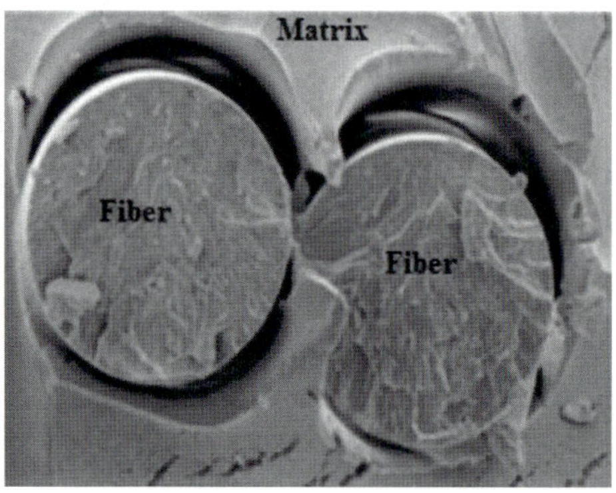

FIGURE 5: Fiber-matrix debonding [12].

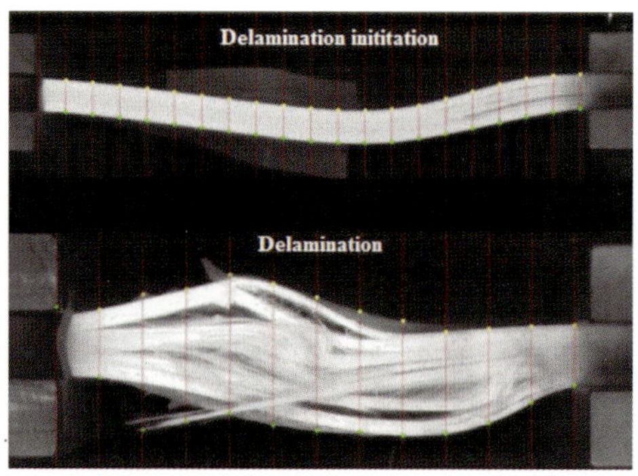

FIGURE 6: Delamination [13].

material with a hole was set up (Figure 9). This model is just a representation of composite structure with a hole that was cut through the fibres. The material properties assigned were E-glass and vinylester (Tables 3 and 4). The diameter of the fibres were 0.014 mm, and size of the micromechanical model was 0.06 mm × 0.06 mm × 0.06 mm. Fibre volume fraction is considered as around 66% (pultrusion). Fibres and matrix were created separately and assembled together. Cohesive surface contact (interface) properties were assigned between fibres and matrix.

2.3. Theoretical Considerations of Interaction Properties. Surface-based cohesive behaviour allows the specification of generalized traction-separation behaviour for surfaces. This behaviour was offering capabilities that are very similar to the cohesive elements, which were defined using traction-separation law, and it requires only less computational time compared with cohesive element in ABAQUS. Surface-based cohesive behaviour was very easy to define, and it allows the simulation of a wider range of cohesive interactions. This method was very useful in this analysis because the surfaces

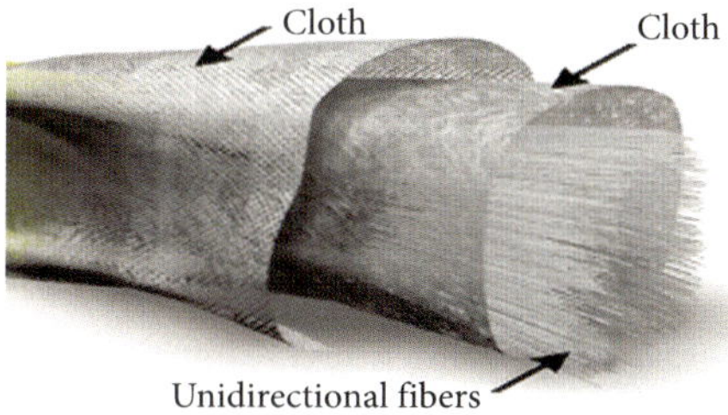

Figure 7: Bidirectional E-glass cloth and unidirectional E-fiberglass [14].

Table 1: Properties of E-glass vinylester bidirectional fiberglass composite [15].

Parameter (unit)	E-glass vinylester bidirectional
Density (kg/m^3)	1988
E_1 (MPa)	20,769
E_2 (MPa)	20,769
ϑ_{12}	0.09
G_{12} (MPa)	4133
Thickness (mm)	0.1

Table 2: Parameters for damage analysis of E-glass vinylester bidirectional fiberglass composite [15].

Parameter (unit)	Notation	Value
Longitudinal tensile strength (MPa)	X_T	395
Longitudinal compressive strength (MPa)	X_C	260
Transverse tensile strength (MPa)	Y_T	395
Transverse compressive strength (MPa)	Y_C	260
Longitudinal shear strength (MPa)	S_L	75
Transverse shear strength (MPa)	S_T	75
Cross product term coefficient		−0.5
Stress limit		0

were in contact (fibre and matrix), and there were no interface thicknesses.

Nominal traction stress vector t is a combination of three components: t_n, t_s, and t_t which represent the normal and two shear tractions, respectively. The corresponding separations are denoted by δ_n, δ_s, and δ_t. And cohesive element thickness is defined as T_0. Nominal strain can be written as

$$\epsilon_n = \frac{\delta_n}{T_0}\quad \epsilon_s = \frac{\delta_s}{T_0}\quad \epsilon_t = \frac{\delta_t}{T_0}. \tag{1}$$

Elastic behaviour can be represented by

$$t = \begin{Bmatrix} t_n \\ t_s \\ t_t \end{Bmatrix} = \begin{bmatrix} k_{nn} & k_{ns} & k_{nt} \\ k_{sn} & k_{ss} & k_{st} \\ k_{tn} & k_{ts} & k_{tt} \end{bmatrix} \begin{Bmatrix} \delta_n \\ \delta_s \\ \delta_t \end{Bmatrix}. \tag{2}$$

Damage modelling was helpful to simulate degradation and eventual failure of the bond between two cohesive surfaces. The failure mechanism in surface-based cohesive behaviour consists of two parameters:

(i) Damage initialization criterion
(ii) Damage evolution law

Damage will evolve according to the damage evolution law after damage initiation criterion is met. A quadratic stress damage initiation criterion and an energy damage evolution law are defined. The graph (Figure 10) shows the typical traction-separation response with a failure mechanism. The damage was assumed to initiate when a quadratic interaction function involving the separation ratios reaches a value of one. This criterion can be represented as

$$\left\{\frac{\langle t_n\rangle}{t_n^0}\right\}^2 + \left\{\frac{\langle t_s\rangle}{t_s^0}\right\}^2 + \left\{\frac{\langle t_t\rangle}{t_t^0}\right\}^2 = 1, \tag{3}$$

where t_n, t_s, and t_t are the contact stress normal to the interface along first and second shear directions, respectively; t_n^0, t_s^0, and t_t^0 are the peak values of contact stress when separation is either purely normal to interface or purely in the first or second shear direction, respectively; and $\langle\rangle$ is the Macaulay bracket indicating that a purely compressive stress state does not initiate damage.

Damage evolution was assigned based on the energy. Simplest way to define the fracture energy was to specify it directly as a function of the mixed mode in tabular form. Fracture energy G_c is equal to the area under the traction-separation curve. The tie contact interaction was used to attach bidirectional fiberglass cloths to micromechanical model of unidirectional pultruded composite [16].

The cohesive elements are modelled as undergoing progressive damage leading to failure and the progressive damage modelling involves softening in the material response, which can lead to convergence difficulties in an implicit solution procedure: ABAQUS/Standard. In order to avoid these convergence issues, ABAQUS/Standard provides a viscous regularization capability that helps in improving the convergence. The use of viscous regularization of the constitutive equations causes the tangent stiffness matrix of the softening material to be positive for sufficiently small time increments [17].

In this study, traction-separation laws are regularized in ABAQUS/Standard using viscosity by permitting stresses to be outside the limits set by the traction-separation law, and this regularization process involves the use of viscous stiffness degradation variable, D_v, which is defined by the following evolution equation:

$$\dot{D}_v = \frac{1}{\mu}(D - D_v), \tag{4}$$

where μ is the viscosity parameter and D is the current damage.

For viscous material, the damage response is

$$t = (1 - D_v)\bar{t}. \tag{5}$$

Using viscous regularization with a small value of the viscosity parameter (small compared to the characteristic time increment) usually helps in improving the rate of convergence of the model in the softening regime, without compromising results. The basic idea is that the solution of the viscous system relaxes into that of the in viscid case as $t/\mu \to \infty$ where t represents time. The use of viscous regularization is a powerful and often necessary tool that enables accurate prediction of delamination, and this approach

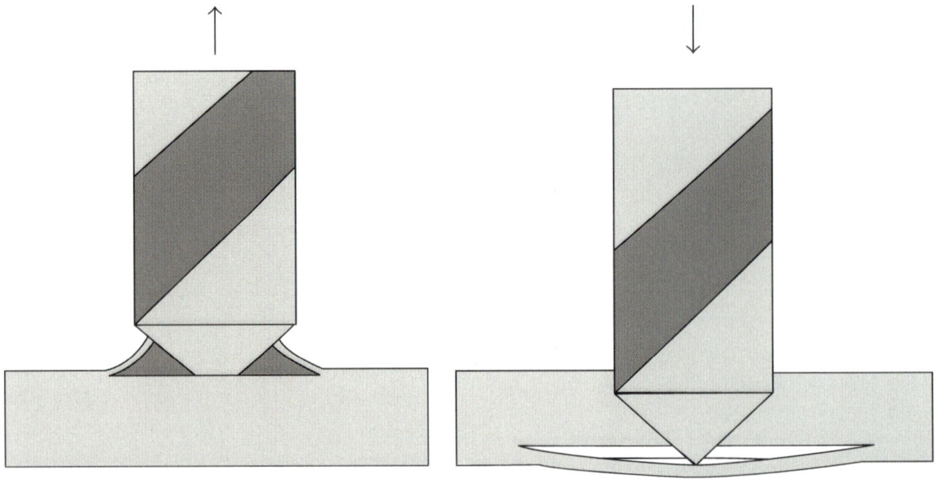

FIGURE 8: Schematic of delamination push-out at exit (a) and peel-up at the entry (b).

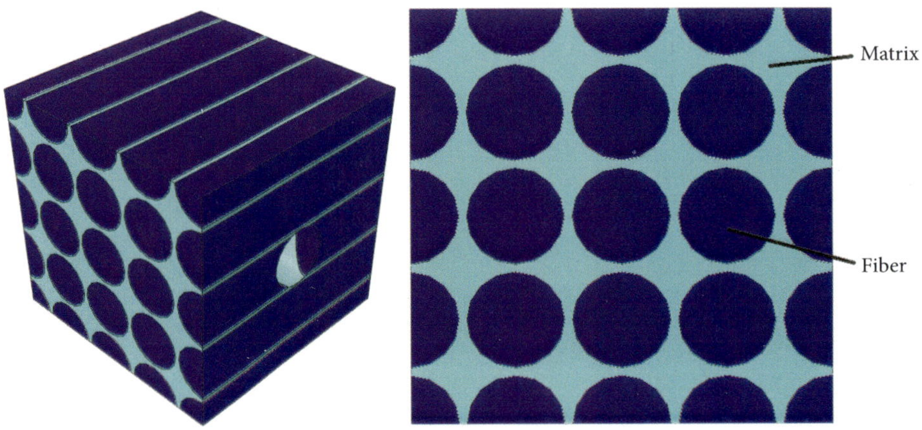

FIGURE 9: Composite micromechanical model.

TABLE 3: Properties of vinylester resin [11].

Parameter (unit)	Vinylester
Density (g/cm^3)	1.05
Poisson coefficient, ϑ	0.35
Shear modulus (GPa)	1.24
Tension	
Modulus, E (GPa)	3.2
Tensile strength, σ_{ult} (MPa)	86
Compression	
Modulus, E (GPa)	3.35
Yield strength, σ_y (MPa)	103

TABLE 4: Properties of E-fiberglass [15].

Parameter (unit)	E-glass
Density (g/cm^3)	2.54
Young's modulus in tension (GPa)	73
Tensile strength (MPa)	2400
Compressive strength (MPa)	1450
Shear modulus (GPa)	33.3
Poisson coefficient, ϑ	0.22
Diameter (μm)	14

is used here in order to improve the rate of convergence of the model [17, 18].

In this analysis, a global stabilization factor (dissipated energy fraction) of 2×10^{-4} was used to perform numerical analysis [19]. By adopting a weakening factor of f_w 0.3, the interfacial strength properties were reduced from the matrix material [20–22] which indicate a relatively weak interface in pultruded glass fibre-reinforced material. The elastic stiffness (per unit area) of the interface was assumed to be equal to 10^6 N/mm^3 [23]. The critical normal interface tractions of the cohesive zone elements were equal to the normal tensile strength of the unidirectional layers in the direction normal to the fibres (Table 5), that is, 13.5 N/mm^2, and the critical shear interface tractions were equal to the unidirectional shear strength (Table 5), that is, 13.5 N/mm^2. For the pultruded material mode I, mode II, and mode III critical energy release rates G_{Ic} and $G_{\mathrm{IIc}} = G_{\mathrm{IIIc}}$ were taken as 0.2 N/mm and 0.5 N/mm, respectively [24]. The bidirectional fibreglass cloth was attached to the unidirectional micromechanical model using tie contact constraint in the four surfaces.

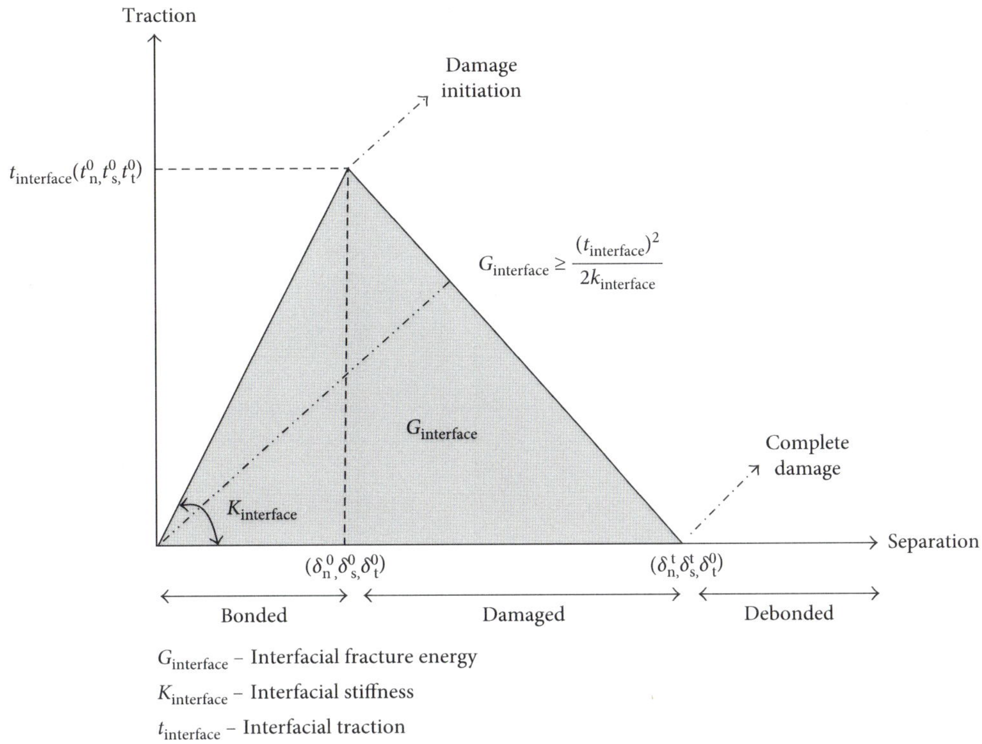

FIGURE 10: Typical traction-separation response [18].

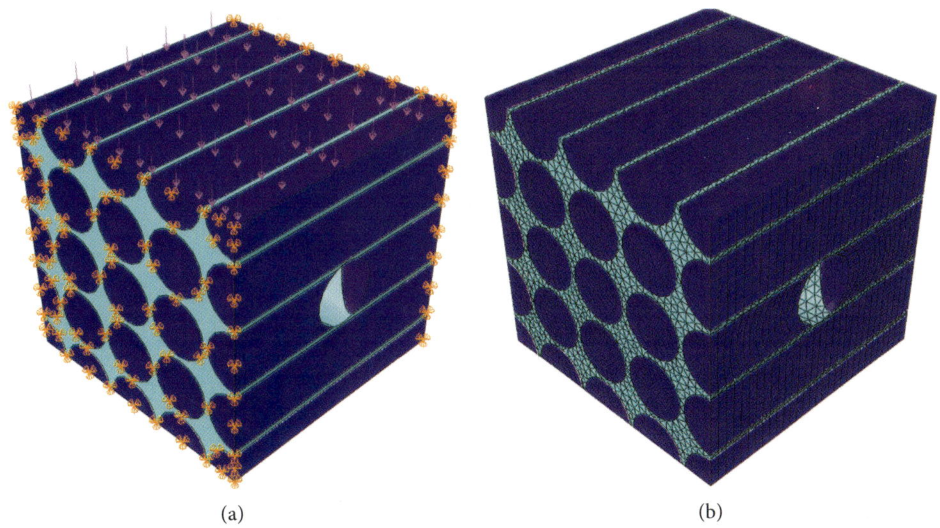

FIGURE 11: Load and boundary conditions (a) and mesh (b) on the composite micromechanical model.

2.4. Load and Boundary Conditions and Mesh. The micromechanical model was subjected to pressure force from the top. As the interest was to study delamination, left and right surfaces are constrained using the pinned join boundary condition that was allowing the material to delaminate easier and was comparatively easy for simulation convergence (Figure 11). This method of load and boundary conditions and mesh was maintained similar in all simulations that were carried out.

The parts of the model were meshed separately. Fibre was meshed using 26,664 C3D8R elements and matrix using 65,290 C3D10 elements. Cloth layer 1 was meshed using

TABLE 5: Parameters for damage analysis of E-glass vinylester unidirectional composite [16].

Parameter (unit)	Notation	Value
Longitudinal tensile strength (MPa)	X_T	1200
Longitudinal compressive strength (MPa)	X_C	620
Transverse tensile strength (MPa)	Y_T	45
Transverse compressive strength (MPa)	Y_C	100
Longitudinal shear strength (MPa)	S_L	45
Transverse shear strength (MPa)	S_T	45
Cross product term coefficient		−0.5
Stress limit		0

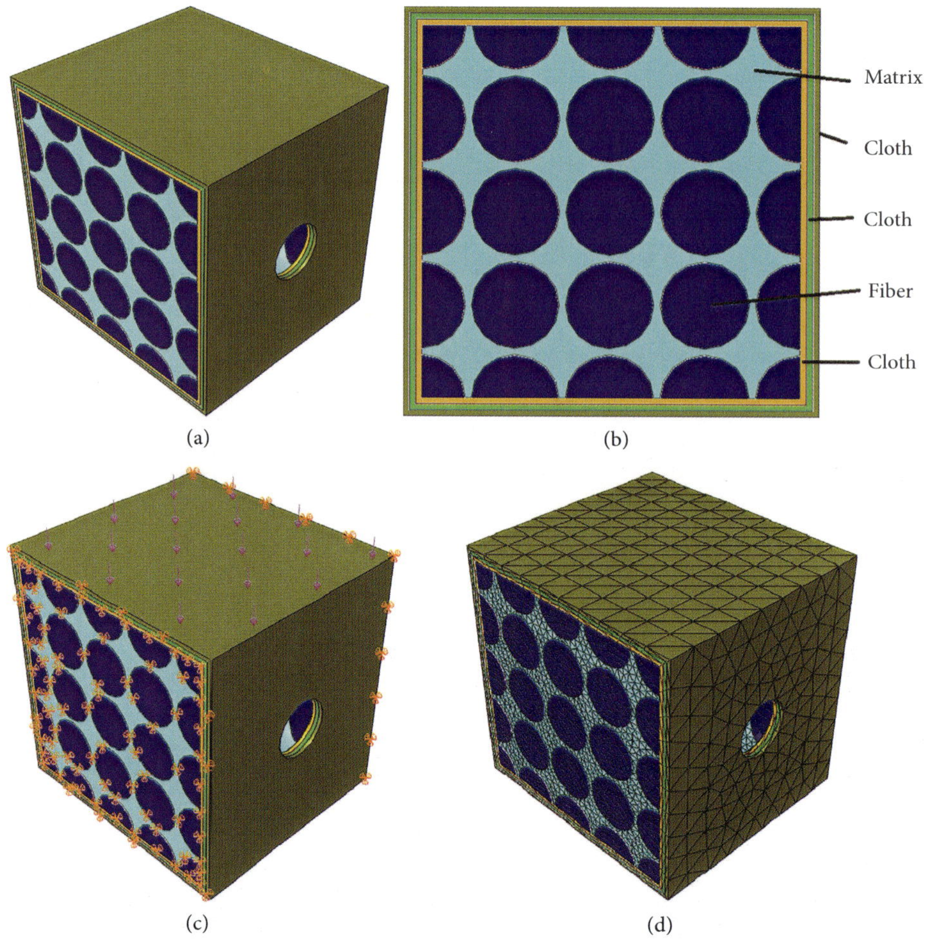

FIGURE 12: Composite micromechanical model: (a) isometric view; (b) front view; (c) load and boundary conditions; (d) mesh of composite micromechanical model with 3-layer cloth.

C3D10 2643 elements, cloth layer 2 was meshed using C3D10 2423 elements, and cloth layer 3 was meshed using 2494 C3D10 elements (Figure 12). Different simulations were performed using these composite micromechanical models to study the damage. (C3D8R: continuum 3-D, 8-node, reduced integration; C3D10: continuum 3-D, 10-node).

The bidirectional fiberglass cloths with different number of layers and densities were wrapped on the composite micromechanical model, and different simulations were performed. These simulations were carried out to select a better bidirectional fiberglass cloth, which can decrease stress concentration and damage. Due to the cost-effectiveness considerations and small thicknesses of beams, the appropriate number of layers of wraps and the better material selection were evident. Weight and thickness of the bidirectional fiberglass cloth chosen are shown in Table 6.

3. Results and Discussion

For comparison, numerical simulations were carried out without wrapping fibreglass cloths and with wrapping different layers of bidirectional fiberglass cloths. Damage initiation stress and interface damage occurring between fibre and matrix of each model were used for comparison.

TABLE 6: Weight and thickness of the bidirectional fiberglass cloth [25].

Weight (g/m^2)	Thickness (mm)
186	0.14
295	0.23
318	0.28

3.1. Micromechanical Model. Stress contour plots obtained from different numerical simulations were plotted to visualize the performance of the model before and after wrapping different layers of bidirectional fibreglass cloths. Micromechanical models were subjected to high loads in order to study damage. Figure 13(a) shows stress contour plot of the micromechanical model before wrapping bidirectional fibreglass cloths. A maximum stress of 5344 MPa and fibre-matrix separation are observed in this model. Debonding is decreased in Figure 13(b) after addition of 1 layer of bidirectional fibreglass cloth. Bidirectional fibreglass cloth wrapping reinforced the model and decreased maximum stress to 3674 MPa. Similarly, maximum stress is decreased for 2 layer and 3 layer as shown in Figure 13(c), and maximum stress is also decreased to 3175 MPa and 2433 MPa, respectively, as shown in Figure 13(d). Maximum

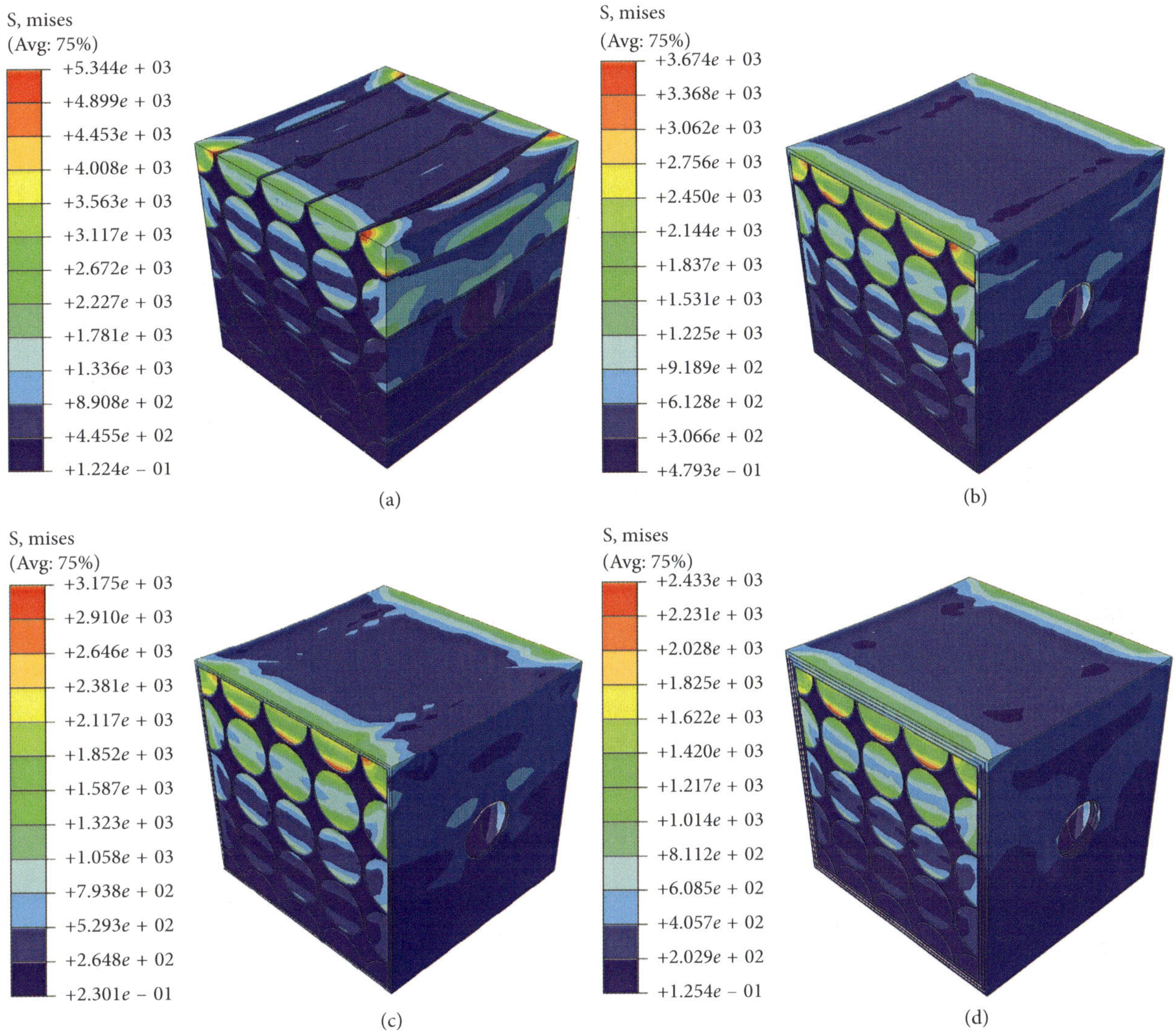

FIGURE 13: Stress on the composite micromechanical model: (a) without cloth; (b) with 1 layer cloth; (c) with 2 layer cloth; (d) with 3 layer cloth.

stresses were found near the fibre-matrix interface in all models, and addition of bidirectional fiberglass cloths results in decreasing of these stresses.

The result of stress versus damage obtained from different models was plotted together, and Figure 14 shows the behaviour of stress versus damage for each model. In the micromechanical model without cloth, damage initiation stress was estimated as 18 MPa. High interface damage and fibre-matrix debonding are observed in this model. In order to increase damage initiation stress, the micromechanical model is wrapped with one layer of bidirectional fiberglass cloth. After wrapping, fibre-matrix debonding is restricted and interface damage is decreased. As a result, damage initiation stress is increased to 23 MPa. Bidirectional fiberglass cloth wrapping has reduced fibre-matrix debonding by holding them together. Similarly, more bidirectional fiberglass cloths were wrapped, and damage initiation stress is increased to 28 MPa and 36 MPa for 2 layer and 3 layer of bidirectional fibre cloth wraps, respectively. It was observed that bidirectional fiberglass cloth wrapping assists fibre-matrix bonding to be stronger and reduces delamination, which occurs around the bolt hole of the unidirectional pultruded model.

Considering Figure 14 into 2 areas, in Area 1, damage is initiated and stress is incrementing steadily from 16 MPa to 22 MPa for the model without cloth, 23 MPa to 51 MPa for 1-layer cloth, 28 MPa to 60 MPa for 2-layer cloth, and 34 MPa to 83 MPa for 3-layer cloth in the damage coefficient limit 0 to 0.8. The increase in the stress value near to damage coefficient 0.8 is due to starting of complete damage. In Area 2, huge variation in stress can be observed due to the complete damage to the models, and the difference in variation of the stress value for each model shows their restriction to damage. A model without cloth is fully damaged in less stress limit comparatively, and after adding bidirectional fibreglass cloths, damage is restricted and stress required for complete damage is extended for each layer.

3.2. Effect of Bidirectional Fiberglass on Platform Model. The micromechanical analysis provided an evident result of the impact of wrapping bidirectional fiberglass cloths on

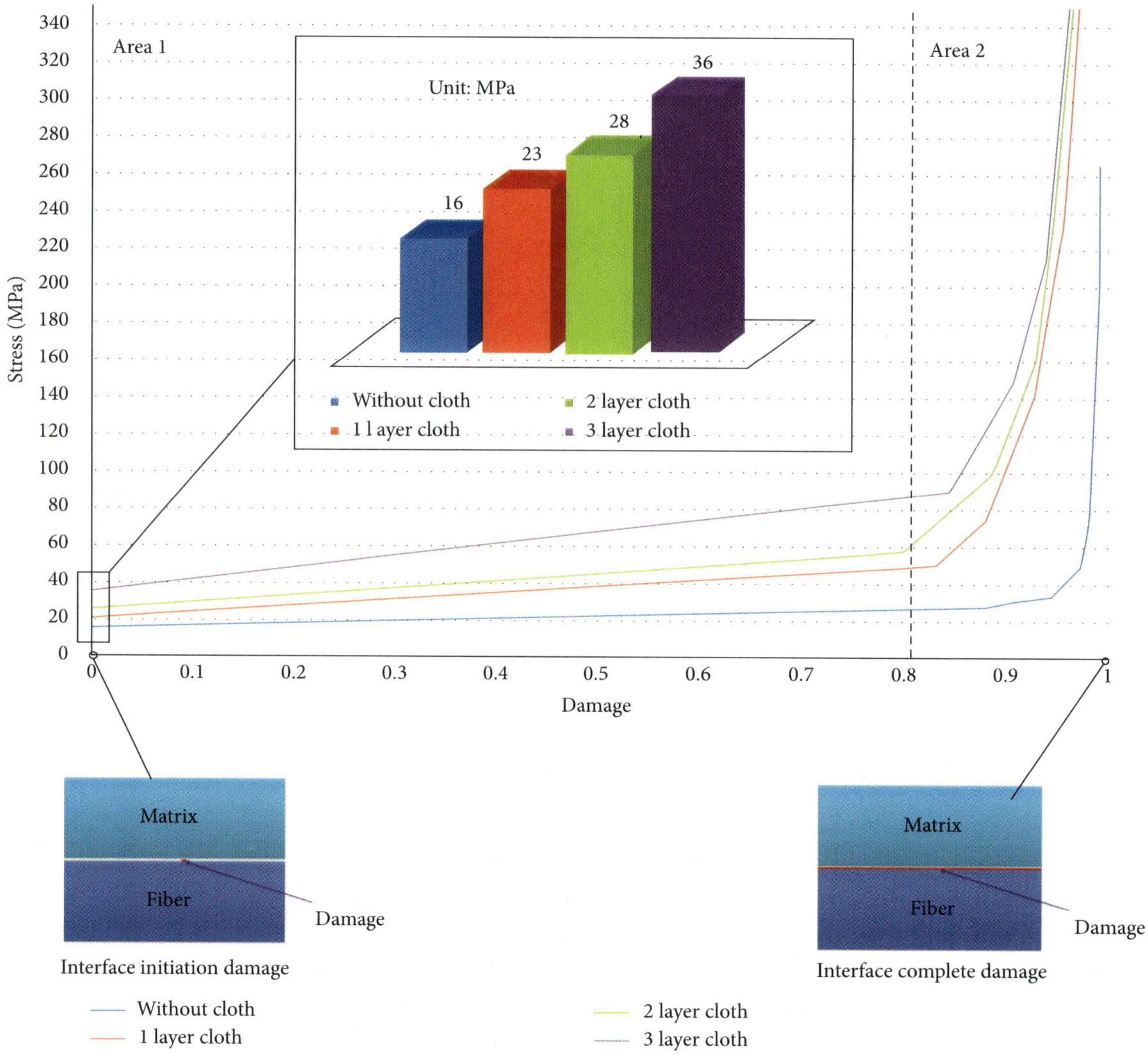

Figure 14: Comparison of stress versus damage.

unidirectional pultruded structure, and this method is implemented on the large smart composite platform. From the previous study [4–7], maximum stress near the bolt hole was around 75 MPa which was on ply-1 (Figure 15). After adding bidirectional fiberglass cloths to the large platform shell model, it is observed that the effect of stress concentration near the bolt hole had decreased to around 67 MPa in ply-1 (Figure 16). In this model, a maximum stress of 82 MPa was on bidirectional fiberglass cloth ply. Here, cloths act as a protective layer to the unidirectional fibreglass material from damage and decreased the stress from 75 MPa to 67 MPa. This result shows that there is an improvement in reduction of stress concentration effect on large smart composite platform, and this can enhance delamination strength.

3.3. Comparison of Different Bidirectional Fibreglass Cloths. To select an appropriate bidirectional fibreglass cloth material which can enhance delamination strength, numerical simulations were performed with three different densities and results were plotted (Figure 17). They were selected considering the beam thickness of smart composite platform and industrial availability. The bidirectional fiberglass cloth with a weight of 186 g/m^2 and thickness of 0.14 mm has damage initiation stress around 41 MPa; the bidirectional fiberglass cloth with a weight of 295 g/m^2 and thickness of 0.23 mm has damage initiation stress around 49 MPa; and the bidirectional fiberglass cloth with a weight 318 g/m^2 and thickness 0.28 mm has damage initiation stress around 59 MPa. The bidirectional fiberglass cloth with a weight of 318 g/m^2 and thickness of 0.28 mm which showed better performance in reducing damage has been selected for large smart composite platform.

In order to study the effect of number of wraps, different layers of bidirectional fiberglass cloths were wrapped on the micromechanical model and numerical simulations were performed. The results were plotted (Figure 18), and the damage initiation stress value after adding each layer was estimated. It was observed that after 3 layers of wrapping, damage initiation stress was saturating. Simulations were carried out up to 7 layers in order to estimate the variation in

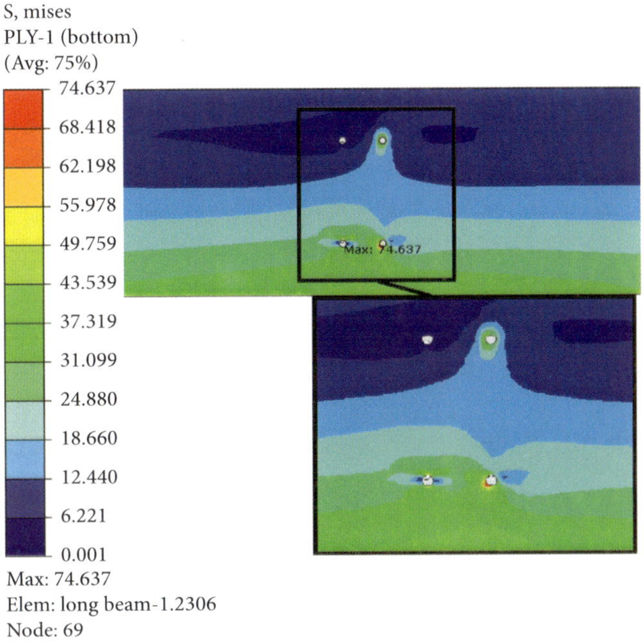

FIGURE 15: Stress contour plot around bolt for unidirectional pultruded composite platform before adding cloth.

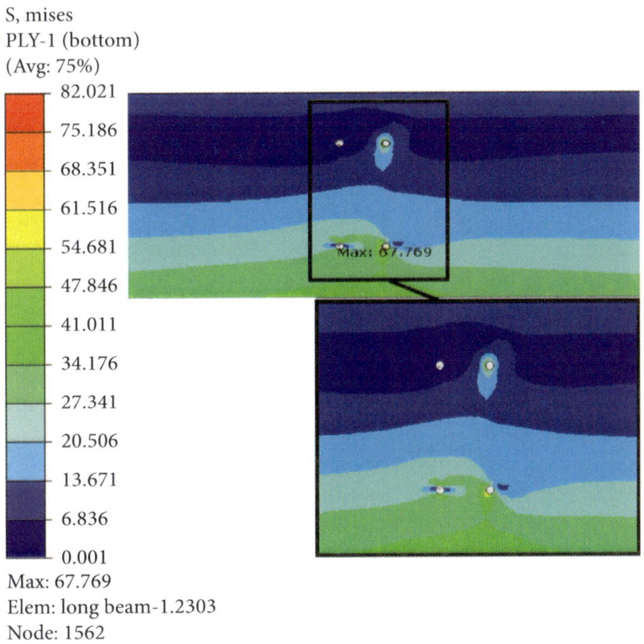

FIGURE 16: Stress contour plot around bolt for unidirectional pultruded composite platform after adding cloth.

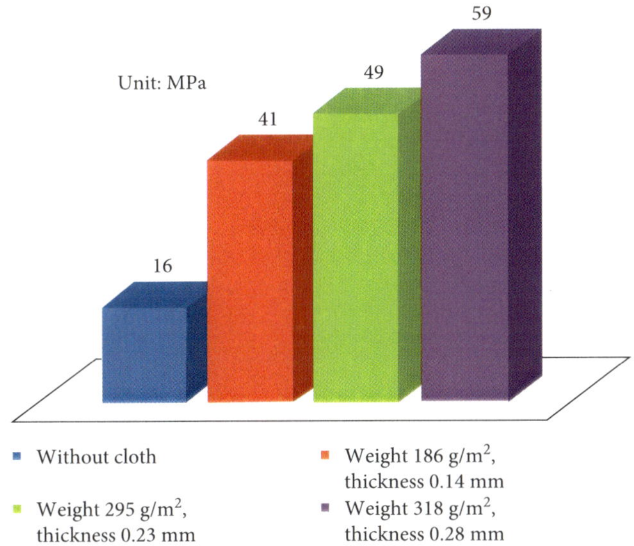

FIGURE 17: Comparison of damage initiation stress for different bidirectional fiberglass cloths with 3 layers.

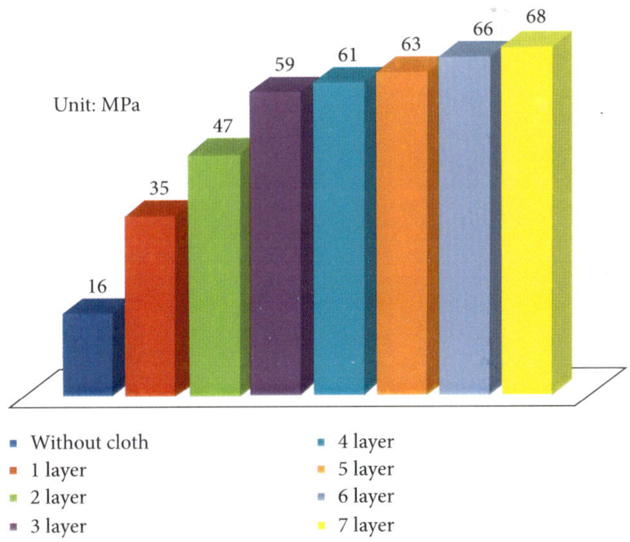

FIGURE 18: Comparison of damage initiation stress for 7 layers.

stress values. Damage initiation stress was 35 MPa, 47 MPa, 59 MPa, 61 MPa, 63 MPa, 66 MPa, and 68 MPa from layer 1 to layer 7, respectively. From Figure 19, it is evident that layer 4, layer 5, layer 6, and layer 7 have very less improvement in damage initiation stress. The addition of more layers without obtaining better performance will increase the budget and the weight of the structure. Delamination failure will occur when the transverse shear load experienced will exceed interlaminar shear strength (ILSS) [26]. In Handbook of Adhesion [27], interlaminar shear strength (ILSS) for E-glass/vinylester was estimated as 51 MPa. As mentioned above, damage initiation stress for layer 3 is 59 MPa; that is, damage initiation stress for layer 3 is above interlaminar shear strength (ILSS) limit, and it can restrict delamination. So, the number of layers is selected as three.

3.4. Comparison of Bidirectional Fiberglass Cloths and Conventional Continuous Fiberglass Mats. Bidirectional fiberglass cloths have better mechanical properties compared to the conventional continuous fibreglass mat due to the proper orientation of fibres in both directions, whereas fibres are oriented randomly in mats (Figure 20). As the aim of this study is to decrease damage and increase delamination strength, a better material that can reinforce the structure

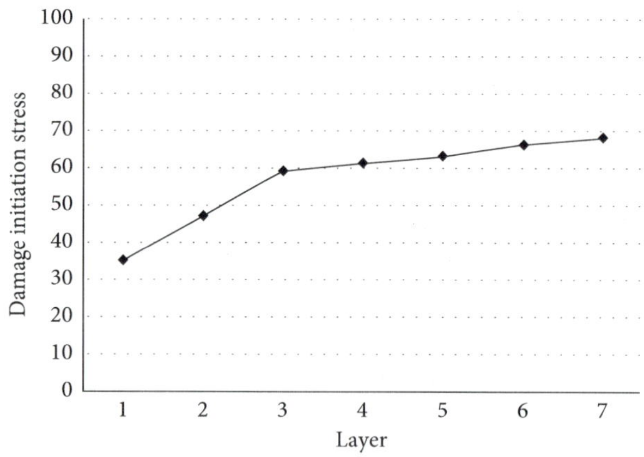

(Unit: MPa)	1	2	3	4	5	6	7
	35	47	59	61	63	66	68

FIGURE 19: Damage initiation stress versus layer.

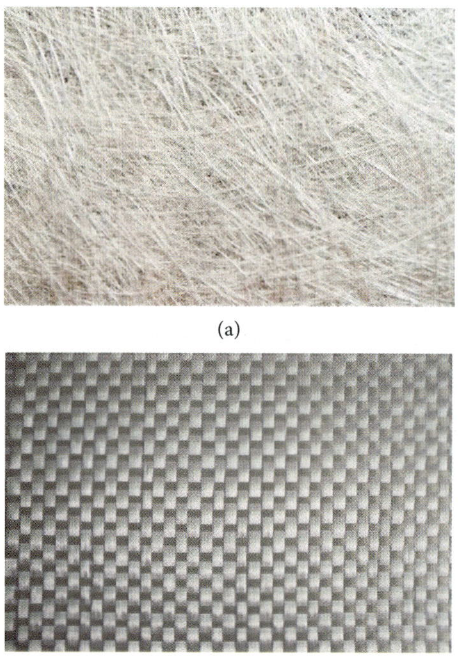

(a)

(b)

FIGURE 20: (a) Continuous fiberglass mat and (b) bidirectional fiberglass cloth.

will be a better choice. The quality of each mat from different industries is dissimilar, and as mentioned earlier, the applications of pultruded structures are mainly in civil engineering. So, selection of good quality material cloth is better compared to that of lower quality mat in order to avoid the risk of human life and increase safety. Moreover, from Figure 21, damage initiation stress for bidirectional fiberglass and continuous fiberglass mat are 59 MPa and 44 MPa, respectively. This difference shows that bidirectional fiberglass cloths have better properties in restricting delamination. Considering all these facts, bidirectional fiberglass cloths have better performance despite their cost compared to continuous fibreglass mats.

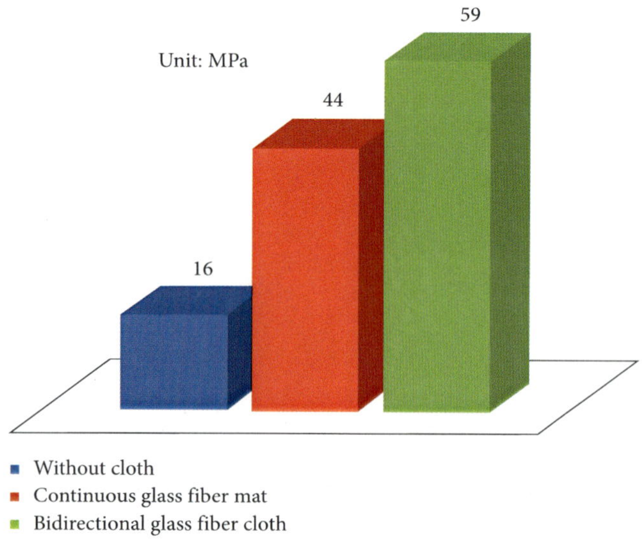

FIGURE 21: Comparison of the damage initiation stress for bidirectional fiberglass cloths and continuous fiberglass mats.

4. Conclusion

Unidirectional composites are subjected to delamination at very low stress, and these issues cannot be tolerated in some demanding applications. Most of the applications of pultruded structures are in civil engineering, which involves human safety. Therefore, it is essential to study the correlation of delamination onset and find a technology to enhance delamination strength as high as possible. The aim of this work was to enhance delamination strength of unidirectional pultruded large smart composite platforms. The micromechanical model of unidirectional pultruded composite was created, and different simulations were carried out in order to investigate defects. From the bibliography, it was found that debonding between fibre and matrix will occur before other defects, and this will lead to delamination. Bidirectional fiberglass cloth wrappings were used to prevent delamination, and different simulations were performed without wrapping cloths and with different layers of cloth wrapping. From a numerical simulation results, it was estimated that the bidirectional fiberglass cloth wrapping method was effective in preventing delamination. Due to the high cost of bidirectional fiberglass cloths, a study was also performed considering the number of layers and density to select an appropriate bidirectional fiberglass cloth material that can be considered as cost effective. After these studies, bidirectional fiberglass cloth with a weight of 318 g/m^2, thickness of 0.28 mm, and 3 layers of wrap were selected. The result of a comparison study shows that bidirectional fiberglass cloths have better ability to restrict delamination compared to continuous fiberglass mats.

Conflicts of Interest

The authors declare that there are no conflicts of interest regarding the publication of this paper.

References

[1] M. Drissi-Habti, "Smart composite materials project DECID2 for structural applications," *JEC Composites*, vol. 42, pp. 28-29, 2008.

[2] L. M. P. Durão, J. M. R. Tavares, V. H. C. De Albuquerque, J. F. S. Marques, and O. N. Andrade, "Drilling damage in composite material," *Materials*, vol. 7, no. 5, pp. 3802–3819, 2014.

[3] L. M. P. Durão, D. J. S. Gonçalves, J. M. R. S. Tavares, V. H. C. de Albuquerque, and A. T. Marques, "Comparative analysis of drills for composite laminates," *Journal of Composite Materials*, vol. 46, no. 14, pp. 1649–1659, 2012.

[4] X. Chapeleau, M. Drissi-Habti, and T. Tomiyama, "Embedded optical fiber sensors for in situ and continuous health monitoring of civil engineering structures in composite materials," *Materials Evaluation*, vol. 68, no. 4, pp. 409–415, 2010.

[5] X. Chapeleau, M. Drissi-Habti, and T. Tomiyama, "Embedded optical fiber sensors for in situ and continuous health monitoring of composite materials in civil engineering structures," *Japan Society for Composite Materials*, vol. 36, no. 1, pp. 227–235, 2010.

[6] A. Cordelle and M. Drissi-Habti, "Nanoindentation characterization of vinylester glass-fiber composites submitted to dense ultraviolet radiation exposure," *Materials Evaluation*, vol. 71, no. 4, 2013.

[7] I. D. Madukauwa-David and M. Drissi-Habti, "Numerical simulation of the mechanical behavior of a large smart composite platform under static loads," *Composites Part B: Engineering*, vol. 88, pp. 19–25, 2016.

[8] P. Carlone, G. Salvatore Palazzo, and R. Pasquino, "Pultrusion manufacturing process development by computational modelling and methods," *Mathematical and Computer Modelling*, vol. 44, no. 7, pp. 701–709, 2006.

[9] Creative Pultrusions, *Composite Sheet Pile Wall System*, Creative Pultrusions, Pleasantville, PA, USA, 2016, http://www.creativepultrusions.com/index.cfm/fiberglass-pultruded-systems/composite-sheet-pile-system.

[10] G. Ginger, *Rail Line Landmark Restoration via Pultrusion*, CompositesWorld, Scottsdale, AZ, USA, 2016, http://www.compositesworld.com/articles/rail-line-landmark-restoration-via-pultrusion-.

[11] M. U. Farooq, L. A. Carlsson, and B. A. Acha, "Determination of fiber/matrix adhesion using the Outwater–Murphy single fiber specimen," *Engineering Fracture Mechanics*, vol. 76, no. 18, pp. 2758–2765, 2009.

[12] M. S. Kumar, K. Raghavendra, M. A. Venkataswamy, and H. V. Ramachandra, "Fractographic analysis of tensile failures of aerospace grade composites," *Materials Research*, vol. 15, no. 6, pp. 990–997, 2012.

[13] Y. Bai, T. Vallée, and T. Keller, "Delamination of pultruded glass fiber-reinforced polymer composites subjected to axial compression," *Composite Structures*, vol. 91, no. 1, pp. 66–73, 2009.

[14] DuraGrid, *DURAGRID® Benefits*, 2016, http://www.prmetals.com/Product/Grating/Fiberglass-Grating/Pultruded-Fiberglass-Grating/DURADEK/DURADEK-I6000-Pultruded-Grating–112-Depth.

[15] R. Karakuzu, T. Gülem, and B. M. İçten, "Failure analysis of woven laminated glass–vinylester composites with pin-loaded hole," *Composite Structures*, vol. 72, no. 1, pp. 27–32, 2006.

[16] A. Shukla, Y. D. Rajapakse, and M. E. Hynes, *Blast Mitigation: Experimental and Numerical Studies*, Springer Science & Business Media, Berlin, Germany, 2013.

[17] ABAQUS, *Abaqus 6.10 Analysis User's Manual, Volume V: Prescribed Conditions, Constraints & Interactions. 33.1.10 Surface-Based Cohesive Behavior*, 2016, http://abaqusdoc.ucalgary.ca/pdf_books/ANALYSIS_5.pdf.

[18] W. G. Jiang, R. Z. Zhong, Q. H. Qin, and Y. G. Tong, "Homogenized finite element analysis on effective elastoplastic mechanical behaviors of composite with imperfect interfaces," *International Journal of Molecular Sciences*, vol. 15, no. 12, pp. 23389–23407, 2014.

[19] A. M. G. Coelho, J. T. Mottram, and N. Matharu, "Virtual characterization of delamination failures in pultruded GFRP angles," *Composites Part B: Engineering*, vol. 90, pp. 212–222, 2016.

[20] M. Knops, *Analysis of Failure in Fiber Polymer Laminates: The Theory of Alfred Puck*, Springer Science & Business Media, Berlin, Germany, 2008.

[21] A. Puck and H. Schürmann, "Failure analysis of FRP laminates by means of physically based phenomenological models," *Composites Science and Technology*, vol. 62, no. 12-13, pp. 1633–1662, 2002.

[22] J. Qureshi, J. T. Mottram, and B. Zafari, "Robustness of simple joints in pultruded FRP frames," *Structures*, vol. 3, pp. 120–129, 2015.

[23] P. P. Camanho, C. G. Davila, and M. F. De Moura, "Numerical simulation of mixed-mode progressive delamination in composite materials," *Journal of Composite Materials*, vol. 37, no. 16, pp. 1415–1438, 2003.

[24] A. M. G. Coelho, "Finite element guidelines for simulation of delamination dominated failures in composite materials validated by case studies," *Archives of Computational Methods in Engineering*, vol. 23, no. 2, pp. 363–388, 2016.

[25] HexForce, *Hexcel Technical Fabrics Handbook*, HexForce, Stamford, CT, USA, 2016, http://www.hexcel.com/Resources/DataSheets/Brochure-Data-Sheets/HexForce_Technical_Fabrics_Handbook.pdf.

[26] Z. Fan, M. H. Santare, and S. G. Advani, "Interlaminar shear strength of glass fiber reinforced epoxy composites enhanced with multi-walled carbon nanotubes," *Composites Part A: Applied Science and Manufacturing*, vol. 39, no. 3, pp. 540–554, 2008.

[27] D. E. Packham, *Handbook of Adhesion*, John Wiley & Sons, Hoboken, NJ, USA, 2006.

15

Effects of Shrinkage Reducing Agent and Expansive Admixture on the Volume Deformation of Ultrahigh Performance Concrete

Su Anshuang,[1] Qin Ling,[2] Zhang Shoujie,[1] Zhang Jiayang,[1] and Li Zhaoyu[1]

[1] *Heilongjiang Provincial Hydraulic Research Institute, Harbin 150080, China*

[2] *School of Civil Engineering, Harbin Institute of Technology, Harbin 150090, China*

Correspondence should be addressed to Su Anshuang; bridgecrete@163.com

Academic Editor: Xiao-Yong Wang

This paper investigated the influences of shrinkage reducing agent and expansive admixture on autogenous and drying shrinkage of ultrahigh performance concrete (UHPC) containing antifoaming admixture. The shrinkage reducing agent was used at dosage of 0.5%, 1%, and 2% and the expansive admixture was used at dosage of 2% to 4% by mass of cementitious material. The results show that the air content of UHPC increases with the higher addition of shrinkage reducing agent and expansive admixtures. However, the fluidity, compressive strength, and shrinkage of UHPC exhibit a declining tendency. The usage of expansive agent at dosage of 4% significantly reduces the shrinkage of UHPC. The 7-day autogenous shrinkage was decreased by 16.0% and 28-day drying shrinkage was decreased by 29.5%, respectively. Shrinkage reducing agent at dosage of 2% reduced the 7-day autogenous shrinkage by 44.3% and 28-day drying shrinkage by 50.2%. Compared with expansive admixture, shrinkage reducing agent exhibits more efficient shrinkage reduction effect on UHPC.

1. Introduction

As a new generation of concrete, ultrahigh performance concrete (UHPC) possesses numerous advantages such as high strength, high durability, high reliability, and high performance [1]. However, there is no unified standard about technical requirements for UHPC. In France, the guide "Ultrahigh Performance Fiber Reinforced Concrete-Interim Recommendations" described ultrahigh performance fiber reinforced concrete (UHPFRC) as a highly viscous cement based composite material with the 28-day compressive strength over 150 MPa, mixed with the steel fiber to increase toughness and mixed with fine quartz sand as aggregate [2]. And two French standards related to UHPFRC have been published in April 2016. Japanese Civil Engineering Society defined UHPC as a cement based material reinforced by steel fiber with 28-day compressive strength over 150 MPa and mixed with cement and highly active volcanic ash as cementitious materials [3]. The diameter of aggregate was less than 2.5 mm and water-to-binder ratio was less than 0.24. The content of steel fiber was less than 2% by volume of concrete. The diameter of steel fiber ranged from 0.10 mm to 0.25 mm and the length of steel fiber was in the range of 10 mm to 20 mm. At present, the research and application of UHPC in China are still in the experimental stage and the accumulation of experimental data and engineering experience is very necessary.

Low water-to-binder ratio, high-efficiency water reducing agent, and highly active volcanic ash were applied in order to achieve the excellent performance of UHPC [4]. Nevertheless, the autogenous and drying shrinkage of UHPC also increased because of the application of the mentioned methods. The shrinkage of concrete would lead to the generation and propagation of internal microcracks, which could greatly decrease the durability of concrete, shorten its service life, and even cause serious engineering accidents [5]. Therefore, control of the shrinkage is crucial for the current research and application of UHPC [6].

Cwirzen et al. systematically studied the mechanical properties, volume stability, and durability of UHPC [7]. Experimental results showed that high curing temperature tended to improve the compressive strength of reactive powder concrete (RPC). The strength of the samples cured at high temperature could reach 200 MPa, while the highest compressive strength of the specimens cured at room temperature was only 150 MPa. The shrinkage of UHPC was high; furthermore, the shrinkage of the ultrahigh strength mortar (UHSM) was approximately twice as that of UHPC. However, UHPC and UHSM possessed excellent durability. Liu et al. investigated the influences of mineral powders on volume stability of RPC. The research results indicated that high content of mineral powders could decrease early-stage shrinkage value of RPC to 300×10^{-6} and the development of shrinkage at later age was negligible [8]. Tam et al. investigated the influences of mixing proportions and curing conditions on the shrinkage of RPC [9]. The results showed that shrinkage increased with the decrease of water-to-cement ratio and the increase of dosage of high-efficiency water reducing agents. However, the steel fiber and high-temperature autoclave curing could reduce the shrinkage of RPC effectively. Wong et al. investigated the shrinkage of RPC cured at room temperature [10]. The results showed that 1-day shrinkage accounted for 77% of the 7-day total shrinkage when curing temperature was 20°C and the relative humidity was 50%. Moreover, the size of specimen also had certain influence on the shrinkage value. The shrinkage of large specimen was lower than that of small one. Han concluded that the fly ash and other mineral powders had significant effects on the early-age shrinkage of RPC. Shrinkage reducing admixture and expansive agent could effectively decrease the early-age shrinkage of RPC, while excessive expansive agent could significantly increase the plastic shrinkage [11].

Shrinkage had negative impacts on the durability of UHPC, which needed to be controlled [12]. The incorporation of shrinkage reducing admixture and expansive agent was the effective method to improve the volume stability of UHPC. The introduction of antifoaming admixture could efficiently reduce internal bubbles of UHPC and make the microstructure more dense [13]. What is more, the addition of antifoaming admixture could greatly improve the mechanical and durability properties of UHPC. For UHPC with the addition of antifoaming admixture, few studies focused on the influence of shrinkage reducing admixture and expansive agent on the volume stability. This paper investigated the effects of shrinkage reducing agent and expansive agent on the shrinkage of UHPC mixed with antifoaming admixture and proposed some suggestions for future research work.

2. Raw Materials and Test Methods

2.1. Raw Materials. The cement used was ordinary Portland cement with the strength grade of 52.5 complying with the Chinese National Standard GB175-2007. The particle size distribution of cement was shown in Figure 1. The specific surface area of silica fume (SF) was 21000 $m^2 kg^{-1}$ and the pozzolanic activity index was 104%. Two types of silica sand (SS) with bulk density of 1.405 g/m^3 and 1.329 g/m^3, respectively, were used. According to the theory of high packing, the mass ratio of the coarse and fine silica sand was 3.9 : 4.9. A polycarboxylate-based superplasticizer (SP) from Harbin Qiang Shi Company was used. It has a water reducing efficiency greater than 30% and a solid content

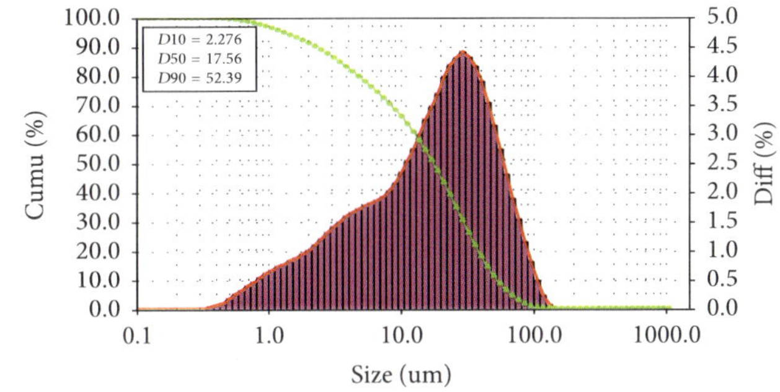

FIGURE 1: Particle size distribution of cement.

FIGURE 2: Autogenous shrinkage measuring device.

of about 40%. The antifoaming admixture (AF) used was produced by Japanese Toho Chemical Industry Company. The shrinkage reducing agent (SRA) used was produced by Germany Evonik Industries. The expansive agent (EA) used was calcium magnesium compound expansive agent produced by Jiangsu Heitman Company Limited.

2.2. Test Methods. The mixing proportions of UHPC were shown in Table 1. The water-to-binder ratio was 0.2 for all the mixtures.

The fluidity of UHPC was tested according to the Chinese standard GB/T 2419-2005. The mixture was cast into a minicone mold uniformly. The mini-slump cone has a base diameter of 60 mm, a top diameter of 36 mm, and a height of 60 mm. Then the mold was lifted vertically and two diameters perpendicular to each other were measured. The mean value was recorded as slump flow to evaluate the fluidity. The air content was tested according to the Chinese standard GB/T 50080-2002. The mixture was cast into a cylinder container uniformly and the entrapped air content was recorded through a pressure gauge. The compressive strength was tested according to the Chinese standard GB/T 50081-2002.

Autogenous shrinkage of UHPC was tested by the self-designed measuring device shown in Figure 2. UHPC was mixed and cast into plastic corrugated pipe. After 6 hours, its initial length was measured. Then the dial gauge readings were recorded every 24 hours until the age of 7 days.

Drying shrinkage of UHPC was tested by the self-designed measuring device shown in Figure 3. UHPC was mixed and cast into the mold with the size of 25 mm ×

TABLE 1: Mixing proportions of UHPC.

Sample	Cement	SF	SS	Water	SP	AF	SRA	EA
JP0	100	25	110	25	1.8	0.5	0	0
J1	100	25	110	25	1.8	0.5	0.5	0
J2	100	25	110	25	1.8	0.5	1	0
J3	100	25	110	25	1.8	0.5	2	0
P1	100	25	110	25	1.8	0.5	0	2
P2	100	25	110	25	1.8	0.5	0	3
P3	100	25	110	25	1.8	0.5	0	4

FIGURE 3: Drying shrinkage measuring device.

25 mm × 285 mm. After curing under the temperature of 20 ± 3°C and relative humidity beyond 90% for 2 days, the UHPC specimens were demolded and the initial length of the specimen was measured. Then the dial gauge readings were recorded every 24 hours until the age of 28 days.

The shrinkage of UHPC was calculated using the following equation:

$$\varepsilon_{st} = \frac{L_0 - L_t}{L_b}, \tag{1}$$

where ε_{st} was the shrinkage of UHPC, L_0 was the initial length of the specimen, measured value, L_t is the measured value, L_b was the standard length of the specimen.

3. Results and Discussion

3.1. Fluidity and Air Content. Figure 4 exhibited the effects of shrinkage reducing agent on the air content and fluidity of UHPC. It could be found that the fluidity of UHPC decreased with the increase of dosage of shrinkage reducing agent. It showed that the highest fluidity value reached 256 mm when UHPC was prepared without mixing the shrinkage reducing agent. But the fluidity of UHPC was decreased by 15.8% to 221 mm and by 24.2% to 201 mm when shrinkage reducing agent was used at the dosage of 0.5% and 2.0%, respectively.

The incorporation of shrinkage reducing agent also had a significant effect on the air content of UHPC mortar. The air content of UHPC without the shrinkage reducing agent was 1.5%. With the increase of the dosage of shrinkage reducing agent, the air content of UHPC showed sharp increase. The air content of UHPC was increased by 106.7% to 3.1% and by 180% to 4.2% when the shrinkage reducing agent was used at the dosage of 0.5% and 2.0%, respectively. As mentioned above, the fluidity of UHPC exhibited declining tendency with the introduction of the shrinkage reducing agent, which might hinder the release of the air in UHPC during the vibration procedure and resulted in the increase of the air content in UHPC.

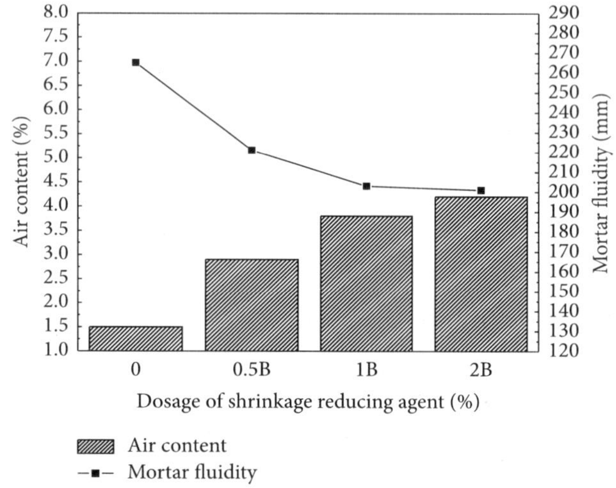

FIGURE 4: Effects of shrinkage reducing agent on air content and fluidity of UHPC.

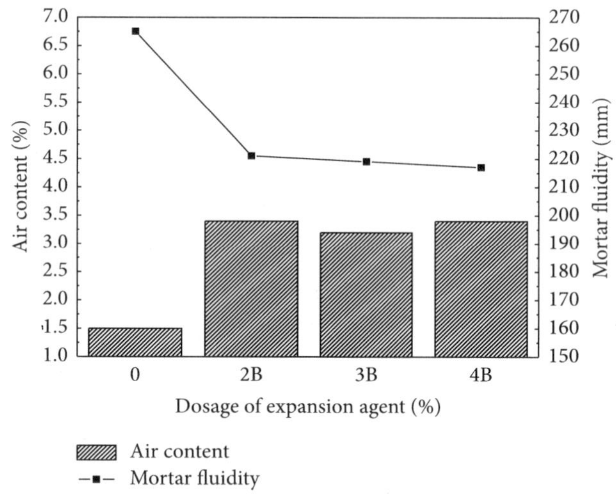

FIGURE 5: Effects of expansive agent on air content and fluidity of UHPC.

Figure 5 showed the effects of expansive agent on the air content and fluidity of UHPC. The results indicated that the

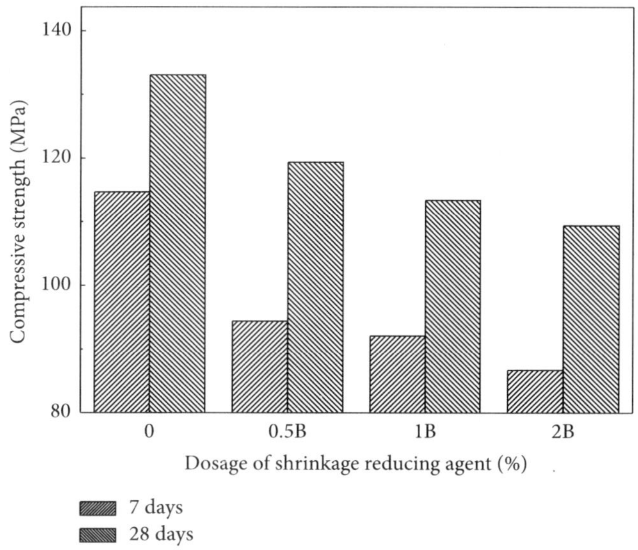

FIGURE 6: Effects of shrinkage reducing agent on compressive strength of UHPC.

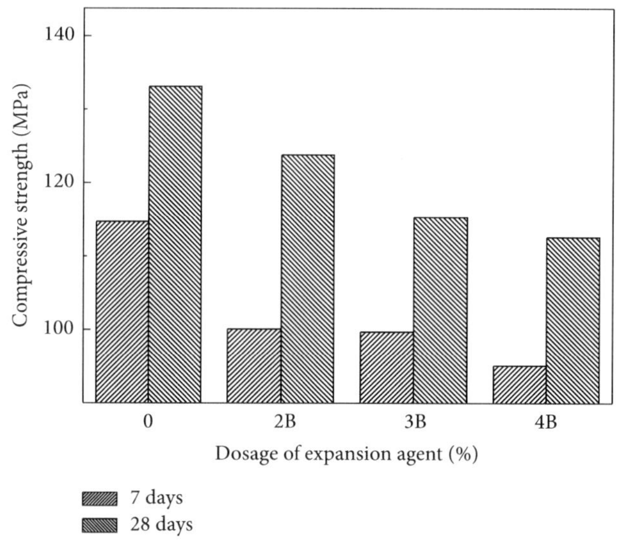

FIGURE 7: Effects of expansive agent on compressive strength of UHPC.

fluidity decreased and air content increased obviously with the addition of expansive agent. Since the main chemical compositions of the expansive agent were MgO, CaO, and sulphoaluminate, much of mixing water was consumed due to the physical absorption and chemical reaction for UHPC with the introduction of expansive agent. The consumption of free water decreased the fluidity and hindered the release of air, which led to the increase of the air content. Some published research work also indicated that the addition of the expansive agent decreased the fluidity of concrete [14, 15].

3.2. Compressive Strength. Figure 6 showed the effects of shrinkage reducing agent on the compressive strength of UHPC. The results indicated that the compressive strength of UHPC decreased with the increase of dosage of shrinkage reducing agent. Some published studies also drew the same conclusion that the addition of shrinkage reducing agent decreased the compressive strength of mortar sample [16–18]. For the samples without mixing with shrinkage reducing agent, 7-day and 28-day compressive strength reached 114.7 MPa and 133.1 MPa. However, when the shrinkage reducing agent was introduced at dosage of 2%, 7-day and 28-day compressive strength was decreased by 24.4% to 86.7 MPa and by 17.8% to 109.4 MPa, respectively. The reduction caused by shrinkage reducing agent was more significant for the 7-day strength of UHPC, which indicated that shrinkage reducing agent had a more obvious effect on the early-age strength of UHPC.

Figure 7 exhibited the effects of expansive agent on the compressive strength of UHPC. The results showed that the compressive strength of UHPC decreased with the increase of dosage of expansive agent. The published research work obtained similar conclusion that the compressive strength of concrete decreased when the dosage of expansive agent exceeded 40 $kg{\cdot}m^{-3}$ [19].

For the samples without mixing with expansive agent, 7-day and 28-day compressive strength reached 114.7 MPa and 133.1 MPa. However, when the expansive agent was introduced at dosage of 2%, 7-day and 28-day compressive strength was decreased by 12.7% to 100.1 MPa and by 7.0% to 123.8 MPa, respectively. The reduction caused by expansive agent was more significant for the 7-day strength of UHPC, which indicated that expansive agent had a more obvious effect on the early-age strength of UHPC.

3.3. Autogenous Shrinkage. Figure 8 exhibited the effects of shrinkage reducing agent on the autogenous shrinkage of UHPC. The results showed that the autogenous shrinkage of UHPC decreased with the introduction of shrinkage reducing agent. A published study obtained similar result that the use of shrinkage reducing agent was favorable for improving the restrained autogenous shrinkage behaviors of UHPFRC [20]. The autogenous shrinkage developed in the first 3 days was dominant and then the growth rate of shrinkage tended to be slow. For the sample without mixing with shrinkage reducing agent, 7-day autogenous shrinkage reached 1080×10^{-6}. However, 7-day autogenous shrinkage decreased to 718×10^{-6}, 649×10^{-6}, and 602×10^{-6} when the shrinkage reducing agent was used at dosage of 0.5%, 1.0%, and 2.0%, respectively. That is to say, 7-day autogenous shrinkage of samples with shrinkage reducing agent at dosage of 0.5%, 1.0%, and 2.0% was decreased by 33.5%, 39.9%, and 44.3%, respectively.

Figure 9 showed the effects of expansive agent on the autogenous shrinkage of UHPC. The results indicated that the autogenous shrinkage of UHPC decreased with the introduction of expansive agent. For the sample without mixing with expansive agent, 7-day autogenous shrinkage reached 1080×10^{-6}. However, 7-day autogenous shrinkage decreased to 907×10^{-6}, 920×10^{-6}, and 937×10^{-6} when the expansive agent was introduced at dosage of 4.0%, 3.0%, and 2.0%, respectively. That is to say, 7-day autogenous shrinkage of samples with expansive agent at dosage of 4.0%, 3.0%, and 2.0% was decreased by 16.0%, 15.2%, and 13.2%, respectively.

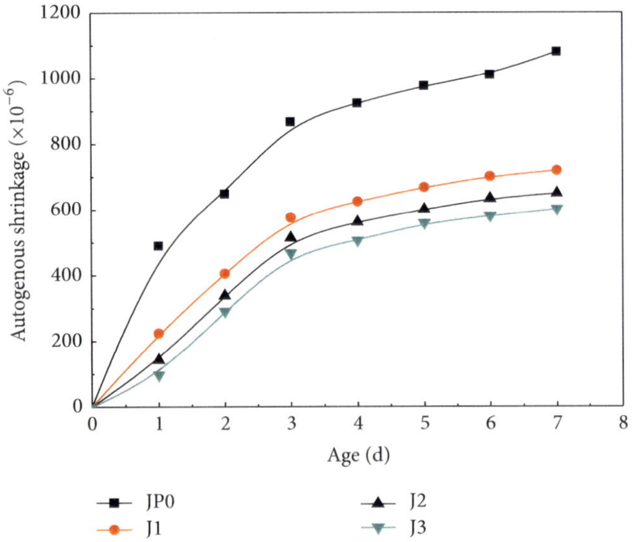

FIGURE 8: Effects of shrinkage reducing agent on autogenous shrinkage of UHPC.

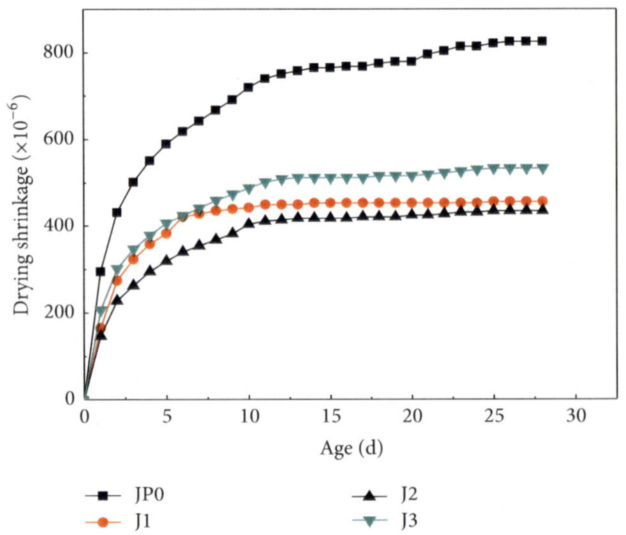

FIGURE 10: Effects of shrinkage reducing agent on drying shrinkage of UHPC.

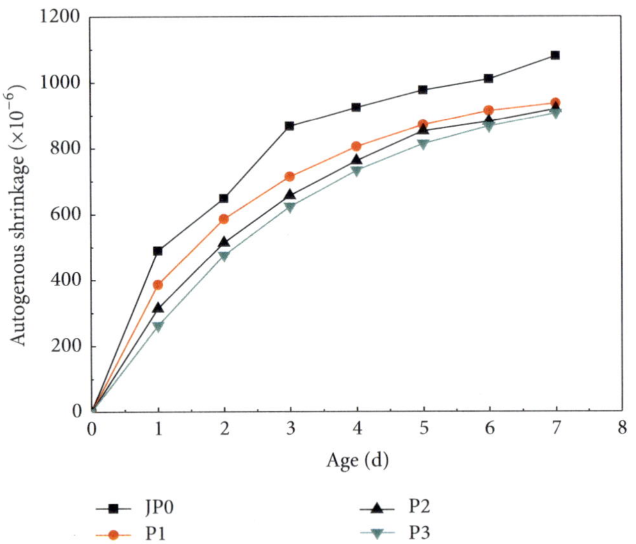

FIGURE 9: Effects of expansive agent on autogenous shrinkage of UHPC.

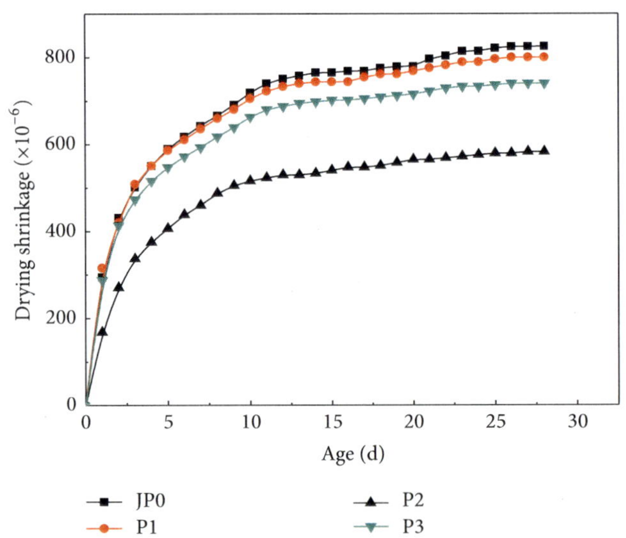

FIGURE 11: Effects of expansive agent on the drying shrinkage of UHPC.

Compared with the results in Figure 8, shrinkage reducing agent had more significant shrinkage reducing effect than expansive agent for autogenous shrinkage of UHPC.

3.4. Drying Shrinkage. Figure 10 exhibited the effects of shrinkage reducing agent on the drying shrinkage of UHPC. The results showed that the drying shrinkage developed in the first five days was dominant and then the growth rate of shrinkage tended to be slow. The results also indicated that the variation of drying shrinkage after 10 days was relatively low. The 10-day drying shrinkage of samples JP0, J1, J2, and J3 was 719×10^{-6}, 442×10^{-6}, 404×10^{-6}, and 371×10^{-6}, which accounted for 87.2%, 96.9%, 92.9%, and 90.3% of 28-day drying shrinkage, respectively.

It could also be found that the drying shrinkage of UHPC decreased obviously when the shrinkage reducing agent was introduced. A published research work obtained similar conclusion [21]. Shrinkage reducing agent was regarded as one of the important measures to prevent shrinkage cracking of concrete [22]. 28-day drying shrinkage decreased to 411×10^{-6}, 435×10^{-6}, and 456×10^{-6} when the shrinkage reducing agent was introduced at dosage of 2.0%, 1.0%, and 0.5%, respectively. That is to say, 28-day drying shrinkage of samples with shrinkage reducing agent at dosage of 2.0%, 1.0%, and 0.5% was decreased by 50.2%, 47.3%, and 44.7%, respectively. With the increase of the dosage, the shrinkage reducing agent exhibited more significant reducing effect for the drying shrinkage.

Figure 11 exhibited the effects of expansive agent on the drying shrinkage of UHPC. The results showed that the drying shrinkage of UHPC decreased with the introduction of expansive agent. It was also reported that expansive agent decreased the drying shrinkage of UHPC [15]. The main chemical compositions of expansive agent were MgO, CaO, and sulphoaluminate; the reaction mechanism of those crystals are as follows:

$$MgO + H_2O \longrightarrow Mg(OH)_2 \quad (2)$$

$$CaO + H_2O \longrightarrow Ca(OH)_2 \quad (3)$$

$$C_4A_3S + H_2O \longrightarrow AFt \quad (4)$$

The generation and growth of these crystals filled the pore and increased the density of UHPC, which could help in the expansion of cement paste and decrease the shrinkage of concrete. For the sample without mixing with expansive agent, 28-day drying shrinkage reached 825×10^{-6}. However, 28-day drying shrinkage decreased to 582×10^{-6}, 740×10^{-6}, and 800×10^{-6} when the expansive agent was introduced at dosage of 4.0%, 3.0%, and 2.0%, respectively. That is to say, 28-day drying shrinkage of samples with expansive agent at dosage of 4.0%, 3.0%, and 2.0% was decreased by 29.5%, 10.3%, and 3.0%, respectively. Compared with the results in Figure 10, shrinkage reducing agent had more significant reducing effect than expansive agent for drying shrinkage of UHPC.

4. Conclusion

From the above experimental results, the following conclusions can be obtained.

(1) The air content of UHPC increases with the increasing dosage of shrinkage reducing agent and expansive admixture and the fluidity of fresh UHPC shows a declining tendency.

(2) The higher addition of shrinkage reducing agent and expansive admixture results in the lower compressive strength of UHPC. This negative effect on compressive strength is more obvious at early ages.

(3) Both autogenous and drying shrinkage of UHPC decrease obviously with the higher dosage of shrinkage reducing agent and expansive admixture. Shrinkage reducing agent presents a more significant reduction effect on shrinkage of UHPC than expansive agent.

Acknowledgments

The support of Heilongjiang Provincial Major Scientific and Technological Project of Applied Technology Research and Development Plan (GA14A501) is acknowledged.

References

[1] F. de Larrard and T. Sedran, "Optimization of ultra-high-performance concrete by the use of a packing model," *Cement and Concrete Research*, vol. 24, no. 6, pp. 997–1009, 1994.

[2] Association Francaise de Genie Civil, *UltRa High Performance Fiber-Reinforced Concretes-Interim Recommendations*, Association Francaise de Genie Civil, Paris, France, 2002.

[3] Japan Society of Civil Engineers, *Recommendations forfbfbDesignfbfbandfbfbConstruction of Ultra High Strength Fiber Reinforced Concrete Structures (DRAFT)*, Japan Society of Civil Engineers, Tokyo, Japan, 2006.

[4] P. Richard and M. Cheyrezy, "Composition of reactive powder concretes," *Cement and Concrete Research*, vol. 25, no. 7, pp. 1501–1511, 1995.

[5] E. Shaheen, N. G. Shrive, S. Allena, and C. M. Newtson, "Optimization of mechanical properties and durability of reactive powder concrete," *ACI Materials Journal*, vol. 104, no. 5, p. 547, 2007.

[6] S. Collepardi, L. Coppola, R. Troli et al., "Mechanisms of actions of different superplasticizers for high-performance concrete," in *Proceedings of the International Conference on High Performance Concrete & Performance & Quality of Concrete Structures*, 1999.

[7] A. Cwirzen, V. Penttala, and C. Vornanen, "Reactive powder based concretes: mechanical properties, durability and hybrid use with OPC," *Cement and Concrete Research*, vol. 38, no. 10, pp. 1217–1226, 2008.

[8] J. H. Liu, D. M. Wang, and S. M. Song, "Research on durability and micro structure of high volume fine mineral mixture of reactive powder concrete," *Journal of Wuhan University of Technology*, vol. 30, no. 11, pp. 54-55, 2008.

[9] C. M. Tam, V. W. Y. Tam, and K. M. Ng, "Assessing drying shrinkage and water permeability of reactive powder concrete produced in Hong Kong," *Construction and Building Materials*, vol. 26, no. 1, pp. 79–89, 2012.

[10] A. C. L. Wong, P. A. Childs, R. Berndt, T. Macken, G. Peng, and N. Gowripalan, "Simultaneous measurement of shrinkage and temperature of reactive powder concrete at early-age using fibre Bragg grating sensors," *Cement and Concrete Composites*, vol. 29, no. 6, pp. 490–497, 2007.

[11] S. Han, Y. Tu, M. An et al., "The shrinkage of reactive powder concrete in early age and its control methods," *China Railway Science*, vol. 36, no. 1, pp. 40–47, 2015 (Chinese).

[12] D.-Y. Yoo, J.-J. Park, S.-W. Kim, and Y.-S. Yoon, "Early age setting, shrinkage and tensile characteristics of ultra high performance fiber reinforced concrete," *Construction and Building Materials*, vol. 41, pp. 427–438, 2013.

[13] H. Yazıcı, "The effect of curing conditions on compressive strength of ultra high strength concrete with high volume mineral admixtures," *Building and Environment*, vol. 42, no. 5, pp. 2083–2089, 2007.

[14] B. You and N. Li, *Expansive Agent and Its Compensation Shrinkage Concrete*, China Building Industry Press, 2005.

[15] Z. Huang, Y. Liu, and C. Li, "Performance research of high performance concrete incorporating HCSA expansion agent," *Materials Review*, vol. 4, pp. 116–121, 2015 (Chinese).

[16] Z. He, Z. J. Li, M. Z. Chen, and W. Q. Liang, "Properties of shrinkage-reducing admixture-modified pastes and mortar," *Materials and Structures*, vol. 39, no. 4, pp. 445–453, 2006.

[17] C. Maltese, C. Pistolesi, A. Lolli, A. Bravo, T. Cerulli, and D. Salvioni, "Combined effect of expansive and shrinkage reducing admixtures to obtain stable and durable mortars," *Cement and Concrete Research*, vol. 35, no. 12, pp. 2244–2251, 2005.

[18] D. P. Bentz, "Influence of shrinkage-reducing admixtures on early-age properties of cement pastes," *Journal of Advanced Concrete Technology*, vol. 4, no. 3, pp. 423–429, 2006.

[19] S. Zhao and B. You, *The Crack Control Technology of Compensation Shrinkage Concrete and Its Application*, China Building Industry Press, 2010.

[20] D.-Y. Yoo, N. Banthia, and Y.-S. Yoon, "Effectiveness of shrinkage-reducing admixture in reducing autogenous shrinkage stress of ultra-high-performance fiber-reinforced concrete," *Cement and Concrete Composites*, vol. 64, pp. 27–36, 2015.

[21] B. L. Cope and G. E. Ramey, "Reducing drying shrinkage of bridge deck concrete," *Concrete International*, vol. 23, no. 8, pp. 76–82.

[22] P. S. Shah, E. M. Karagular, and M. Sarigaphuti, "Effects of shrinkage reducing admixture on rest rained shrinkage cracking of concrete," *ACI Materials Journal*, vol. 89, no. 3, pp. 88–90, 1992.

16

Improving Electromagnetic Shielding Ability of Plaster-Based Composites by Addition of Carbon Fibers

A. Samkova,[1] **P. Kulhavy**,[2] **V. Tunakova**,[1] **and M. Petru**[2]

[1]*Faculty of Textile Engineering, Technical University of Liberec, Studentska 2, 461 17 Liberec, Czech Republic*

[2]*Institute for Nanomaterials, Advanced Technologies and Innovation, Technical University of Liberec, Studentska 2, 461 17 Liberec, Czech Republic*

Correspondence should be addressed to A. Samkova; alzbeta.samkova@tul.cz

Academic Editor: Bing Zhang

The size of electromagnetic shielding in plaster composites by the means of different volume fractions of carbon fibers was studied in this paper. Conventional types of plaster, which are commonly used in industry, that is, cement, lime, gypsum, and lime cement (Thermo UM), were the base materials of the created composites. The fundamental idea of improving the electromagnetic shielding properties was verified based on a numerical simulation conducted by means of electromagnetic module in Comsol Multiphysics. The carbon microfibers with the above-critical length of 8 mm were added as the reinforcing and simultaneously shielding element into the plaster samples. From the viewpoint of the mechanical properties, fibers shorter than the critical length do not provide sufficient reinforcement. The samples were created at three different volume fractions of the dispersion and one without any reinforcement for the possibilities of their mutual comparison. The results of the carried measurement show that the electromagnetic shielding in the plaster composite grows with the increase of fiber content within the tested ratio proportionately. Also, the dependency of shielding ability on the inner material moisture has been studied. Any measureable influence of the moisture content on to the total shielding effect has not been found. Only in the lime plaster reinforced with fibers, the increased moisture could significantly decrease the shielding effect.

1. Introduction

Over the last few years, there has been mounting concern about the possibility of adverse health effects resulting from exposure to radiofrequency electromagnetic (EM) fields, such as those emitted by wireless communication devices [1]. This phenomenon is connected with extensive development of electronic systems and telecommunications including generation and transmission of electricity, domestic appliances and industrial equipment, and telecommunications and broadcasting, whereas electrical and EM fields in certain frequency bands have wholly beneficial effects, which are applied in medicine, other nonionizing frequencies sourced from extremely low frequencies, power lines or certain high-frequency waves used in the fields of radar, telecommunications, and mobile communications, and they appear to have more or less potentially harmful, nonthermal, biological effects on plants, insects, and animals as well as the human body [2–4].

For the reasons given above, a demand for protection of human beings against undesirable influence of EM signals and troublesome charges (especially for professional use) has been increasing. The best material for shielding or absorption of EM wave must have both high electric conductivity (imparts higher reflection of EM wave especially at lower frequencies) and high magnetic permeability (provides higher absorption of EM wave especially at lower frequencies); therefore, shields based on the use of metals, magnetic alloys, and carbon-based materials are the best ones [5]. The materials used in the shielding of radiofrequency fields are usually copper, aluminum, or silver based (in the form of wires, metal plating, or conductive fillers) [6–10]. These shields work by reflecting the waves because these metals have a high electric conductivity. To shield a magnetic field, materials must have a high magnetic permeability to be able to absorb the fields [5]. Carbon-based materials such as carbon fibers, carbon particles, carbon nanotubes, and so on fully fulfill this requirement [11–13].

One way to protect human beings is to use protective clothing made of textile materials with increased electrical conductivity [14, 15]. These fabrics are usually based on the use of very thin metal fibers incorporated to the structure of yarn from which woven or knitted fabric is made [10]. By these fabrics, sufficient electrical conductivity connected with certain level of electromagnetic shielding protection is achieved, but on the other hand, this method of protection brings some disadvantages. For example, by addition of conductive components (metal fiber), thermophysiological comfort properties of clothes are usually deteriorated [16]. In addition, even the appearance (color) of these special fabrics is also very often influenced [17].

It seems that more robust solution offers controlled insulation of whole buildings. However, it is clear that just the passive instruments, such as wall cladding by damping panels and the use of electro conductive curtains or special wallpapers solve only the consequences and not the causes of electromagnetic radiation. Since in many cases the passive instruments are the only possible solution, they are sometimes necessary and therefore must be addressed.

The most important material in the building industry is cement. Dried cement composites do not contain any free electrical charge carriers. Their specific electrical resistance in the fully dried state is around 10^9 Ωm, making it one of the best insulators, which means they are fully transparent to electromagnetic radiation of higher frequencies [18]. The increased moisture content of cement composites results in a reduction of specific electrical resistance. This is due to salt ions contained in the pore solution. Significant reduction of electrical resistivity accompanied with the increase of electromagnetic shielding ability of cement composites occurs after an addition of a sufficient amount of electrically conductive additives to form conductive paths. The dominant charge carrier in this case is the electron [18].

There are many references in the literature concerning the addition of a varying amount of conductive filler as a reinforcing element to a cement matrix in order to produce an electrically conductive, electromagnetic shielding material with favorable mechanical properties such as tensile strength, compressive strength, or impact strength. For example, Wang et al. [19] prepared multiwalled carbon nanotube (MWCNT)/cement composites and studied the effect of sample thickness and the nanodispersion content on the shielding effectiveness. It was found that at MWCNT content 0.6 wt.%, the cement mortar sample with a thickness of 25 mm remarkably absorbs electromagnetic waves close to the absorbing peaks in the frequency range of 2–8 GHz. Yawen et al. [20] described investigation of main properties of high-structure carbon-black (CB) cement-based composites (CBCCs). It was confirmed that these composites exhibit good performance in the absorption of electromagnetic waves in the frequencies higher than 8 GHz. For CBCC containing 2.5 wt.%, the minimum reflectivity reaches about 20 dB, but compressive strength significantly decreases with the increase of CB content. Unfortunately, both mentioned papers analyze only reflection part of electromagnetic shielding effectiveness, whereas it is expected from theory [5] that for carbon-based materials, main portion of electromagnetic shielding effectiveness will be created by absorption coefficient. Khushnood et al. [21] focused their work on the usage of cost-effective material (pyrolyzed agricultural wastes such as peanut and hazelnut shells) for enhancing the electromagnetic shielding effectiveness of cement composites. Numerically evaluated electromagnetic shielding effectiveness showed that shielding ability of composite is poor at low frequencies (about 0.2 GHz). The shielding ability is close to 10 dB at the highest of evaluated frequency (10 GHz) using the highest portion of conductive component that is 5 wt.% of carbonized peanut shells or 0.5 wt.% of carbonized hazelnut shells.

As mentioned above, the use of fibrous materials as reinforcing elements is favorable especially for its high aspect ratio (length/diameter ratio). It was proven that fibrous materials contribute to improvements in their mechanical properties [22]. In the authors' previous studies [23], it was reported that usage of optimal volume ratio of fibers (1% to 2% of reinforcement) shows significant improvement in the main mechanical properties of plaster composites. With the higher volume of fibers, the improvement is not proportional, and the demanded effect is just minimal.

The aim of the carried work was to find out whether the addition of fibrous reinforcement could improve the electromagnetic shielding ability. To achieve the required effect, some conductive components must be added to the plaster. Carbon fiber shows very good electrical and thermal conductivity, good strength properties, and improved environmental resistance of the final plaster [24].

2. Materials and Preparation

In the previous works of the author [23, 25], all realistic combinations of plasters and fibers have been tested with respect to their mechanical properties (Figure 1). Some glass, carbon, basalt fibers, and even polyvinyl butyral nanofibrous dispersion have been used as the reinforcing element of composites. The thickness of plasters has been chosen based on the real values used in the construction industry. The results show that the mechanical properties (flexural tensile strength, compressive strength, and impact strength) of plasters could be significantly improved by using the weight ratio of fiber reinforcement of 1% to 2%. Higher content of carbon reinforcement does not have any equivalent effect or could even worsen the mechanical parameters [23]. The content of the fibrous reinforcement in concrete materials is usually 6%. However, a mixture of the plasters and fibers with such high contents usually causes problems in their bonding [23, 26]. For this reason, the content of the fibrous reinforcement has been set to 1%, 2%, and 3%. The calculation (1) of the mixing ratio [27] is based on the percentage of the components in the composite and their densities.

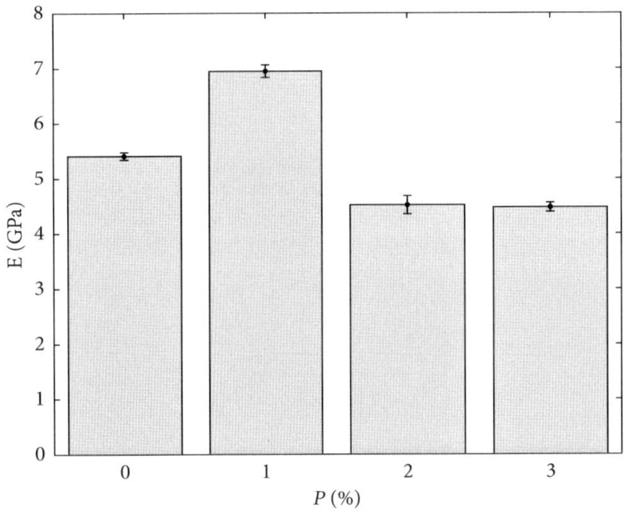

FIGURE 1: Comparison of the modulus of elasticity in the tension E (GPa) of the Thermo UM Xtra plaster with different volume fractions (P (%)) of the carbon reinforcement.

The mutual conversion between the mass and volume fraction is in Table 1.

$$w_i = \frac{V_i \times \rho_i}{\sum V_i \rho_i}, \quad (1)$$

where w_i (%) is the mass fraction of the ith component, V_i (%) is the volume fraction of the ith component, and ρ_i (kg/m^3) is the density of the ith component.

With the higher volume fraction, the mechanical properties suddenly start falling down because the reinforcement is transformed into a bundle of stress concentrators and the rupture initiators. Based on these results, four types of plasters, whose mechanical properties could be improved or at least not worsened, have been chosen. For the mutual comparison and consistency of the results, pure plasters without any fiber dispersion were tested. Because of the natural climatic loading of the plasters, the samples were also tested at different humidities (0.8%, 2.2%, and "wet"-not physically possible to measure). The individual plasters were prepared according to the recommended manufacturer's technical standards. The inner moisture content was measured after acclimatization of the samples in the standardized ambient conditions, and the so called "wet" was achieved by simultaneous spraying of the samples by water. The carbon fibers with the declared mean diameter 1.76 μm were selected to achieve the shielding effect. For plaster composites with microfiber reinforcement, it was necessary to compute the critical fiber length: the final length of carbon fibers is 8 mm. The fiber distribution in the tested parts was assessed by image analysis on randomly selected samples. However, it is possible to say, that after approx. 20 min of mixing, the distribution is equal. The main physical properties of the used materials are given in Table 2.

2.1. Plaster. The type and amount of the used binder influences the resulting plaster properties, so it is one of the main criteria based on which the plasters could be divided. The first plaster selected for the study was the lime-cement. This plaster belongs among extremely effective heat-insulating plasters with the trade name Thermo UM (this designation is used in this study). Another significant feature of this material is low bulk density; compared to commonly used plasters, it could be up to 4 times lighter. Due to its low bulk density and low modulus of elasticity, the plaster eliminates the volume or shape changes in the base wall material, thus creating no cracks. Above all, the Thermo UM plaster is characterized by a high value of the thermal conductivity coefficient, and the chemical composition of the plaster eliminates growth of microorganisms, algae, and fungi on its surface [28].

TABLE 1: Volume fraction of the individual components.

Volume fraction w_i	1%	2%	3%
	Mass fraction (%)		
Thermo UM	3.98	7.74	11.28
Cement	0.75	1.51	2.27
Lime	1.79	3.55	5.29
Gypsum	1.96	3.87	5.75

TABLE 2: Mechanical properties of the base used materials.

Material	Length (mm)	Diameter (μm)	Density (kg·m^{-3})	E (GPa)	υ (–)
Carbon fiber	8	1.6	1800	241	0.27
Thermo UM	—	—	551	5.4	0.18
Gypsum	—	—	1100	14	0.27
Cement	—	—	3020	39	0.2
Lime	—	—	1250	8.8	0.21

Another type, the gypsum plaster, is generally cheaper material characterized by good thermal and acoustic resistance. Due to its thermal properties, the plaster is perceived as a hygroscopic material, which allows the regulation of the microclimate in the room [29, 30]. This plaster is vapor-permeable, does not close the water in the construction, and at the same time allows the natural drying of residual water from the blocks. The gypsum plaster is suitable for antibacterial environments because the smoothness of its surface prevents dust from settling on. At high temperatures, gypsum is released by water vapor, which is often used to increase fire resistance [31].

The cement plaster [32] exceeds especially with the high strength, resistance to mechanical damage, and very good adhesion to the basic substrate. This kind of plaster is therefore commonly used for highly loaded areas and areas where frost and weather-resistant is also required.

The lime plaster is characterized by very good water vapor permeability and ability to absorb moisture from surrounding bricks. It does not retain capillary water and it also dries very well. These properties allow lime plasters to regulate indoor climate. Lime plaster is very flexible, making it suitable for uneven surfaces. Another advantage is its ability to absorb all odors [33]. The microimages of the individual types of plaster with the carbon dispersion are in Figure 2.

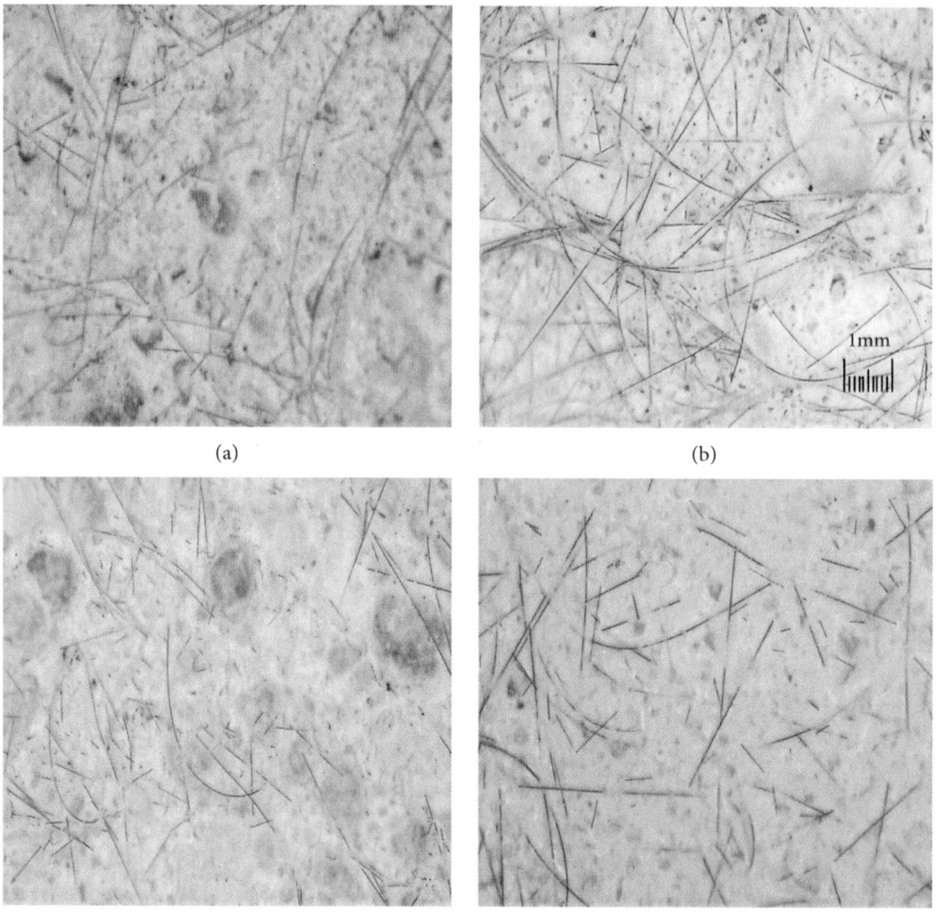

(a) (b) (c) (d)

FIGURE 2: The studied plasters reinforced with carbon fibers: (a) cement, (b) gypsum (c) Thermo UM, and (d) lime.

2.2. Fiber Dispersion. The carbon *"An old but still new material"* and its natural or synthetic compounds exceed the number of all the compounds of all other elements together, which is about a million. This is possible even though the carbon materials are composed of only one element. The properties of these materials could not be predicted or estimated based on the known composition, as is the case of, for example, conventional metal alloys. The fibers are commercially available on fiber coils of several hundred meters in length, and each bundle consists of 1000 to 10000 filaments with a diameter of 5–12 μm. Standard carbon fibers have a tensile strength between 3 and 4 GPa, and the modulus of elasticity of 230–300 GPa [34, 35]. Carbon fibers have the largest range of properties. High strength, modulus, and heat resistance with low density are associated with carbon fibers. Very good thermal and electrical conductivity were the important criterion for selecting this fiber for our study [36]. The importance of carbon fibers is growing steadily despite the fact that compared to glass fibers, their price is ten to hundred times more expensive. However, due to their characteristics, customers' constant interest could be expected. In recent years, the carbon fibers have become an integral part of modern civil engineering. They are primarily used to increase the load-bearing capacity of building components and to perform building reconstructions. In such cases, not only the simple carbon fiber is used, but carbon reinforcement in the form of rods, lamellas, and carbon fabric is used as well.

The potential of using those fibres also in the civil engineering could lay in utilization of some wasted carbon fibres, remaining in a big amounts from manufacturing of composite parts for vehicles and airplanes. Due to the fact that once the fibers are cut or split, there is no possibility to merge or dissolve them again, and they are usually just energetically utilized by burning in the heating plants. To add them into building materials is significantly a better idea.

The conductance (2) is a physical parameter that describes the material ability to lead an electrical current.

$$G = \frac{I}{U} = \sigma \frac{S}{l} = \frac{1}{R}, \quad (2)$$

where I (A) is the electric current flowing through the conductor, U (V) is the voltage at the ends of the conductor, σ (SI) is the conductivity of the substance, S (m^2) is the cross-sectional area, l (m) is the length of the conductor, and R (Ω) indicates the electrical resistance.

3. Modeling

The created model of the electromagnetic absorption has been created in the Comsol Multiphysics. The model uses

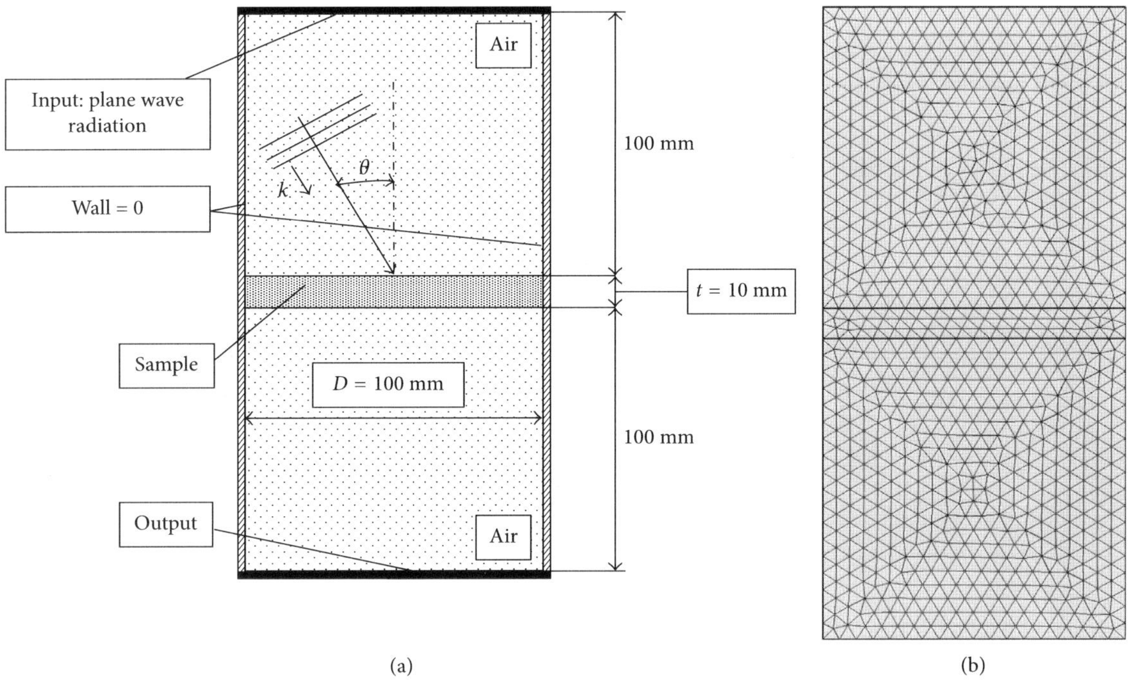

FIGURE 3: Scheme of the created FEM model and boundary conditions (a) and FEM model (b).

a 2D geometry of such a system (Figure 3). The aim of the model is to characterize the absorption properties—more specifically, the specific surface impedance with and without the conductive reinforcement. The base models of the four samples (Thermo UM, gypsum, cement, and lime) were created using the module Acoustic-Electromagnetic Interaction and Radiation, with the frequency domain. Layers are also assumed to be infinite in the lateral dimensions; the only geometric information is then the sample. The surrounding fluid is air. The incident angle is given as (3). The task was solved with two directions of acting waves 0° and 45° as could be seen in Figure 4. As shown in Figure 5, the only solid body in the created model was the desk with diameter $D = 100$ mm, thickness $t = 10$ mm, and material properties of the individual materials (density, conductivity, moisture, and resistivity). The maximal frequency in the model is 1000 Hz.

$$k = k_0 (\sin \theta, \cos \theta), \quad (3)$$

where θ is the incidence angle and k_0 is the wave number in the free field.

The orthogonal impact of the wave to the partition is the worst case in terms of shielding. Therefore, it is not necessary to calculate any other impacting angles for comparison of the numerically obtained results with the experiments. In order to determine the shielding efficiency, it is necessary to determine the value of the incident electromagnetic waves H_t transmitted through the insulated sample described by the parameters μ, G, and the sample thickness t. Where μ is the permeability of vacuum ($\text{H·m}^{-1} = \text{N·A}^{-2}$). Parameters of the free continuum filled with air around the simulated sample are the permeability constant $\mu_0 = 4\pi \times 10^{-7}\,\text{H·m}^{-1}$ and the vacuum permittivity $\varepsilon_0 = 8854 \times 10^{-12}\,\text{F·m}^{-1}$. The characteristic impedance of the free space on both sides of the sample is Z_0 (4), and the characteristic impedance of the sample is $Z_S\,je$ (5). Subsequently, the shielding efficiency can be expressed by impedance according to the following:

$$Z_0 = \sqrt{\frac{\mu_0}{\varepsilon_0}}, \quad (4)$$

$$Z_S = \sqrt{\frac{j\mu\omega}{G}}, \quad (5)$$

$$SE = 20 \cdot \log\left|\frac{(Z_0 + Z_S)^2}{4Z_0 Z_S} \cdot e^{\gamma t} \cdot \left[1 - \left(\frac{Z_0 - Z_S}{Z_0 + Z_S}\right)^2\right] \cdot e^{-2\gamma t}\right|, \quad (6)$$

where γ denotes the number of wavelengths per unit length (propagation constant) of the electromagnetic wave in space $\gamma = \sqrt{j\omega\mu(G + j\omega\mu)} = \alpha + j\beta$, where α (dB·m^{-1}) is the attenuation constant and β (rad·m^{-1}) is the phase constant. The overall shielding efficiency can be broken down into several components. The reflection of electromagnetic wave occurs when the wave arrives from the free space to the sample surface and then again when passing through the sample into the free space. The total attenuation caused by reflection R (the electrical resistance—already defined in (2)) defined according to (7) could be subsequently described by 2 parameters R_1 and R_2. The characteristic reflection on the first (input) and on the second partition interface, where a part of the unattenuated wave and a part of the attenuated wave, is shown in Figure 5.

$$R = R_1 + R_2 = 20 \cdot \log\left|\frac{Z_0 + Z_S}{2Z_S} \cdot \frac{Z_0 + Z_S}{2Z_0}\right|. \quad (7)$$

When some shielding material is used, the greatest number of the reflected waves belongs to the first interface. This phenomenon could be easily observed on some metallic

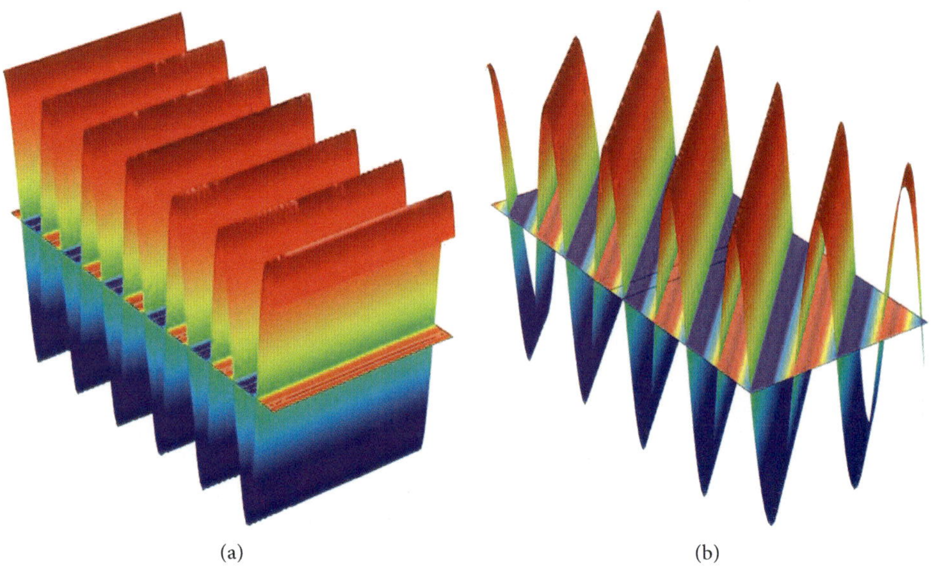

FIGURE 4: FEM model: examples of the incident waves in the two computed directions $\theta = 0^\circ$ (a), $\theta = 45^\circ$ (b).

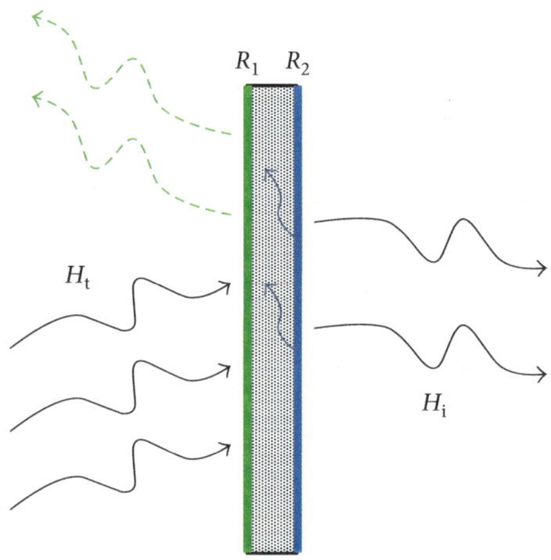

FIGURE 5: Model scheme of propagation, reflection, and attenuation of an electromagnetic wave in a solid partition.

materials where the best possible conductivity is necessary to obtain the highest shielding efficiency. When the electromagnetic wave passes through a metal wall of the enclosure, portion of its energy is dissipated to heat. The size of the absorbed portion of the energy depends on the wall material, its thickness, and frequency of the waves. Therefore, in a case of modeling of nonmetallic materials with just a minimal primary conductivity, such are the tested materials: Thermo UM, gypsum, cement and lime, it is necessary to use conductive dispersion that could affect their permeability because the greatest absorption $\vartheta(\mathrm{dB})$ can be achieved by a composite material with high permeability, whose thickness exceeds the wavelength of the penetrating waves

$$\vartheta = 20 \cdot \log e^{t/\delta}, \tag{8}$$

where t is the material thickness and δ is the depth of the impacting wave penetration, which is dependent on field frequency, permeability, and material resistance $\delta = \sqrt{(2\rho/\omega\mu)}$, where ρ is the specific resistance $(\Omega{\cdot}\mathrm{mm}^2{\cdot}\mathrm{m}^{-1})$.

The dispersion used in the form of added fibers affects the resulting permeability and the overall absorption shielding. However, the problem could be to determine the optimal amount of the fiber dispersion. The optimal added fiber ratio could cause positive increase, whereas in the case of nonoptimal amount, the reflected and pulsed waves within the structure may have even a negative effect on the overall shielding efficiency. The shielding efficiency caused by multiple reflections $\vartheta_{\mathrm{M}}\,(\mathrm{dB})$ could be then expressed as

$$\vartheta_{\mathrm{M}} = 20 \cdot \log\left\| \left[1 - \left(\frac{Z_0 - Z_{\mathrm{S}}}{Z_0 + Z_{\mathrm{S}}} \right)^2 \cdot e^{-(2t/\delta)} \cdot e^{-j(2t/\delta)} \right] \right\|. \tag{9}$$

When the shielding structure is designed from the material where $Z_0 \gg Z_{\mathrm{S}}$ and when the wall thickness is greater than the wave's penetration depth $t \gg \delta$, then the value of ϑ_{M} is approaching zero, and therefore, the effect of multiple reflections of electromagnetic waves on the overall shielding could be neglected. The reflected wave is easily converted to heat by the means of absorption attenuation. However, if the wall thickness is less than the penetration depth, of the ratio $(t/\delta) \leq 1$, then ϑ_M achieves negative values, and thereby, the overall shielding efficiency is reduced, as stated by Mardiguian or Chatterton et.al. [37, 38].

4. Experiment and Testing

Electromagnetic shielding effectiveness (SE) of composites was measured according to ASTM D 4935-10 [39] for planar materials using a plane-wave, far-field EM wave. SE of samples was measured over the frequency range of 30 MHz to 3 GHz. The most common wireless technologies such as the broadcasting television and radio, cell phones, satellite radio and TV, wireless computer networks, Bluetooth, GPS, and so on use radio frequencies, which extend from about 10 kHz to 300 GHz and match wavelengths within the range 30 km to 0.1 mm. Majority of practical applications is working at frequencies below 30 GHz. Table 3 summarizes the approximate frequency bands used by the most present wireless technologies [2], which means that the measured frequency range includes working frequencies of the most common equipment for public communication and data transfer.

TABLE 3: Frequency range for different applications [40].

Application	Frequency or frequency range
AM broadcast	530 kHz~1.7 MHz
Broadcast TV	54~88, 174~261, 470~698 MHz
FM broadcast	88~108 MHz
Cell phones	~850, ~900, ~1800, ~1900 MHz
GPS	~1.5 GHz
Satellite radio	~2.3 GHz
Wireless Comp. Network	2.4 & ~5.8 GHz
Satellite TV	~12 GHz

The setup consisted of a sample holder with its input and output connected to the network analyzer (Figure 6). A shielding effectiveness test fixture (Electro-Metrics, Inc., model EM-2107A) was used to hold the sample. The design and dimension of the sample holder follows the ASTM method mentioned above. Network analyzer Rohde & Schwarz ZN3 was used to generate and receive the electromagnetic signals. Sinusoidal signal was generated over the mentioned frequency range with the use of 50 Ω output impedance to minimize reflections caused by mismatches. The standard mentioned above determines the electromagnetic shielding effectiveness of the fabric using the insertion-loss method. A reference measurement for the empty cell was required for the shielding effectiveness assessment.

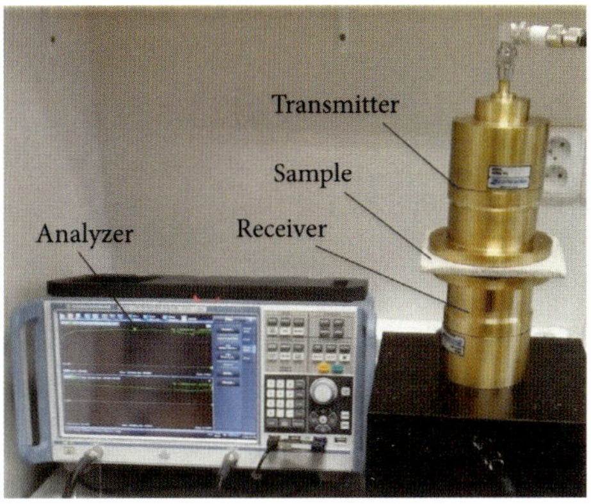

FIGURE 6: The testing equipment.

Samples were air-conditioned before testing ($T = 22^{\circ}\mathrm{C} \pm 3$, RH = 50% ± 10%). In addition, effect of moisture content was studied. Samples of dimensions 10 × 150 × 150 mm were produced for this testing. Five samples were produced for each conductive component content in order to the conduct subsequent statistical analysis.

The total shielding of a solid material with no apertures is equal to the sum of the absorption loss (A) plus the reflection loss (R) plus a correction factor (B) to account for multiple reflections in thin shields. Total electromagnetic shielding effectiveness therefore can be written as (10) [5]:

$$\mathrm{SE} = A + R + B \ (\mathrm{dB}). \tag{10}$$

The reflection is usually the primary mechanism of electromagnetic interference shielding. The wave, which is incident to the boundary with the second medium, is partially reflected back and partially transmitted to second medium. The same situation occurs at the interface between the second and third material. Reflection loss for plane waves is greater at low frequencies and for high conductivity materials [5].

Secondary mechanism of EMI shielding is usually absorption. When an electromagnetic wave passes through a medium, its amplitude decreases exponentially. This decay occurs because currents induced in the shield produce ohmic losses and heating of material. General expression for absorption loss can be written as (11) [20]:

$$A = 0.0848t\sqrt{f\mu_r\sigma_r} \ (\mathrm{dB}), \tag{11}$$

where t is the thickness of the shield in meters. For significant absorption of the radiation by the shield, the shield should have electric and/or magnetic dipoles, which interact with the electromagnetic fields in the radiation. The electric dipoles may be provided by materials having a high value of the dielectric constant. The magnetic dipoles may be provided by materials having a high value of the magnetic permeability.

The absorption loss is a function of the product $\sigma_r \cdot \mu_r$, whereas the reflection loss is a function of the ratio (σ_r/μ_r), where σ_r is the electrical conductivity relative to copper and μ_r is the relative magnetic permeability. Silver, copper, gold, and aluminum are excellent for reflection due to their high conductivity. Super-malloy and mu-metal are excellent for absorption due to their high magnetic permeability. The reflection loss decreases with increasing frequency, whereas absorption loss increases with increasing frequency [5].

Other than reflection and absorption, a mechanism of shielding is based on multiple reflections (B), which refer to the reflections at various surfaces or interfaces in the shield. This mechanism requires the presence of a large surface area or interface area in the shield. An example of a shield with a large surface area is a porous or foam material. The loss due to multiple reflections can be neglected when the distance between the reflecting surfaces or interfaces is large compared to the skin depth.

Operation of the electromagnetic shield could be characterized by the so called shielding attenuation coefficient (dimensionless) defined as a ratio between electromagnetic field power density in a specific place of shielded space P_t and incident electromagnetic wave power density P_i (W/m^2)

$$SE = \frac{P_t}{P_i} \ (-). \quad (12)$$

Logarithmic size of this coefficient called electromagnetic shielding effectiveness (SE) is used more frequently:

$$SE = 10 \log \frac{P_t}{P_i} = 20 \log \frac{H_t}{H_i} \ (\text{dB}), \quad (13)$$

where H_t, E_t, and P_t are the electric field strength, magnetic field strength, and electromagnetic field density values measured in the presence of the textile material, whereas H_i, E_i, and P_i are the same values measured without the textile material.

For the analysis of the shielding effectiveness of test samples, a formula expressing absorption loss A_{sheet} (dB) and reflection loss R_{sheet} (dB) can be used expressed as

$$A_{\text{sheet}} = 0.0848 \times t \sqrt{\frac{K}{K_C} f}, \quad (14)$$

$$R_{\text{sheet}} = C + 10 \log \left(\frac{K}{K_C f} \right), \quad (15)$$

where the constant C is listed in Table 4 for plane waves, electric fields, and magnetic fields, respectively. K (S/cm) is the volume conductivity, K_C (S/cm) is cooper conductivity, f (MHz) is the frequency, and t (m) is thickness of the shield [41].

During the experiment, the dependency of inner material moisture has also been measured. The measured moistures were 0.8% for the dried samples, 2.2% for the normalized samples, and for the higher level of the inner humidity, it was not possible to set the value because of the physical limits of the measuring tools.

Table 4: Constants used in (7) and (8).

Type of field	C
Electric field	322
Plane wave	168
Magnetic field	14.6

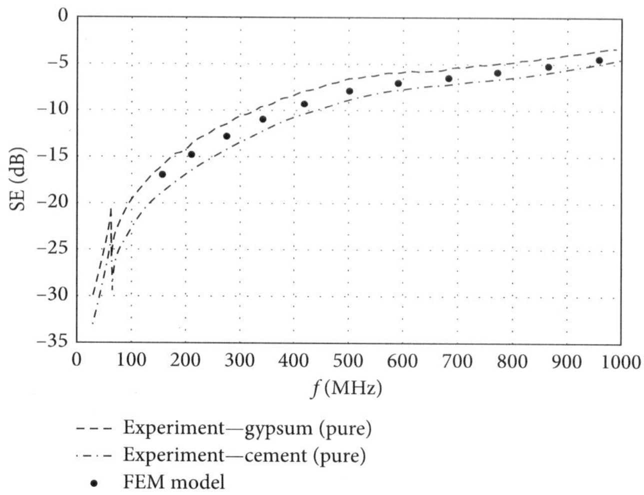

Figure 7: A comparison of the FEM model and experimental measurement.

5. Result and Discussion

First of all, the theoretical model was prepared to study electromagnetic shielding ability of plaster composites with different thicknesses, containing different fiber reinforcements, and applying different types of incident wave. Based on findings gained from modeling, real composite samples were produced, and effect of plaster type, effect of conductive reinforcement content, and effect of moisture content on total electromagnetic shielding effectiveness were studied. A comparison of the FEM model and experimental measurement revealed similar results of electromagnetic shield by testing materials obtained from the measurements; however, Figure 7 shows a uniform course. The FEM model also shows a study of the absorption coefficient where it is evident that in the mixture of the selected plaster types without fiber dispersion the absorption coefficient of the electromagnetic waves is low and remains almost unchanged. Improvement can be achieved by adding conductive carbon fibers with value of about ≤3% in our case. As shown in Figures 8 and 9, it is possible to achieve an increase in the effective plaster thickness by the fiber addition and the final electromagnetic shielding could also be improved. As it was already mentioned in the description of materials, the plaster itself, due to brittleness and porosity, does not have the corresponding elastic properties. However, when reinforcing the plaster with just small ratio of fibers (1-2%), composite material with significantly improved elastic and damping properties can be obtained. This phenomenon is illustrated in Figure 9, where the FEM model compares the structure of the pure and the reinforced plaster. Figure 9 shows the overall shielding efficiency, where the density of intensity of the overall attenuation of the electromagnetic shield R is defined by (7), whereas in case of a nonreinforced sample, the shielding intensity decreases steeply beginning by the first contact of waves and material boundary; the increased absorption is evident in the entire thickness of the reinforced sample; thus, less of undamped waves pass through the reinforced sample.

The electromagnetic shielding ability was measured for three sets of samples having 4 different types of plaster with the various carbon fiber weight ratios (from 0 to 3 wt.%). Figure 10 shows the variation in SE for composites with

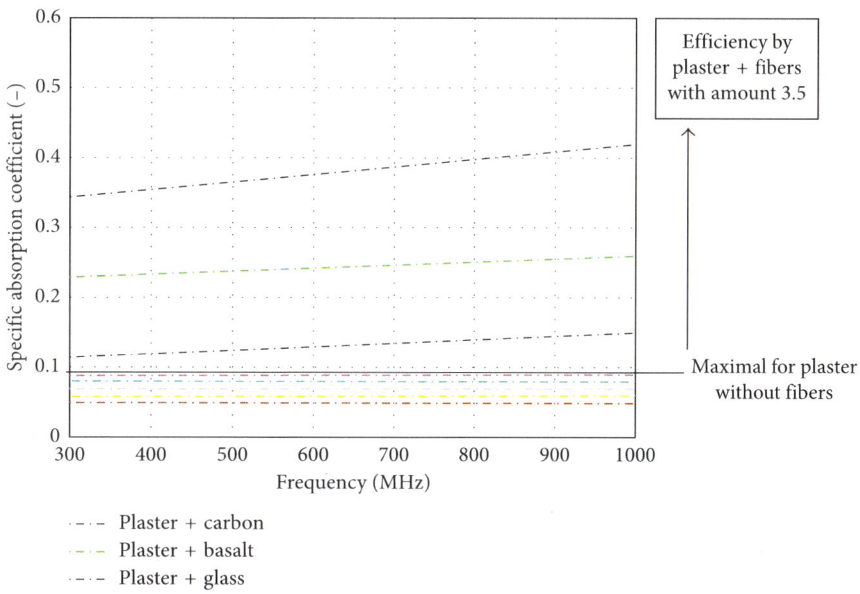

Figure 8: FEM model: study of absorption coefficient for plaster.

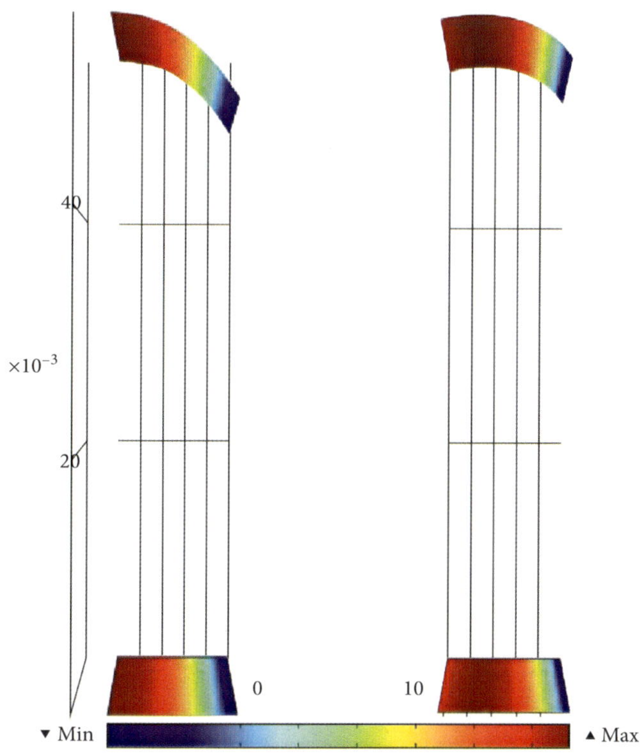

Figure 9: FEM model: comparison of intensity of electromagnetic shielding for the same thickness of sample—the plaster without the fibers (left) and in the plaster with the fibers (right).

incident frequency in the range of 30–3000 MHz. It can be concluded that electromagnetic shielding effectiveness (SE) in (dB) is dependent not only on carbon fiber content but also on the type of plaster. Assumption based on theory saying that addition of conductive component guarantees improved shielding ability of material validated by modeling was also confirmed by experimental evaluation of real samples. It is visible that pure plasters regardless of its type show poor electromagnetic shielding effectiveness higher than −10 dB for frequency range 500–3000 MHz, whereas by addition of carbon fiber, SE can be decreased up to −60 dB for higher frequencies. When studying frequency-dependent behavior, Thermo UM, cement, and lime plaster embody similar behavior with the increase of SE in lower frequencies

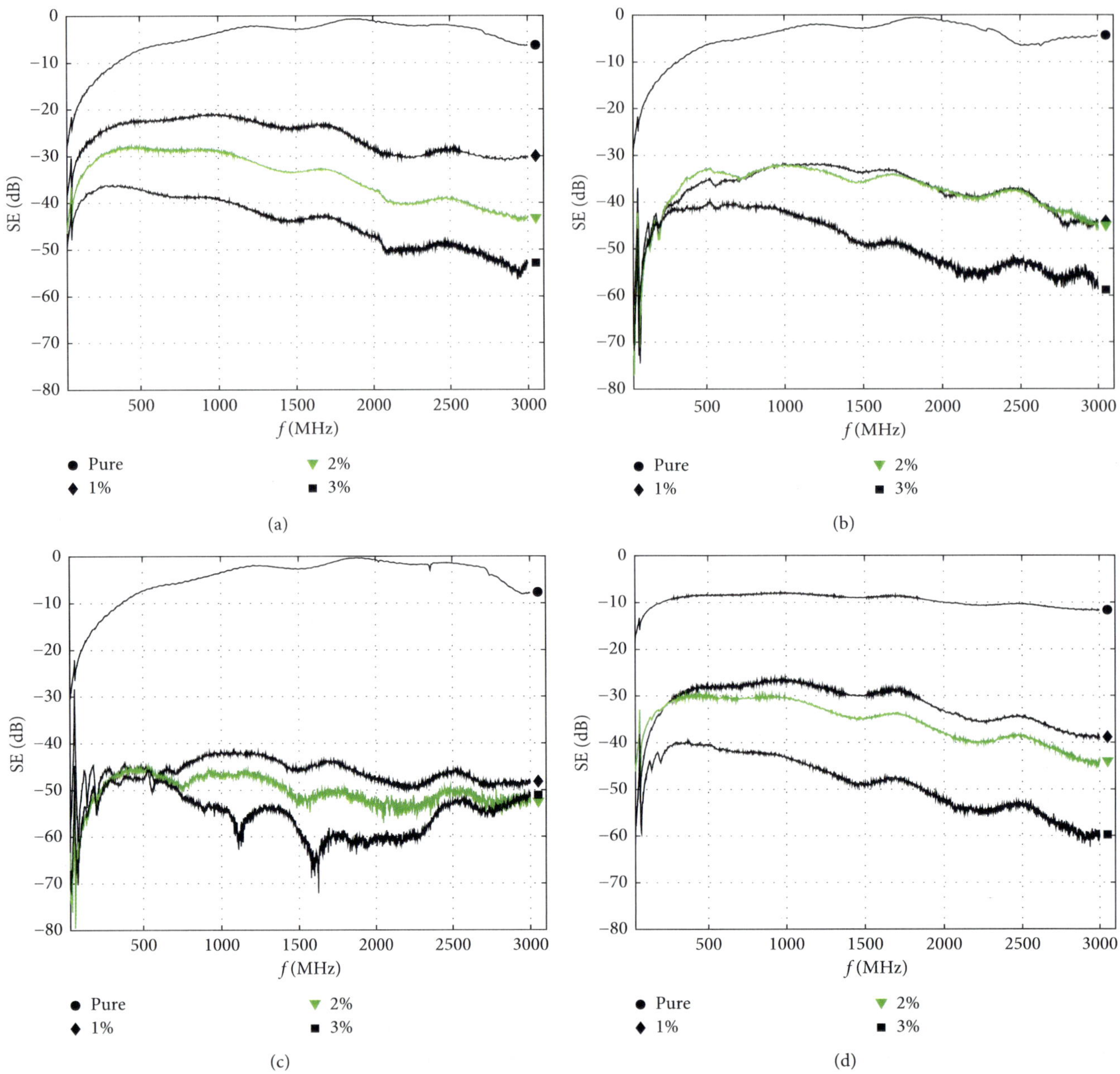

Figure 10: Frequency-dependent electromagnetic shielding ability of samples with different carbon fiber contents and different plaster types: (a) Thermo UM, (b) cement, (c) gypsum, and (d) lime.

($f < 500$ MHz) followed by decrease of SE in a higher frequency range. Gypsum plaster behaves differently having almost constant SE on the whole measured frequency range. In Figure 11, the dependence of SE on carbon fiber weight content for all plaster materials for frequency 1.5 GHz is shown. This frequency was found interesting because many devices use this frequency to operate. It is visible that SE increases with increasing carbon fiber content. Dependence of SE on fiber content can be described using a power function. The solid lines in this graph correspond to the regression models (straight line) with parameters obtained by the minimizing sum of squared errors. High coefficients of determination displayed in the figure indicate the good quality of fit. From this figure, effect of plaster type on SE can also be studied. All pure plasters have very poor shielding ability (SE > −10 dB), whereas gypsum plaster provides the highest SE (SE = 60 dB at 3 wt.% of carbon fiber) compared to other plasters when conductive fibers are added to the composite structure. On the contrast, Thermo UM plaster has the poorest shielding ability accompanied by the highest SE values (SE = 43 dB at 3 wt.% of carbon fiber). This phenomenon is in agreement with electric properties of plasters, namely, its electric conductivity. Thermo UM plaster is thermally insulating plaster having low thermal conductivity and also very low electric conductivity. Gypsum plaster is known for its hygroscopic behavior, which means that this material can attract and hold water molecules from the surrounding

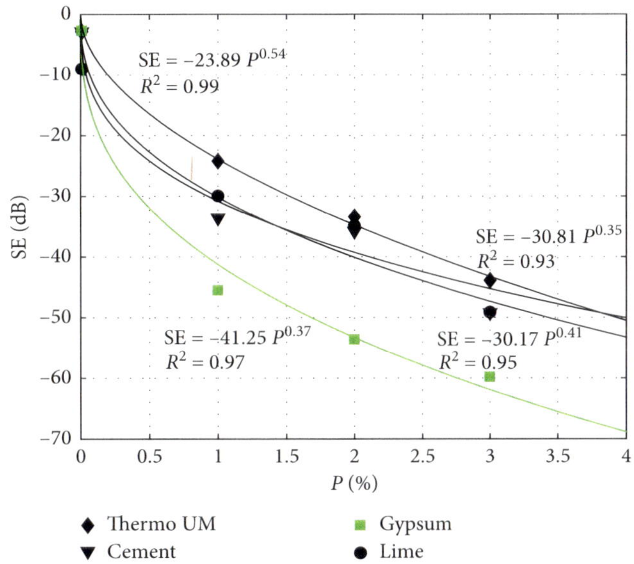

FIGURE 11: The dependence of SE on carbon fiber content P for all plaster types for frequency 1.5 GHz.

environment, which can cause higher electric conductivity accompanied by greater electromagnetic shielding ability. It is known from theory that shielding effectiveness of material is directly proportional to its electric conductivity, see (14) and (15).

Effect of moisture content on electromagnetic shielding ability of carbon fiber-reinforced plaster-based composites was also studied. It has been found that analyzed moisture content has very low impact on output variable. That is why, only the frequency-dependent SE of samples with 0 and 2% of reinforcement for all plaster types are shown in Figure 12. It is known from theory that increased moisture content increases electric conductivity and also positively influences electromagnetic shielding ability because liquid water has a broad absorption spectrum in the microwave region (300 MHz–300 GHz). It is visible from figures that this effect was not uniquely confirmed. Especially for pure samples, the effect of moisture content is neglectable. Difference in shielding effectiveness of samples with 2% of carbon fiber with different moisture contents was also not statistically significant. The observed phenomenon is presumably caused by very low level of moisture in the structure of composites. To notice the increase of electromagnetic shielding ability, the water content in the sample should be much higher as shown in [10].

6. Conclusion

Based on the measured data, it is possible to declare that with increasing content of fibers, the shielding ability of the plaster composites grows significantly. For samples containing 1% of fiber reinforcement, the highest value of the elastic module was measured, and at the same time, it was confirmed that the carbon content had a statistically significant effect on the module. Despite this fact, the content of fibrous reinforcement of more than 3% is less suitable because of arising negative impact to the basic mechanical properties and their sudden degradation, as could be read in the author's previous articles, for example [23, 25]. Also, when we consider the value of improved shielding properties achieved with 1-2% of carbon content, it is not effective to try to use more dispersion because the obtained benefits will only be small. In the future work, it would be advisable to consider the possibility of optimizing shielding capabilities using not only the conductive fibers (1-2%) but also conductive particles. This method could lead to the maintenance of optimal mechanical properties to increase the electromagnetic shielding. This could be beneficial especially when using the mentioned wasted composite fibers and particles. From the global manufacturing processes of composite materials, there are a lot of waste carbon fibers, which are not possible to be used again. However, their usage as the plaster filling could be a way to use them. Another possibility could be the use of conductive grids.

The content 1-2% of carbon dispersion, which according to Table 5 means shielding group: good–very good, is the most suitable. From the measured results, it is obvious that the sample's inner moisture does not have a major impact on the shielding effect. The only noticeable differences in shielding occurred in lime plaster. The reason could be the ability of the plaster to accept moisture from the surrounding environment and its good vapor permeability.

The numerical simulation has been one of the fundamental parts of the work. The main problem is to define the constitute equations or parameters in the material model because it is really difficult to obtain some of the required constants, and often it was necessary to rely just on the literature. However, the main dependency on the variable dispersion content has been verified. It could be stated that it is possible to predict the trends in shielding properties based on the simulation with sufficient accuracy.

Therefore, it is advisable to measure, calculate, or estimate the frequency of unwanted signals and based on that

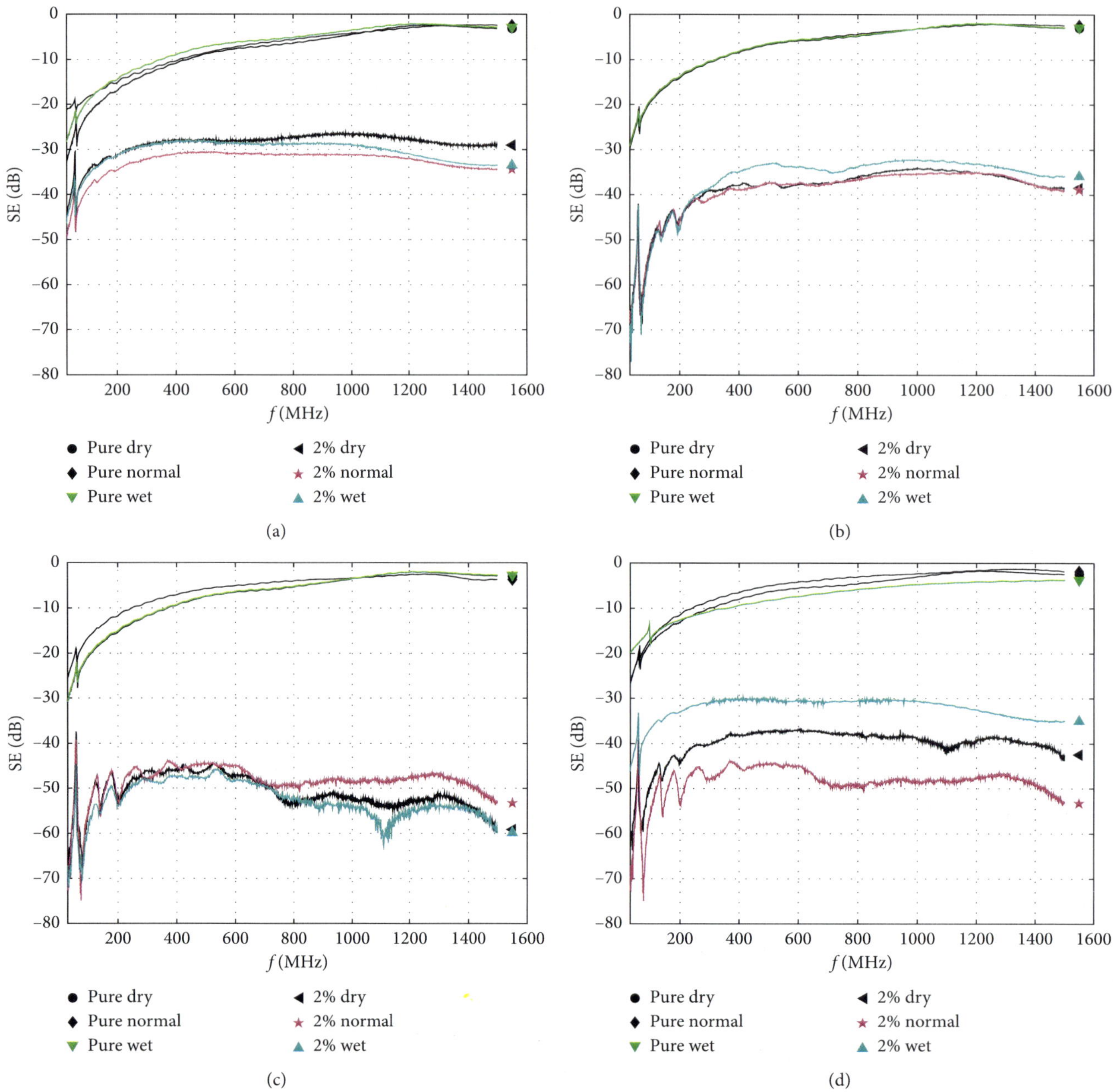

Figure 12: Frequency-dependent electromagnetic shielding ability of samples with different moisture content: (a) Thermo UM, (b) cement, (c) gypsum, and (d) lime.

Table 5: Classes for professional use [42].

Grade	5, excellent	4, very good	3, good	2, moderate	1, fair
Electromagnetic effectiveness range	SE > 60 dB	60 dB ≥ SE > 50 dB	50 dB ≥ SE > 40 dB	40 dB ≥ SE > 30 dB	30 dB ≥ SE > 20 dB

choose the shielding material and thickness of the enclosure, before designing the main shielding element. It is possible to state that at lower frequencies, the reflection of the waves prevails, whereas the absorption effect increases at higher frequencies. Therefore, the conductive nonmagnetic materials are suitable for reflection, and the thickness of the shielding layer is not a critical parameter. This finding corresponds with the probably most common example of electromagnetic shielding in microwave ovens, where just a thin transparent conductive film on the door provides sufficient protection.

The future of our work should be based on the numerical modeling with the use of numerical optimization in order to set the optimal length and content of fibers and thickness of the base materials.

Conflicts of Interest

The authors declare that they have no conflicts of interest.

Acknowledgments

This publication was written at the Technical University of Liberec as part of the project "Assessment of change physical behavior of plaster reinforced by fibre material" (21252) with the support of the Specific University Research Grant, as provided by the Ministry of Education, Youth and Sports of the Czech Republic in the year 2017, and the results of this project (LO1201) were obtained with cofunding from the Ministry of Education, Youth and Sports as part of targeted support from the "Národní program udržitelnosti I" program.

References

[1] A. Zamania and C. Hardiman, "Electromagnetic radiation and human health, a review of sources and effects," *High Frequency Electronics Summit Technical Media*, vol. 4, no. 7, pp. 16–25, 2005.

[2] World Health Organization, *Establishing a Dialogue on Risks from Electromagnetic Fields*, WHO, Geneva, Switzerland, 2002.

[3] Agriculture and Local and Regional Affairs, *The Potential Dangers of Electromagnetic Fields and Their Effect on the Environment*, Committee on the Environment, Strasbourg, France, 2011.

[4] J. F. B. Bolte and M. J. M. Pruppers, *Electromagnetic Fields in the Working Environment*, Netherlands Ministry of Social Affairs and Employment, Hague, Netherlands, 2006.

[5] H. W. Ott, *Electromagnetic Compatibility Engineering*, John Wiley & Sons, Hoboken, NJ, USA, 2009.

[6] M. S. Ersoy and E. Onder, "Electroless silver coating on glass stitched fabrics for electromagnetic shielding applications," *Textile Research Journal*, vol. 84, no. 19, pp. 2103–2114, 2014.

[7] A. Ali, V. Baheti, J. Militky et al., "Copper coated multifunctional cotton fabrics," *Journal of Industrial Textiles*, 2017, In press.

[8] R. Guo, X. Jing, L. Peng et al., "Nickel-catalyzed deposition of Cu film on PET fabric with supercritical fluid," *Journal of Materials Science: Materials in Electronics*, vol. 28, no. 22, pp. 16618–16626, 2017.

[9] I. Su and J. T. Chern, "Effect of stainless steel-containing fabrics on electromagnetic shielding effectiveness," *Textile Research Journal*, vol. 74, no. 1, pp. 51–54, 2004.

[10] V. Safarova and J. Militky, "Multifunctional metal composite textile shields against electromagnetic radiation–effect of various parameters on electromagnetic shielding effectiveness," *Polymer Composites*, vol. 38, no. 2, pp. 309–323, 2017.

[11] S. I. Mistik, E. Sancak, S. Ovali et al., "Investigation of electromagnetic shielding of boron, carbon and boron-carbon fibre hybrid woven fabric and their polymer composites," *Journal of Electromagnetic Waves and Applications*, vol. 31, no. 13, pp. 1289–1303, 2017.

[12] J. M. Keith, N. B. Janda, and J. A. King, "Shielding effectiveness density theory for carbon fiber/nylon 6,6 composites," *Polymer Composites*, vol. 26, no. 5, pp. 671–678, 2005.

[13] J. M. Thomassin, C. Jerome, T. Pardoen et al., "Polymer/carbon based composites as electromagnetic interference (EMI) shielding materials," *Materials Science and Engineering*, vol. 74, no. 7, pp. 211–232, 2013.

[14] B. Saravanja, K. Malric, T. Pusic et al., "Impact of dry cleaning on the electromagnetic shield characteristics of interlining fabric," *Fibres and Textiles in Eastern Europe*, vol. 23, no. 1, pp. 104–108, 2015.

[15] A. Asghar, M. R. Ahmad, M. F. Yahya et al., "An alternative approach to design conductive hybrid cover yarns for effective electromagnetic shielding fabrics," *Journal of Industrial Textiles*, vol. 48, no. 1, pp. 38–57, 2017.

[16] S. Palanisamy, V. Tunakova, D. Karthik et al., "Study on textile comfort properties of polypropylene blended stainless steel woven fabric for the application of electromagnetic shielding effectiveness," *IOP Conference Series: Materials Science and Engineering*, vol. 254, no. 7, article 072018, 2017.

[17] V. Tunakova, Z. Hrubosova, M. Tunak et al., "Laser surface modification of electrically conductive fabrics: material performance improvement and design effects," *Optics and Laser Technology*, vol. 98, pp. 178–189, 2018.

[18] G. E. Monfore, *The Electrical Resistivity of Concrete*, Portland Cement Association, New York, NY, USA, 1968.

[19] B. Wang, Z. Guo, Y. Han, and T. Zhang, "Electromagnetic wave absorption properties of multi- walled carbon nanotube/cement composites," *Construction and Building Materials*, vol. 46, pp. 98–103, 2013.

[20] D. Yawen, S. Mingqing, L. Chenguo et al., "Electromagnetic wave absorbing characteristics of carbon black cement-based composites," *Cement and Concrete Composites*, vol. 32, no. 7, pp. 508–513, 2017.

[21] R. Khushnood, S. Ahmad, P. Savi, J. Tulliani, M. Giorcelli, and G. A. Ferro, "Improvement in electromagnetic interference shielding effectiveness of cement composites using carbonaceous nano/micro inerts," *Construction and Building Materials*, vol. 85, pp. 208–216, 2015.

[22] P. K. Mallick, *Fiber-Reinforced Composites: Materials, Manufacturing, and Design*, CRC Press, Boca Raton, FL, USA, 2007.

[23] A. Samkova, P. Kulhavy, and M. Pechociakova, "Optimization parameters of plaster composites," in *Proceedings of the 54th International Conference on Experimental Stress Analysis*, Pilsen, Czech Republic, June 2016.

[24] C. Bing, W. Ker, and Y. Wu, "Conductivity of carbon fiber reinforced cement-based composites," *Cement and Concrete Composites*, vol. 26, no. 4, pp. 291–297, 2004.

[25] A. Samkova, P. Kulhavy, and M. Pechociakova, "Fibre reinforcement effect on plaster composite properties," in *NART Conference Proceedings*, pp. 189–195, Liberec, Czech Republic, August 2015.

[26] A. Samková, P. Kulhavý, and M. Pechočiaková, "Possibilities to improve electromagnetic shielding of plaster composites adding carbon fibers," *IOP Conference Series: Materials Science and Engineering*, vol. 254, no. 4, 2017.

[27] Standard ČSN EN 1015-14, *Methods of Test for Mortar for Masonry-Part*, Czech Standards Institute, Prague, Czech Republic, 2000.

[28] *Material and Technical List of the Manufacturer: SATSYS*, https://www.maxit.de/produkte/putz/.

[29] D. A. Kontogeorgos and M. A. Founti, "Gypsum board reaction kinetics at elevated temperatures," *Thermochimica Acta*, vol. 529, pp. 6–13, 2012.

[30] S. L. Manzello, R. G. Gann, S. R. Kukuck, and B. L. David,

"Influence of gypsum board type (X or C) on real fire performance of partition assemblies," *Fire and Materials*, vol. 31, no. 7, pp. 425–442, 2007.

[31] P. Kulhavy, T. Martinec, O. Novak, M. Petru, and S. Pavel, "Testing fireproof materials in a combustion chamber," *EPJ Web of Conferences*, vol. 143, article 02058, 2017.

[32] M. Bentchikou, A. Guidoum, K. Scrivener, K. Silhadi, and S. Hanini, "Effect of recycled cellulose fibres on the properties of lightweight cement composite matrix," *Construction and Building Materials*, vol. 34, pp. 451–456, 2012.

[33] *Material and Technical List of the Manufacturer: MAXIT PUTZ*, https://www.maxit.de/produkte/putz/.

[34] S. Chand, "Carbon fibers for composites," *Journal of Materials Science*, vol. 35, 2000.

[35] R. W. Cahn and B. Harris, *Newer Forms of Carbon and Their Uses*, University of Sussex, Brighton, UK, 1969.

[36] E. Fitzer and L. M. Manocha, *Carbon Reinforcements and Carbon/Carbon Composites*, Springer-Verlag, Berlin, Heidelberg, Germany, 1998.

[37] M. Mardiguian, *Controlling Radiated Emissions by Design*, Springer, Berlin, Germany, 2014.

[38] P. A. Chatterton and M. A. Houlden, *EMC-Electromagnetic Theory to Practical Design*, John Wiley & Sons, Hoboken, NJ, USA, 1992.

[39] Standard ASTM D4935–10, *Test Method for Measuring the Electromagnetic Shielding Effectiveness of Planar Materials*, ASTM International, West Conshohocken, PA, USA, 2010.

[40] A. V. Raisanen and A. Lehto, *Radio Engineering for Wireless Communication and Sensor Applications*, Artech House, London, UK, 2003.

[41] V. Šafářová, M. Tunák, and J. Militký, "Prediction of hybrid woven fabric electromagnetic shielding effectiveness," *Textile Research Journal*, vol. 85, no. 7, pp. 673–686, 2015.

[42] Committee for Conformity Assessment on Accreditation and Certification of Functional and Technical Textiles, *Specified Requirements of Electromagnetic Shielding Textiles*, 2010, http://www.ftts.org.tw/images/fa003E.pdf.

Permissions

All chapters in this book were first published in AMSE, by Hindawi Publishing Corporation; hereby published with permission under the Creative Commons Attribution License or equivalent. Every chapter published in this book has been scrutinized by our experts. Their significance has been extensively debated. The topics covered herein carry significant findings which will fuel the growth of the discipline. They may even be implemented as practical applications or may be referred to as a beginning point for another development.

The contributors of this book come from diverse backgrounds, making this book a truly international effort. This book will bring forth new frontiers with its revolutionizing research information and detailed analysis of the nascent developments around the world.

We would like to thank all the contributing authors for lending their expertise to make the book truly unique. They have played a crucial role in the development of this book. Without their invaluable contributions this book wouldn't have been possible. They have made vital efforts to compile up to date information on the varied aspects of this subject to make this book a valuable addition to the collection of many professionals and students.

This book was conceptualized with the vision of imparting up-to-date information and advanced data in this field. To ensure the same, a matchless editorial board was set up. Every individual on the board went through rigorous rounds of assessment to prove their worth. After which they invested a large part of their time researching and compiling the most relevant data for our readers.

The editorial board has been involved in producing this book since its inception. They have spent rigorous hours researching and exploring the diverse topics which have resulted in the successful publishing of this book. They have passed on their knowledge of decades through this book. To expedite this challenging task, the publisher supported the team at every step. A small team of assistant editors was also appointed to further simplify the editing procedure and attain best results for the readers.

Apart from the editorial board, the designing team has also invested a significant amount of their time in understanding the subject and creating the most relevant covers. They scrutinized every image to scout for the most suitable representation of the subject and create an appropriate cover for the book.

The publishing team has been an ardent support to the editorial, designing and production team. Their endless efforts to recruit the best for this project, has resulted in the accomplishment of this book. They are a veteran in the field of academics and their pool of knowledge is as vast as their experience in printing. Their expertise and guidance has proved useful at every step. Their uncompromising quality standards have made this book an exceptional effort. Their encouragement from time to time has been an inspiration for everyone.

The publisher and the editorial board hope that this book will prove to be a valuable piece of knowledge for researchers, students, practitioners and scholars across the globe.

List of Contributors

Vladimir L. Shkuratnik and Petr V. Nikolenko
Mining Institute of National University of Science and Technology MISiS, 119991 Moscow, Russia

Katalin Kopecskó
Department of Engineering Geology and Geotechnics, Budapest University of Technology, Muegyetem Rkp. 3, Budapest 1111, Hungary

György L. Balázs
Department of Construction Materials and Technologies, Budapest University of Technology, Muegyetem Rkp. 3, Budapest 1111, Hungary

Gao Ma, Liang Huang, Hui Wang and Peng Yin
College of Civil Engineering, Hunan University, Changsha 410082, China

Libo Yan
Department of Organic and Wood-Based Construction Materials, Technical University of Braunschweig, Hopfengarten 20, 38102 Braunschweig, Germany
Centre for Light and Environmentally-Friendly Structures, Fraunhofer Wilhelm-Klauditz-Institut WKI, Bienroder Weg 54E, 38108 Braunschweig, Germany

M. P. F. Graça and L. C. Costa
I3N and Physics Department, University of Aveiro, 3810-193 Aveiro, Portugal

K. D. A. Sabóia
Universidade Estadual do Ceará, Fortaleza, Brazil
Laboratório de Telecomunicações e Ciência e Engenharia de Materiais–LOCEM, Departamento de Física, Universidade Federal do Ceará, Fortaleza, Brazil

F. Amaral
I3N and Physics Department, University of Aveiro, 3810-193 Aveiro, Portugal
Coimbra Health School, Polytechnic Institute of Coimbra, Coimbra, Portugal

Ning Zhuang, Honghan Dong, Da Chen and Yeming Ma
Jiangsu Key Laboratory of Coast Ocean Resources Development and Environment Security, Hohai University, Nanjing, China
College of Harbor Coastal and Offshore Engineering, Hohai University, Nanjing, China

Ladislav Reinprecht, Ján Iždinský and Zuzana Vidholdová
Faculty of Wood Sciences and Technology, Technical University in Zvolen, T. G. Masaryka 24, 960 53 Zvolen, Slovakia

Sung-Won Yoo
Department of Civil and Environmental Engineering, Gachon University, Seongnam-si, Gyeonggi-do 13120, Republic of Korea

Young Cheol Choi and Wonchang Choi
Department of Architectural Engineering, Gachon University, Seongnam-si, Gyeonggi-do 13120, Republic of Korea

Fan Wu
Institute of Disaster Management and Reconstruction, Sichuan University and e Hong Kong Polytechnic University, No. 1 Huanghe Road, Chengdu 610065, China

Changwu Liu
Institute of Disaster Management and Reconstruction, Sichuan University and e Hong Kong Polytechnic University, No. 1 Huanghe Road, Chengdu 610065, China
College of Water Resource and Hydropower, Sichuan University, No. 24 South Section 1, Yihuan Road, Chengdu 610065, China

Zhaofeng Diao, Bo Feng, Wei Sun and Xiaolong Li
College of Water Resource and Hydropower, Sichuan University, No. 24 South Section 1, Yihuan Road, Chengdu 610065, China

Shuang Zhao
College of Water Resource and Hydropower, Sichuan University, No. 24 South Section 1, Yihuan Road, Chengdu 610065, China
China Yangtze Power Co., Ltd., China ree Gorges Corporation, No. 1 Construction Road, Yichang, China

Duanyi Wang, Xiayi Liang, Danning Li, Hehao Liang and Huayang Yu
School of Civil Engineering and Transportation, South China University of Technology, Guangzhou 510641, China

Wenchang He
Centre for Advanced Wood Processing, University of British Columbia, 2424-2900 Main Mall, Vancouver, British Columbia, Canada V6T 1Z4

Philip D. Evans
Centre for Advanced Wood Processing, University of British Columbia, 2424-2900 Main Mall, Vancouver, British Columbia, Canada V6T 1Z4
Department of Applied Mathematics, Research School of Physics and Engineering, The Australian National University, Canberra, ACT 0200, Australia

Hongwei Zhang, Zhijun Wan, Yuan Zhang and Dong Wu
Key Laboratory of Deep Coal Resource Mining (CUMT), Ministry of Education of China, School of Mines, China University of Mining & Technology, Xuzhou 221116, China

Pejman Azarsa and Rishi Gupta
Department of Civil Engineering, University of Victoria, Victoria, BC, Canada

Francisco J. Molina and Pierre Pegon
European Commission, Joint Research Centre (JRC), Institute for the Protection and Security of the Citizen (IPSC), 21027 Ispra, Italy

Pierre Labbé
Université Paris-Est, RENON (IRC-ESTP, IFSTTAR), IRC-ESTP, 28 avenue du Président Wilson, 94234 Cachan, France

Sheedev Antony, Monssef Drissi-Habti and Venkadesh Raman
PRES LUNAM IFSTTAR, CS4 Route de Bouaye, 44344 Bouguenais, France

Su Anshuang, Zhang Shoujie, Zhang Jiayang and Li Zhaoyu
Heilongjiang Provincial Hydraulic Research Institute, Harbin 150080, China

Qin Ling
School of Civil Engineering, Harbin Institute of Technology, Harbin 150090, China

A. Samkova and V. Tunakova
Faculty of Textile Engineering, Technical University of Liberec, Studentska 2, 461 17 Liberec, Czech Republic

P. Kulhavy and M. Petru
Institute for Nanomaterials, Advanced Technologies and Innovation, Technical University of Liberec, Studentska 2, 461 17 Liberec, Czech Republic

Index